CURSOR MUNDI

Cursor Mundi is produced under the auspices of the Center for Medieval and Renaissance Studies, University of California, Los Angeles.

Previously published volumes in this series are listed at the back of the book.

VOLUME 25

The Mirror in Medieval and Early Modern Culture

Specular Reflections

Edited by

Nancy M. Frelick

BREPOLS

British Library Cataloguing in Publication Data

A catalogue record for this book is available from the British Library.

D/2016/0095/2
ISBN: 978-2-503-56454-8
e-ISBN: 978-2-503-56564-4
DOI: 10.1484/M.CURSOR-EB.5.108028
Printed in the EU on acid-free paper

Contents

List of Illustrations

Elizabeth Black

Tom Conley

Sergius Kodera

Ulrike Feist

Preface

The essays in this collection are revised versions of selected papers from an international conference entitled *Specular Reflections: The Mirror in Medieval and Early Modern Culture*, held at Green College on the campus of the University of British Columbia, on 16–18 March 2012. This conference — the 40th UBC Medieval Workshop, organized by Nancy M. Frelick, Juliet O'Brien, and Chantal Phan — contributed to a long-standing tradition of interdisciplinary colloquia exploring various aspects of the premodern period, several of which have produced edited volumes.[1] We also followed in the footsteps of some predecessors in expanding the geographical and chronological confines of the Medieval Workshop.

The theme was selected by the Medieval Studies Committee (Chair: Courtney Booker) because of its rich interdisciplinary potential and because of the importance and polyvalence of mirrors as objects or instruments, as well as signs, in the periods in question. Later on, in the modern period, mirrors became more ubiquitous and therefore lost some of the symbolic charge obtained when they were rare, precious commodities — the implications of some of these changes are considered in the introduction.

This multidisciplinary collection was not designed to resemble ordinary conference proceedings; instead, it presents the very best from among submissions written specifically with this thematic volume in mind. It features chapters dealing with 'specularity' through text, image, and various branches of

[1] Recent titles include: Stephen Partridge and Erik Kwakkel, eds, *Author, Reader, Book: Medieval Authorship in Theory and Practice* (Toronto: University of Toronto Press, 2012); Niall Christie and Maya Yazigi, eds, *Noble Ideals and Bloody Realities: Warfare in the Middle Ages* (Leiden: Brill, 2006); and Siân Echard and Stephen Partridge, eds, *The Book Unbound: Editing and Reading Medieval Manuscripts and Texts* (Toronto: University of Toronto Press, 2004).

learning, especially those that engage with the paradoxical ways mirror images are used in the Middle Ages and early modern period. As such, it represents an original contribution on mostly secular (rather than sacred) mirrors in premodern times.

The volume contains articles based on approximately half of the original conference papers, so it does not reflect every nuance of the lively conversations that took place during the colloquium. We hope, however, that this collection offers a valuable cross-section of the wide range of issues that were addressed during the sessions and ensuing discussion periods, thanks to the diversity of participants and disciplinary approaches.

We were fortunate to receive several grants and financial contributions that enabled us, among other things, to secure travel funding for many esteemed colleagues and graduate students from Europe and from all over North America. For this financial assistance, we thank: the Social Sciences and Humanities Research Council of Canada's Aid to Research Workshops and Conferences; the UBC Humanities and Social Sciences Symposium Grant program (Faculty of Arts); the Office of the Provost and Vice President Academic; along with several Departments at the University of British Columbia: History; French, Hispanic, and Italian Studies; Central, Eastern, and Northern European Studies; Art History, Visual Arts, and Theory; English; and Classical, Near Eastern, and Religious Studies.

Most sessions were held at Green College, a beautifully situated, multidisciplinary graduate college at the University of British Columbia. We are grateful to Mark Vessey, Principal of Green College, for his gracious and generous support, and for continuing the annual tradition of welcoming the Medieval Workshop. We also wish to acknowledge the Faculty of Arts for providing large auditoriums for the two plenary lectures, which were open to the general public. The Head of French, Hispanic, and Italian Studies at the time, André Lamontagne, supported this project from the start and we are grateful for his generosity. We also wish to thank Courtney Booker, then Chair of Medieval Studies, for his support and for suggesting we approach Blair Sullivan, Publications Director at the UCLA Center for Medieval and Renaissance Studies, to see whether there might be interest in including this volume in the Cursor Mundi series. Thanks also go to Claire Carlin at the University of Victoria, who shared her experience and expertise and provided invaluable advice at crucial stages of the project.

It is a special privilege to underscore the contributions of our two distinguished plenary speakers, Suzanne Conklin Akbari (University of Toronto) and Tom Conley (Harvard University), who offered very different angles on

Medieval and Renaissance literature. We sincerely thank all the presenters, introducers, and session chairs, as well as other participants who contributed to convivial and academically rich exchanges.[2]

Indeed, among the individuals who planned this meeting over the year that preceded it, special thanks go to Juliet O'Brien and Chantal Phan, co-founders of the Early Romance Studies Research Cluster and co-organizers of the conference. Juliet O'Brien brought her keen appreciation of medieval and early modern literature, as well as her remarkable technological know-how, as designer of our conference website, webmistress, and general technical wizard. Chantal Phan contributed not only her deep commitment to medieval studies but also her considerable organizational skills to making many practical aspects of the conference possible, including room bookings and meals, as well as the coordination of student volunteers. She also proved attentive to various editorial tasks along the way.

The meeting ran smoothly from beginning to end, thanks to our two hired graduate student assistants, Deborah Moniuk and Stephanie Williams, and to a very efficient team of volunteers (present and former graduate students): Monika Edinger, Marina Lushchenko, Margherita Romengo, Dorian Simonneaux, Marilyse Turgeon-Solis, Cristina Uroni, Mike Wilkinson, and Kristyna Zemanova.

We are also indebted to the readers of paper submissions for the conference and to the anonymous referees for the present volume; and we want to express our special appreciation to our very competent and conscientious graduate research assistants, Han Fei and Dorian Simonneaux, who worked with us on the bibliographies and typescript.

Last, but not least, we especially wish to express gratitude to Blair Sullivan, Executive Editor of Cursor Mundi (UCLA CMRS), as well as Guy Carney, Publishing Manager at Brepols, and his team, whose patience, wisdom, encouragement, and editorial assistance have been invaluable from start to finish.

Nancy M. Frelick,
on behalf of the organizing committee

[2] For more information about the 40th UBC Medieval Workshop, including paper abstracts, see the conference website designed and maintained by Juliet O'Brien: <https://ubc2012medieval.wordpress.com/home-2/> [accessed 10 December 2015].

Introduction

Nancy M. Frelick

The mirror's multiple uses as an object translate into highly diversified symbolic functions in the Middle Ages and early modern period.[1] At a very pragmatic level, a looking glass could help one ensure that one's appearance was spotless. It thus functioned as an aid to self-examination, self-knowledge, and 'self-fashioning' (to borrow Stephen Greenblatt's term). By extension, mirrors were therefore associated with both self-improvement (moral edification or spiritual purification) and vanity (excessive pride and preoccupation with the self or worldly goods). In addition, while they were linked with scientific exploration, knowledge, and contemplation (owing largely to analogies with their instrumental uses), mirrors also functioned as metaphors for the illusory nature of representation.[2] Sometimes guarantors of clarity, divine

[1] See Schmidt, 'Miroir', p. 1290.

[2] As Laurie Schneider reminds us in 'Mirrors in Art', p. 286: 'In a rather unflattering commentary on painters, Plato (*Republic* X, 596–98) compares painting to a mirror. The first creator, according to Plato, is God, who conceives of a thing; next is the carpenter who makes it; and third is the painter who merely repeats (i.e., mirrors) what the carpenter has made. If the artist is good, concludes Plato, his picture will deceive the simple-minded and the child.' Mirrors were not only linked to painting because art was seen as mimetic or illusionistic (in antiquity and in the early modern period) and because Renaissance artists like Brunelleschi used mirrors in experiments on geometrical perspective, but also because artists like Leonardo da Vinci recommended 'using mirrors [as instruments] to judge the accuracy of one's work and to make the painting look like a reflected scene' (p. 287). Mirrors were also essential for the self-portrait.

Nancy M. Frelick (nancy.frelick@ubc.ca) is Associate Professor of French at the University of British Columbia. She is the author of *Délie as Other: Toward a Poetics of Desire in Scève's 'Délie'*. Her current book project explores mirror imagery and discourses of desire in early modern texts and visual culture.

The Mirror in Medieval and Early Modern Culture: Specular Reflections, ed. by Nancy M. Frelick, CURSOR 25 (Turnhout: Brepols, 2016), pp. 1–29 10.1484/M.CURSOR-EB.5.111040
BREPOLS PUBLISHERS

truth, or mimetic accuracy, reflective surfaces could also be opaque and distorting, creating false, shadowy, or deformed images of earthly reality, as suggested by the ubiquitous Ovidian theme of Narcissus at the fountain and the Pauline *per speculum in aenigmate*, for example.[3] The contradictory uses of mirrors in iconography mean they could stand as figures of either virtue or vice, depending on whether they accompanied personifications of Prudence or Pride, Wisdom or Idleness, Lust and the enticements of Venus (one of whose attributes is the mirror), or whether they represented the immaculate Virgin Mary — the *speculum sine macula* — or the repentant sinner Mary Magdalene.[4] Mirrors were not only ambivalent but also Janus-like: whether examined as objects — in their instrumental, decorative, or other functions — or as visual or textual figures, mirrors have fascinated humankind, not least because they seem to serve as a kind of threshold phenomenon allowing for the contemplation of inner and outer worlds, as well as the otherworldly (including astrology and magic, catoptromancy, or divination by mirrors, as used by the *specularii* in ancient Rome, or by the renowned magician and alchemist John Dee in the court of Elizabeth I of England, for instance).[5] While these thresholds promise access to other realms

[3] See Hugedé, *La Métaphore du miroir*, and Nolan, *Now Through a Glass Darkly*, for more on Paul's famous epistles.

[4] See Santore, 'The Tools of Venus', for more on images of Venus or Aphrodite and Renaissance courtesans or prostitutes represented with mirrors in early modern iconography or paintings playing on the themes of *Vanity* and of the *Lady at Her Toilet*, all of which Santore links in significant ways. For more on Mary as the *speculum sine macula*, see Goldberg, *The Mirror and Man*, p. 115. See Schwarz's 'The Mirror in Art', on specular representations of Mary and Christ, as well as portrayals of Prudence and Wisdom on the one hand, and Vanity, Venus, Lust, Pride, and so on, on the other hand.

[5] As Crystal Addey explains, in 'Mirrors and Divination', p. 32: 'The practice of divination using mirrors was called Catoptromancy, a name derived from the Greek word *katoptron*, which means "mirror", or "reflection"; since one looks into a mirror as one would look into a reflective pool. Catoptromancy involved divination by images seen reflected on a shining surface, usually a mirror, but sometimes water or another analogous reflective medium.' For more information on various modes of 'scrying' (gazing into reflective or translucent substances to foretell the future, see into the past, or even communicate at a distance or with the otherworldly), including 'catoptromancy' (divination with mirrors), 'crystallomancy' (gazing into a crystal ball), 'hydromancy' (using water as an all-seeing mirror), or 'cyclicomancy' (divination with a scrying cup), see Goldberg, *The Mirror and Man*, pp. 7–23. See also Gregory, *Mirrors in Mind*, pp. 65–67. For more on John Dee, whose famous obsidian mirror is now on display in the British Museum, see Goldberg, *The Mirror and Man*, pp. 15–19. For the uses of mirrors in the Ancient world, see Frontisi-Ducroux and Vernant, *Dans l'œil du miroir*, as well as Taylor, *The Moral Mirror of Roman Art*. For an analysis of divination with a cup in Persian Sufi literature,

— earthly, imaginary, or divine — they are also suggestive of the limitations of human perception, knowledge, and wisdom.

The importance accorded to mirrors and optical phenomena in natural philosophy and science seems to stem from the fact that knowledge was primarily viewed or explained in visual terms, perhaps because vision was generally considered the noblest and thus the most dependable of senses.[6] Physicians, scientists, and natural philosophers often used the mirror as a simile or metaphor for sight or vision. For Avicenna and Averroes the eye itself is like a mirror.[7] Indeed, not only was vision an important area of study in its own right, but the cognitive perception of ideas (insight) was also related to visual sense perception (sight). Similarly, mirrors or *specula* were linked to thought, *speculation*, or inner reflection.[8]

In ancient philosophical and sacred traditions, knowledge of abstract ideas, of truth, or of the divine was generally translated in visual terms, using analogies borrowed from the two areas studied in premodern optics: light and visual perception.[9] The discernment of knowledge and truth have long been explained through such visual metaphors, drawing on comparisons with light, shadow, and reflectivity, as can be seen in the allegory of the cave in Plato's *Republic*, where sunlight came to represent the world of Forms, whereas ignorance was associated with obscurity.[10] Shadows and reflections became symbols of indirect knowledge, letting one see what could not be viewed directly or what could

see Leila Rahimi Bahmany's contribution in this volume. For more on catoptromancy and magic in Della Porta, see Sergius Kodera's article in this volume.

[6] Lindberg, *Theories of Vision*, p. x. For more on vision as the noblest of senses, and on the importance of the eye and the gaze, see also Havelange, *De l'œil et du monde*.

[7] See Lindberg, *Theories of Vision*, pp. 49 and 55. As David Summers explains in *Vision, Reflection, and Desire in Western Painting*, p. 60: 'The metaphor of the eye as a mirror assumed new importance with the resurgence and final triumph of Aristotle's intromissive theory of vision.' For more on intromission and extramission theories of vision, see Akbari's *Seeing through the Veil*. See also Berthold Hub's contribution in this volume.

[8] According to Mark Pendergrast in *Mirror, Mirror*, p. 121: 'Thomas Aquinas, a contemporary of Bonaventura, offered one of the first etymological links between mirrors as *specula* and the modern meaning of *speculation*: "to see something by means of a mirror is to see a cause in its effect wherein its likeness is reflected", Aquinas wrote. "From this we see that 'speculation' leads back to meditation."'

[9] Lindberg, *Theories of Vision*, p. x.

[10] See the allegory of the cave at the beginning of Book VII of the *Republic*, 514–20.

be seen only in the mind's eye.[11] While such *eidola* or *phantasmata* pointed to human existence in a second-order reality — a realm of projections or appearances inferior to the real world of Forms or Ideas in Platonic theory —, indirect vision was also protective, allowing mortals to catch a glimpse of what would otherwise be blinding or deadly. This is the case in Plato's allegory, which likens the human incapacity to see the world of Forms without mediation to our inability to gaze directly at the sun without harm. Reflective surfaces also provide such mediation and protection in other myths, where gazing directly upon powerful divinities or otherworldly beings is forbidden and could prove fatal.[12] Thus, according to ancient mythology, it is with the aid of a mirrored shield that Perseus was able to behead the snake-haired Gorgon Medusa, whose petrifying gaze turned onlookers to stone when they beheld her directly.[13] In medieval and early modern folklore, it was also believed that the most effective way to slay the deadly basilisk was with the use of a mirror: the venomous creature would kill itself by glancing at its own lethal gaze.[14]

In the Christian Middle Ages, the mirror generally stands as a symbol for the idea of indirect knowledge of the divine, for the Pauline idea of the mediation of knowledge through faith and through signs *per speculum in aenigmate*.[15]

[11] As Stoichita explains in *A Short History of the Shadow*, p. 24: '[A]part from their "degree of clarity and obscurity" [which suggests a progressive itinerary from darkness to light]– there is very little difference between shadows and specular reflections' in Plato's myth. According to Stoichita, mimetic theory later privileged the 'specular paradigm, and the projection of the shadow was relegated to a marginal role. But this did not lead to the shadow being completely eliminated from the representation's arsenal: it became for all time the poor relation of all reflection, the forgotten origin of all representation' (p. 25).

[12] Some Greco-Roman myths suggest that it can be fatal for mortals to look directly at deities: see most notably the story of Zeus/Jupiter and Semele, the daughter of King Cadmus of Thebes who was burned up by the god's lightning and thunderbolts when (goaded on by the jealous Hera/Juno to test his divinity) the young woman convinced him to visit her in his true form. In a very different kind of story, Acteon is punished for having caught a glimpse of the goddess Diana naked at her bath. The young hunter is transformed into prey: he is turned into a stag by the virginal goddess of the hunt and torn apart by his own hounds, who no longer recognize their master. As Nolan explains in *Now through a Glass Darkly*, p. 297 n. 2, the Judeo-Christian tradition also has such an interdiction: its 'doctrine forbidding unmediated access to divinity is dramatized in the colloquy between God and Moses in Exodus 34.20–23, in which God, while denying access to His full being, nevertheless does allow Moses to stand behind a tree and, while God passes by, view His "hinder parts"'.

[13] For more on this myth, see *The Medusa Reader*, ed. by Garber and Vickers.

[14] See Berthold Hub's contribution in this volume for a discussion of the basilisk.

[15] See Colish, *The Mirror of Language*, p. ix.

This idea is related to the Platonic notion (discussed above) that the Real is elsewhere, and that all we can see are pale reflections or shadows of that eternal realm here in the changeable, temporal world.[16] As Benjamin Goldberg explains:

> Paul was no doubt aware of Plato's idea that reality is only a poor reflection, replacing the sun as the life source with the God figure. Using this theme, Paul expressed the imperfect nature of human knowledge in this life. Man may see the glory of God only as in a mirror, that is, dimly, never directly. Man must await his translation to, and acceptance by, the spiritual realm before he can feast his astonished eyes upon the true nature of God in heaven, which he now sees as an imperfect image.[17]

Paul's reference to the mirror (probably made of polished bronze[18]) emphasizes its poor reflective quality, which (like the shield of Perseus, also made of polished metal) provides protection as well as indirect vision. Such mediation imposes constraints (the ability to see only dimly, darkly, enigmatically) but proves as necessary in Judeo-Christian doctrine as it did in the Greco-Roman tradition, for, as Edward Peter Nolan puts it, 'We cannot see God and live'.[19]

For medieval Christians, verbal signs were also likened to mirrors, because they mediated cognition and the knowledge of God (often through verbal images).[20] As Suzanne Conklin Akbari suggests, second-order sign systems,

[16] As Jean Frappier explains in 'Variations sur le thème du miroir', pp. 137–38, just as the shadows on the walls of Plato's cave evoke other realms, mirrors seem to capture reflections of otherworldly ideas. He also plays with the fanciful notion of an etymological link between mirror and cave (*speculum* and *specus*), reminding readers that the same word designated both shadow and reflection in Latin (*umbra*) and in Old French (*ombre*).

[17] Goldberg, *The Mirror and Man*, p. 115. As Anna Torti explains in *The Glass of Form*, p. 5, Paul's Second Epistle to the Corinthians glosses the First ('Videmus nunc per speculum in aenigmate: tunc autem facie ad faciem. Nunc cognosco ex parte: tunc autem cognoscam sicut et cognitus sum') where he uses the mirror to explain that in this world we see things 'as in a glass darkly' (according to the King James translation) but that later, we will see God as if face to face (see Nolan): 'Paul makes a distinction between the mediating function of the *speculum* when this is viewed in its material and imperfect nature, and its ameliorating function when it displays an image of the splendour of our Lord in which man can see his own likeness. The *imaginem* of the Second Epistle is thus a reflection of divine glory [...], and at the same time a new image that takes form gradually, as little by little man continues to grow in God's image and likeness. The Holy Ghost makes possible this transition from obscurity (*in aenigmate*) to clarity (*facie ad faciem*), from what we are to what we ought to be.'

[18] Kauntze, 'Seeing through a Glass Darkly', p. 61.

[19] Nolan, *Now through a Glass Darkly*, p. 2.

[20] Colish, *The Mirror of Language*, p. ix.

like allegories (which often use visual metaphors), were also deemed to function like the Pauline mirror because they convey veiled meanings indirectly or obliquely and conjure up mental images through metaphors, symbols, or enigmatic signs that serve both to hide and to reveal meaning 'that cannot be expressed directly through ordinary language'.[21] Yet, like early mirrors, such mediations are not without limitations or flaws. As Akbari explains, 'In a world after the Fall and after Babel, both sense perception and language are imperfect mediators': things can only be known imperfectly, 'indirectly or, as Paul puts it, "in a mirror, dimly" (I Corinthians 13:12)'.[22] Signs and language are therefore as unfaithful as early mirrors, with their poor reflecting qualities and distortions: 'the mirror reproduces the image, but not perfectly, just as the allegorical text conveys meaning, but never perfect truth'.[23]

Akbari also specifies that in a postlapsarian world, 'both vision and language are imperfect mediators, open to deception. Yet, paradoxically, as pseudo-Dionysius recognized, they represent the only possible approaches to knowledge'. Moreover, 'vision is both the truest and the falsest of the senses: it has the greatest capacity to reveal the truth, and the greatest capacity to deceive'. This has important implications for medieval mirrors, which can represent truth or falsehood, virtue or vice, revelation or deception:

[21] Akbari, *Seeing through the Veil*, p. 9. Further on, Akbari cites Augustine's influential commentary on the Pauline *speculum in aenigmate*, in which he sees allegory as a kind of 'specular enigma': 'He associates *aenigma* not with metaphor but with allegory, stating that "an enigma is an obscure allegory." He discusses the term at length in order to gloss Paul's famous statement that, as Augustine puts it, "we see now through a mirror in an enigma." By declaring that enigma and mirror are identical, and that enigma is a variety of allegory, Augustine implies that at least some allegories are, figuratively, mirrors. They are not deceptive mirrors but revealing ones, which allow the viewer to glimpse things ordinarily hidden from human sight; to put it another way, these allegories allow the reader to apprehend meanings normally inaccessible through language' (p. 10). In *The Glass of Form*, Torti explains that 'Allegory establishes a kind of mirror-relationship between the literal and the secondary (allegorical, tropological, anagogical) levels of discourse, thus revealing the inseparable connection between the world of phenomena and the universal truths which lie behind. Writing allegory is a way of functioning as the mirror does: allegorical narrative relates a fictive story and at the same time is strongly conscious of the story as a fiction of which the validity depends on the nonfictional, *true* relationship with the ideal, with God's Word', p. 2. For more on the subject of allegory and literature of the Middle Ages, see also Wimsatt, *Allegory and Mirror*.

[22] Akbari, *Seeing through the Veil*, p. 44.

[23] Akbari, *Seeing through the Veil*, p. 44. As Summers explains in *The Judgment of Sense*: 'In the visible world, images, shadows, and reflections are in the same relation to the objects of which they are images as opinion is to truth in the realm of the intelligible' (p. 40).

> In medieval literature, the double nature of vision is often figured in the two opposing properties of mirrors, the good mirror which makes visible what could otherwise never be perceived, and the bad mirror which inverts the true image before it. In medieval iconography, these appear as the helpful mirror of foresight held by Prudence, and the dangerous mirror of narcissistic self-absorption held by Laziness.[24]

While some see mirrors as granting access to hidden truths, or as a 'manifestation of the transcendent in the immanent', others portray reflective surfaces as deceptive.[25] As we have seen, mirror symbolism is based on analogy, so divine mirrors are generally depicted as immutable and inalterable, whereas worldly mirrors tend to be imperfect and subject to change. As Margot Schmidt explains, God is the pure, elevated mirror to which mystics aspire. According to some writers, the Creator contains all the mirrors that reflect him and must be distinguished from the creations that serve to mirror him. Thus, while man is also a mirror, created in God's image, his spiritual nature is derivative; it is but a copy, or likeness emanating from the original divine Source, and reflected in the mirror of his soul.[26] The pure soul or heart is a living mirror of God. Yet man is generally an imperfect mirror.

For medieval Christians, only Christ can come close to being a perfect reflection of God, the Father. Christ, the angels, the Virgin Mary, and the saints, frequently depicted as clear, stainless, or unblemished (*specula sine macula*), all serve as exemplary mirrors that reflect the truth.[27] They are sources of reflection, contemplation, and wisdom for the faithful. They are ideal mirrors proffered for self-knowledge and emulation, for they provide not only reflections

[24] Akbari, *Seeing through the Veil*, p. 7.

[25] Schmidt, 'Miroir', p. 1295: 'le miroir évoque *la manifestation du transcendant dans l'immanent*'.

[26] As Torti reminds us in *The Glass of Form*, p. 1–2: 'This fundamental analogy goes back to Genesis 1: 26, where God, mirror of all Creation, creates man in His image and likeness. As man is similar to and different from God, so is the image reflected in and by the mirror in relation to the real thing of which it is a "similar copy." In the mirror-image relationship, the mirror has the active role as the means by which the ideal is seen in a transient image. That image is therefore both to be praised for its similarity to the ideal, and to be treated with caution on account of its temporariness. The mirror's essentially ambivalent role, of reflecting appearances (sensible phenomena) and of revealing an intimation of the invisible, the divine, behind them, has to be borne in mind for an understanding of the frequent use of the mirror metaphor in literature.'

[27] Schmidt, 'Miroir', pp. 1295–1301.

of what one is (flaws and all), but of what one should be. As Ritamary Bradley explains, St Augustine describes Holy Scriptures as a mirror with a two-fold function in his commentary on Psalm 103: 'In its resplendence it shows you what you should be, that is, pure of heart; and it also shows you what you are, that you may confess your deformity and begin to adorn yourself.'[28]

According to Diogenes Laertius, the connection between mirrors and self-improvement goes at least as far back as Socrates, who apparently 'advised his young disciples to look frequently at themselves in mirrors to make themselves worthy of their beauty, or if they were ugly to compensate and to cover their imperfection through education'.[29] Studying one's reflection is related to the injunction to 'Know Thyself' (*nosce teipsum* or *gnōthi seauton*, from the Delphic oracle), a theme explored extensively in didactic or moralizing iconography and literature, and used ironically in Ovid's myth of Narcissus (in *Metamorphoses* III, 316–510), where the blind seer Tiresias predicts that the youth will enjoy a long life provided he does not know himself ('Si se non noverit').[30] In edifying literature and iconography, the mirror or fountain of Narcissus frequently represents *philautia*, *vanitas*, excessive pride, or folly, whereas in Neoplatonic traditions (in Plotinus and Ficino, for example) and courtly literature (like the *Roman de la Rose*) such images can serve as a warning of the dangers of erotic or earthly love, with its seductive and deadly illusions that lead away from the

[28] Bradley, 'Backgrounds of the Title *Speculum*', p. 103. See her notes for the Latin original. As Bradley writes in another paragraph on p. 103, Augustine's 'commentaries on other psalms repeat the comparison of *Enarratio in Psalmum 103*, wherein Holy Scripture is said to be a mirror: a mirror of knowledge, when it is said that all which has been written is our mirror; and a paragon for right living, when it is said that God's commands, whether read or recalled to memory, are seen as in a mirror, with a reference to the man in St James' epistle who beheld his face in a mirror and went off forgetful of his state'. Bradley's article gives an outline of these comparisons in the Middle Ages.

[29] Schwarz, 'The Mirror in Art', p. 110.

[30] As Torti points out, in *The Glass of Form*: 'The curse of one of [those] spurned by the young man ("sic amet ipse licet, sic non potiatur amato" — So may he himself love, and not gain the thing he loves; 405) is also added to Tiresias' terrible prophecy: self-knowledge is dangerous, because it implies aspiration towards the unattainable' (p. 9). Medieval revisions of the myth 'insist on the connection between self-love and earthly *vanitas*. Narcissus becomes either the emblem of those who refuse to submit to the power of love, or the archetype of those who love to excess [...] The central issue is no longer self-knowledge and its risks [...] but the recovery, in a negative sense, of an *exemplum* of pagan folly. Narcissus mistakes his own image for something different, and for this reason his impossible love is insane. The pool-mirror has now become the sign of a fatal confusion between illusion and reality' (p. 10).

true path of divine or spiritual love.[31] As is the case with Narcissus, who longs to embrace the simulacrum he sees in the silvery pool even after he recognizes the image as his own reflection, and even after death (for according to Ovid, he still tries to grasp the unattainable object of his love in the Stygian waters of the underworld), this kind of unquenchable passion for an inaccessible other is generally the result of a fatal glance that can lead to madness and perdition.[32] The mirror, which relies on the noblest and yet most deceptive of senses (sight), thus 'has a double but opposite function: positive in the sense that it provides an image of what we ought to be, and negative in that it displays transitory, and consequently vain images'.[33] Although this appears contradictory, Bradley summarizes the logic of these different mirrors in the hierarchy outlined by Alain de Lille (Alanus de Insulis), who explains that a three-fold mirror arises from human nature: the true mirror of reason 'wherein right things appear right and left things appear to be left'; 'the false mirror of sensuality', where 'things are reversed'; and the 'most false mirror of the flesh', where 'they appear as completely inverted. However, all three mirrors are to be regarded: reason, that you may obey it; sensuality, that you may subject it to reason; and carnality, that you may chastise it'.[34]

[31] For more on Narcissus in Plotinus and Ficino, see Kodera's *Disreputable Bodies*. For more on the *Roman de la Rose*, see Akbari's *Seeing through the Veil*, as well as her contribution in this volume.

[32] See Donaldson-Evans on the dangers of the 'aggressive eye topos' in *Love's Fatal Glance*, as well as Stewart's *The Arrow of Love*. For more on the myth of Narcissus in Ovid, see: Bartsch, 'The Philosopher as Narcissus'; Brenkman, 'Narcissus in the Text'; Hardie, *Ovid's Poetics of Illusion*; and Nouvet, 'An Impossible Response'. For later uses of the myth, see: Frappier, 'Variations sur le thème du miroir'; Goldin, *The Mirror of Narcissus in the Courtly Love Lyric*; Nouvet, 'An Allegorical Mirror'; Vinge, *The Narcissus Theme in Western European Literature*. For more on lovesickness and contagion through the eyes, see: Beecher, 'Windows on Contagion'; Frelick, 'Contagions of Love'; and Wack, *Lovesickness in the Middle Ages*, among others. See Kodera for a Neoplatonic reading of the myth in 'Narcissus, Divine Gazes, and Bloody Mirrors', which is also a chapter in *Disreputable Bodies*.

[33] Torti, *The Glass of Form*, p. 4. Indeed, as Nouvet points out, in 'An Allegorical Mirror', p. 357: '[V]ision is not an immediate perception but a complex and indirect "speculation," since it is condemned to take place in a mirror, the "speculum" of the eye. The medieval conception of the eye as a mirror shatters any illusion of a direct, immediate vision that we may have entertained. We never look directly at the world; we always look at it in a mirror, and as a mirror image. We never see the world itself; we only see the images and phantasms of the world reflected in our eyes.'

[34] Bradley, 'Backgrounds of the Title *Speculum*', p. 112.

Such ideas translate into different kinds of iconography, wherein the mirror sometimes seems to represent one thing and then stand for its opposite, depending on the context, thus requiring thought or careful consideration from the reader or viewer, for as Heinrich Schwarz explains:

> in medieval and later symbolism, the mirror assumes quite opposite roles: it stands for good and for evil; for sacred and for profane; for the spiritual and for the worldly. It was the symbol of the Virgin's chastity and of the repentant sinners St. Magdalen [...] and St. Thais, but it was also the attribute of the two Deadly Sins; Superbia or Pride and Luxuria or Lust, who are frequently represented with a mirror in medieval miniature paintings, as well as in the sculptures [...] and stained glass windows [...] of the great French cathedrals.[35]

According to Schmidt, the Latin word *speculum* also connotes representation, and thus designates a painting, portrait, or description.[36] As Laurie Schneider explains: 'Metaphorically speaking, art *is* a mirror, a reflection of society with all its customs, beliefs, folklore, superstition, religion, even of the artist himself.'[37] A work of art can thus be a mirror of people (self or other), of the world (temporal or celestial), or of the mind, with both 'mirror and picture' seen as 'analogous to the internal images of the imagination and memory'.[38] The fascination with mirrors and reflectivity also led many artists to portray them in paintings (see Jan van Eyck, Robert Campin, Petrus Christus, Quentin Massys, Savoldo, Furtenagel, Von Aachen, or Velázquez, for instance).[39] Mirrors were also used

[35] Schwarz, 'The Mirror in Art', pp. 105–06.

[36] Schmidt, 'Miroir', p. 1291.

[37] Schneider, 'Mirrors in Art', p. 283. As Schneider states later on, on p. 287: 'The very concept of self-portraiture implies the literal use of mirrors even if they are not visible in the final product. But there is also an inevitable metaphorical aspect to the self-portrait, which is a mirror of the artist.'

[38] Dundas, '*Speculum mentis*', p. 373. In 'Man and Mirror in Painting', p. 62, Jan Białostocki cites (Gombrich's translation of) Marsilio Ficino's *Theologia Platonica*: '"In paintings and buildings the skill of the artist shines forth. Moreover, we can see in them the attitude and the image, as it were, of his mind; for in these works the mind expresses and reflects itself not otherwise than a mirror reflects the face of a man who looks into it."'

[39] Some of the most renowned paintings with a convex mirror include: the *Arnolfini Portrait*, by Jan van Eyck (1434); the *Werl Alterpiece*, by Robert Campin (*c.* 1438); *Saint Eligius* or *A Goldsmith in his Shop*, by Petrus Christus (1449); the *Moneylender and his Wife*, by Quentin Metsys or Massys (1514); and Furtenagel's portrait *Hans Burgkmair and his Wife Anna* (1529). Some famous paintings with a flat or plane mirror (including some self-portraits): Savoldo's *Self-Portrait* or *Portrait of a Man in Armour* (*c.* 1525); Von Aachen's *Couple with Mirror* (*c.* 1596); Johannes Gumpp's *Self-Portrait* (1646); Velázquez's *Las Meninas* (1656)

directly or indirectly in the development of the self-portrait, which became more prevalent in the Renaissance and early modern period, and increasingly reflexive, or 'metapictorial', to use Stoichita's term (like the 'self-aware images' of Dürer and Parmigianino, as well as Gumpp, Rembrandt, and Murillo).[40] This also influenced the concept of the literary self-portrait, as seen in Montaigne's *Essais*, for example.[41]

In the Middle Ages and early modern period, analogies based on the instrumental functions of mirrors, along with the Augustinian commentary describing the Scriptures as mirrors (mentioned above), also led to the idea of the book as *speculum*, a convention explored by Ritamary Bradley and Herbert Grabes in their respective studies of mirror-titles.[42] The Latin *speculum* ('mirror', *miroir*,

and *The Toilet of Venus*, also know as *The Rokeby Venus* (*c.* 1647–51). There are also countless other images of Venus or of the *Lady at her Toilet* with a mirror (by Bellini, Titian, Vasari or the Fontainebleau school, etc.) analyzed by Santore in 'The Tools of Venus', as well as images of Narcissus contemplating his image in a reflecting pond or fountain (from antiquity and beyond, in paintings, emblems, and tapestries), some of the most famous of which include pictures for the emblem by Alciato (entitled 'Philautia') and the painting by Caravaggio (*c.* 1559–1600). For more on these artists and other types of specular reflections (there are too many to mention here), see Białostocki, Panofsky, Schwarz, Stoichita, and Summers, among others, as well as Jonathan Miller's exhibition catalogue entitled *On Reflection*.

[40] I also borrow the expression 'self-aware image' from the English title of Stoichita's masterful *L'Instauration du tableau*, translated as *The Self-Aware Image*. For more on this, see the famous series of self-portraits by Dürer and Rembrandt, for example, as well as Parmigianino's *Self-Portrait in a Convex Mirror* (1524) and Murillo's *Self-Portrait*, which uses *trompe l'œil* to extend out of the painting by resting a hand on a frame inside the frame of the painting, thus becoming a kind of self-reflexive image of a self-portrait. For a discussion of this and other images that Stoichita views not just as self-portraits but as representations of self-portraits, see *The Self-Aware Image*, pp. 207–15. I am grateful to Rose Marie San Juan for sharing her passion and wealth of knowledge in this area.

[41] For the relationship between mirrors and literary self-portraits, see Beaujour's *Miroirs d'encre*. Some have intuited that the mirror influenced the rise of modern subjectivity in the Renaissance. Debora Shuger questions this idea in her thought-provoking article 'The "I" of the Beholder', where she says that her 'essay began as an attempt to document [such] an hypothesis that turned out to be false' (p. 21). She states that 'the preponderance of evidence suggests that the Renaissance self lacks reflexivity, self-consciousness, and individuation, and hence differs fundamentally from what we usually think of as the modern self', though she notes that Montaigne is an 'obvious' exception (pp. 34–35). For Shuger, 'early modern selfhood was not experienced reflexively but, as it were, relationally' (p. 37). She writes that 'Only in Montaigne does one find selfhood imagined as a recessed space (*arrière-boutique*) where one goes to be alone' (p. 38).

[42] See Bradley, 'Backgrounds of the Title *Speculum*'; and Grabes, *The Mutable Glass*. In his study, which is mostly concerned with English literature from the thirteenth to the seventeenth

specchio, or *spiegel*, depending on the vernacular) was related to the mediation and presentation of knowledge, the world, and the self in informative or normative ways in works with mirror-titles that use images and/or words to instruct readers about the physical or moral universe (such as the encyclopaedic *Speculum maius* of Vincent of Beauvais, a compendium that attempts to summarize the different branches of knowledge in several parts — the *Speculum naturale*, the *Speculum doctrinale*, the *Speculum historiale*, and the *Speculum morale*) or the anonymous *Speculum humanae salvationis* that illustrates the biblical story of salvation through images.[43] Some mirror-texts exhort readers to lead more virtuous lives through sermons, homilies, and *exampla*, which could be positive (as in the *Speculum virginum*, a kind of manual for monastic life, for instance) or negative (see Nigel de Longchamps's satirical *Speculum stultorum* or *Mirror for Fools*, for example).[44] Mirrors with a spiritual aim also include more unusual mystical texts like Marguerite Porete's *Miroir des simples âmes* (*The Mirror of Simple Souls*) or Marguerite de Navarre's *Miroir de l'âme pécheresse* (translated by a young Elizabeth Tudor as *The Glass of the Sinful Soul*,

century, Grabes gives a synoptic list of 398 works with mirror or *Speculum* titles between 1500 and 1700. See also Einar Már Jónsson, *Le Miroir*. For mirrors of melancholy from antiquity to early modern times, see Cazes and others, *Miroirs de la Mélancolie/Mirrors of Melancholy*, including, among others, a contribution by Frelick, '*Per speculum in aenigmate*', which provides an introduction to and facsimile of a little book entitled *Le Miroir des melancholicques* (1543).

[43] For more on Vincent of Beauvais's *Speculum maius*, as well as alchemical works with mirror-titles, see Anna Dysert's contribution in this volume.

[44] According to Schmidt, there are two sorts of literary mirrors: instructive mirrors and exemplary or normative mirrors — depending on whether their aim was to enhance knowledge or to provide moral and spiritual guidance. In 'Reflections: Spenser, Elizabeth I, and Mirror Literature', Rebecca Dark (quoting Grabes) states that there are four basic types of didactic literary mirrors: 'encyclopedic mirror literature that "reflects things as they are," fantastic mirror literature that reflects what exists only in the writer's imagination, prognostic mirror literature that reflects "the way things will be," and didactic mirror literature that "shows the way things should or should not be" through both exemplary and admonitory examples, this genre remained extremely popular well into the seventeenth century. Herbert Grabes asserts that in this genre "the idealizing function of the mirror (as metaphor) is closely connected with the function of the real toilet-mirror," exhorting its intended audience to use its content to improve his or her "thoughts and conduct" much as one would use a toilet-mirror to improve one's outward appearance. Although the earliest didactic mirror literature was largely religious or contemplative often directed at specific ages or classes of people, religious mirrors were eventually "displaced in quantity and importance" in favor of "exemplary political mirrors," including the well-studied mirror-for-princes sub-genre' (p. 3).

long before she became Elizabeth I, queen of England).[45] There was also a subgenre of the *speculum* book (that did not always have a mirror-title): the *speculum principis* or 'mirror for princes'. Following on Augustine's model — the portrayal of a Christian prince and his duties in the *City of God* —, these treatises offer advice to the ruling elite regarding their spiritual, social, and political responsibilities, and include such works as Christine de Pizan's *Livre du corps de policie*, the *Institutio principis christiani* of Erasmus, and Guillaume Budé's *L'Institution du prince*, among others.[46] Perhaps the most famous and controversial of these is Niccolò Machiavelli's *Il Principe*, which is considered atypical, as most manuals for rulers stress the need for Christian virtues. Also included in this genre are instruction and conduct books for noble children, sometimes authored by a parent, like the *Enseignements de la duchesse de Bourbon à sa fille Suzanne*, written by Anne de France, or King James's *Basilikon Doron*.[47]

Specularity has served not only to inspire titles, themes, and ways of seeing, along with enduring analogies, allegories, and metaphors (as we have seen), but also the very form or structure of certain works in art, architecture, literature, and music, through the use of mirroring, symmetry, inversion, 'parallelism, chiasmus, and other rhetorical devices', as well as *mise en abyme*, such as Shakespeare's famous 'play within a play' in *Hamlet*, or other forms of recursiveness.[48] Mirrors and reflecting surfaces of various kinds have inspired

[45] Both Marguerites saw their mirror-texts condemned by the Church, though the Queen of Navarre fared better than the other Marguerite, as she had the protection of her brother, Francis I of France. Marguerite Porete was burned at the stake, as the ideas in her text (with the full title *Le Mirouer des simples ames anienties et qui seulement demourent en vouloir et desir d'amour*) were considered heretical. See Cottrell, 'Marguerite Porete's Heretical Discourse', and Sommers, '*Le Miroir de l'âme pécheresse* Revisited'. For more on Elizabeth's text, see Shell, *Elizabeth's Glass*.

[46] See Darricau, 'Miroirs des princes'. See David Napolitano's contribution in this volume for a more republican type of *speculum principis*: the *specchio del podestà* or 'mirror for magistrates'.

[47] For more on conduct books for women, see the chapter entitled 'The Mirror, the Distaff, the Pen' in Jones, *The Currency of Eros*, who quotes the words of Anne de France reminding her daughter of the duties of 'noblewomen, who "in all things [...] are and ought to be the mirror, model and example for others"' (p. 15).

[48] Torti, *The Glass of Form*, p. 3. For more on such literary devices, see contributions by Akbari and Rahimi Bahmany, as well as Jean-Philippe Beaulieu, Alison Beringer, Elizabeth Black, Hélène Cazes, and Tom Conley, in this volume. According to Haley, in *Shakespeare's Courtly Mirror*: 'all of Shakespeare's plays contain "mirrors" for the prince, the gentlemen, and the aristocrats who populate them. At times the play also "mirrors" the reflecting hero to *us*

not only devotional works (as we have seen) but also secular poetry, in verse addressed to the Beloved's looking glass, or to speak of her eyes as mirrors, for example.[49] Mirrors, fountains, and related imagery (like allusions to Medusa or Narcissus) have also appeared as conceits (*concetti*) in Petrarchan, metaphysical, and baroque poetry.[50] Indeed, mirrors have influenced texts in a more material way through the very medium of print. As Miranda Anderson explains: 'The closeness of the relationship between mirrors and texts is reflected by the fact that the [European] printing press was invented by Johannes Gutenberg, a man who previously made mirrors and came up with the idea, from the way that they invert images.'[51]

Reflection and refraction were tied to the mathematical or geometrical portion of optics (or catoptrics), as studied in Hero of Alexandria's *Catoptrica* and Alhazen's treatise on light.[52] Such geometrical calculations and reconstructions later led to the idea of linear perspective in Brunelleschi and Alberti, probably with the aid of a flat or plane mirror, as Samuel Edgerton suggests.[53] Some Renaissance artists (like Leonardo da Vinci, following Alberti and Filarete) rec-

with unmistakable contemporaneity. The effect is to heighten our sense of historical existence and make it seem more or less continuous with the immediate theatrical experience. We are drawn into reflexive participation with the subject' (p. 38). For more on Shakespeare's mirrors, see also Kelly, 'Surpassing Glass: Shakespeare's Mirrors'. See Dällenbach's *The Mirror in the Text*, for uses of the *mise en abyme* in the modern novel and *nouveau roman*.

[49] See Dante, Petrarch, Scève, Ronsard, Donne, and Spenser, to name just a few. For studies of the specular emblems and poems in Maurice Scève, see Frelick, 'Lire en abyme' and 'La Passion au miroir'.

[50] As Rayna Kalas explains in 'The Technology of Reflection' (which is also a chapter in her book, *Frame, Glass, Verse*): '*Conceit*, from the Italian *concetto*, a word that has been so central in describing the fanciful inventions distinctive to the sonnet tradition, simultaneously refers to precious and persuasive trifles, objects that have too much significance attached to them' (p. 527). Kalas argues for the influence of technology on poetry and sees a distinction between the use of steel and glass mirrors in Gascoigne's *The Steele Glas*, 'wherein the crystal glass is a rarified conceit': 'Unlike the steel glass, which offers a decorously imperfect reflection of divine intent in the causation of the material world, the glassy mirror proves to be not only a false reflection of reality, but also a false instrument within it.' She argues that Gascoigne prefers the more modest and more English metallic mirror that requires polishing to be true, as opposed to the vain seductions of the precious import (pp. 527–28).

[51] Anderson, 'Early Modern Mirrors', p. 117.

[52] On mirrors in Hero of Alexandria, see Lindberg, *Theories of Vision*, pp. 14–15; and in Alhazen, see pp. 79–80.

[53] See Edgerton, *The Mirror, the Window, and the Telescope*. For more on the development of glass and mirror-making technologies, see Melchior-Bonnet, *The Mirror: A History*.

ommend using (plane or flat) mirrors to check one's work.[54] The more accurate plane glass mirrors were enabled by advances in glass and mirror making (especially through the invention of *cristallo* by Renaissance artisans in Murano), which also allowed for the construction of revolutionary scientific tools such as the telescope. Technological advances in glass and mirror making also led to architectural marvels such as the *Galerie des Glaces* (or Hall of Mirrors, also known as *la Grande galerie*) in Versailles, which was a kind of showcase for new manufacturing techniques for large-scale production of sizeable plane mirrors of superior quality (thanks in large part to the perfection of methods allowing for a move away from blown glass to cast glass). This signalled the French victory over the Venetian quasi-monopoly on glass mirrors (owing to the purity of its *cristallo*) along with the birth of a new national industry in France, but not before a veritable 'mirror craze' that threatened financial disaster, owing to the exorbitant cost of Venetian crystal mirrors.[55] Ironically, the apogee of reflective glass as a symbol of wealth and power in the Hall of Mirrors at Versailles also signaled its ultimate decline as a luxury item, as the nascent *Manufacture royale des glaces* showcased in the palace would make mirrors of all sizes increasingly available to all.

According to Sabine Melchior-Bonnet, the glorification of the mirror and the proliferation of *cabinets de glaces* in the seventeenth century encouraged an interest in appearances and artifice, as well as a more theatrical view of life as a kind of spectacle. Large mirrors and windows created a dazzling light show, reflecting sunlight or the luster of candlelight, opening up interior spaces, sometimes appearing to erase boundaries between inside and outside spaces, and even creating apparently infinite vistas, especially in galleries such as the Hall of Mirrors at Versailles, which also boasted other types of 'specular magic'.[56] No

[54] See especially 'How the Mirror is the Master of Painters' and 'Judging Your Painting' in *Leonardo on Painting*, ed. by Kemp, pp. 202–03.

[55] Melchior-Bonnet, *The Mirror: A History*, p. 36. See Melchior-Bonnet (especially Chapter 2) for a fascinating account of the rivalry between Italy and France over glass and mirror making. The passion for mirrors leads to all sorts of intrigue, as well as huge investments (and often huge losses) of capital. French competition with Venice and methods used to coax glassmakers from Murano, as well as protectionist strategies from Italians who wanted to keep their lucrative monopoly and secrets intact, sometimes reads like a murder mystery, or at least like early modern 'industrial espionage'. For more on the telescope, see Edgerton, *The Mirror, the Window, and the Telescope*. For more on Della Porta's telescope (as opposed to Galileo's), see Kodera's contribution in this volume.

[56] Melchior-Bonnet, *The Mirror: A History*: 'Everything at Versailles was specular magic, not only the castle reflected in the water of the canal, nor the symmetry of an unfolding

doubt, the *Galerie des Glaces* also served as a reflection of the infinite power of the Sun King, Louis XIV, reminding courtiers of the absolute monarch's omniscient gaze and creating in them a kind of self-consciousness (or internalization of that gaze), as well as an awareness of the need for self-fashioning — for keeping up appearances in a world where they were constantly on display and scrutinized by an ever-present social gaze (turning the court into a kind of fishbowl or prefiguration of Bentham's Panopticon). Mirrors have perhaps always been (in the service of) 'technologies of power' of one kind or another (in a Foucauldian sense).[57] Yet, the moral mirror of the Middle Ages that encouraged an introspective gaze to care for the soul (through 'technologies of the self') increasingly seemed to give way to more external modes of social control based on outward appearances, with more superficial conventions and forms of etiquette being used to regulate public behaviours through an ever-present series of gazes and mutual glances (encouraging self-conscious simulation and dissimulation) in an endless courtly spectacle. Indeed, according to Melchior-Bonnet, the technical advances in mirrors ultimately lead to their 'metaphysical decline'.[58]

architecture, nor the repetition of gestures in the mirror, but first and foremost all the rules of etiquette by which the audience of the courtiers bowed in unison, exchanging curtsey for curtsey, glance for glance. The court thought itself a spectacle. Each person wanted to see, to see himself and to be seen, to be narcissistically bedazzled as all gazes converged upon the eye of the Sun King, from whom all light radiated' (p. 147).

[57] In his seminar 'Technologies of the Self' (originally presented in Vermont in 1982), in which he examines pagan and early Christian (monastic) practices, Michel Foucault says: 'technologies of power [...] determine the conduct of individuals and submit them to certain ends or domination, an objectivizing of the subject', and 'technologies of the self [...] permit individuals to effect by their own means or with the help of others a certain number of operations on their own bodies and souls, thoughts, conduct, and way of being, so as to transform themselves in order to attain a certain state of happiness, purity, wisdom, perfection, or immortality'. These generally function along with 'technologies of production, which permit us to produce, transform, or manipulate things' and 'technologies of sign systems, which permit us to use signs, meanings, symbols, or signification'. He stipulates that these 'four types of technologies hardly ever function separately, although each is associated with a certain type of domination' (p. 18).

[58] Melchior-Bonnet, *The Mirror: A History*, pp. 270–71: 'The technical advances in glassmaking in the fifteenth and sixteenth centuries and the greater understanding of the mechanics of vision in the seventeenth century are what led to the "metaphysical decline" of the mirror. By shedding its mystery, the mirror (by then perfectly flawless but ordinary) became an instrument of social conformity and offered man the freedom of a solitary face-to-face encounter. The clarity and distance of the reflection offered a space of performance, a theatre lending itself to disguise and show. All games, all illusions are now possible since the transparent mirror makes

It is also interesting to note that the technological advances that allow for a more accurate type of looking glass (as opposed to the relatively flawed and deforming early mirrors made of polished stone, metal, or convex glass) seem to lead to a new desire to experiment with distorting mirrors, frequently for the sake of entertainment, aesthetic pleasure, and the production of marvels. Similarly, the focus on linear perspective and mimetic accuracy (as seen in Brunelleschi, Alberti, and their followers) also seems to lead to a fashion for what Jurgis Baltrušaitis terms 'depraved perspectives' as the instruments for geometrical perspective devised in the Renaissance enabled the construction of optical games and anamorphoses that played with deformations of perspectival lines.[59] These became especially popular in the sixteenth and seventeenth centuries, along with perspective boxes, catoptric devices, machines, and installations such as those seen in spectacles, cabinets of curiosities, or *Wunderkammern*, such as the collection of Athanasius Kircher, for example.[60] Anamorphoses or optical glasses have also inspired oblique perspectives in literature and have been likened to 'wit'.[61]

one forget its physical presence, and the man manipulating space in this fashion delights in his power.' Similarly, in *L'Art et l'Autre*, Yvonne Neyrat (borrowing Stoichita's terminology to speak of art history) says that painting moves progressively away from its didactic and religious function and becomes more reflexive in the mid-sixteenth century, going from being 'metaphysical' to being 'metapictorial' (*métapicturale*) (p. 138).

[59] The third edition of Jurgis Baltrušaitis's magisterial study, originally entitled *Anamorphoses ou perspectives curieuses* (1955) — revised, with significant additions (including two new chapters on mirrors), and renamed *Anamorphoses ou magie artificielle des effets merveilleux* (1969) in the second edition (translated as *Anamorphic Art* by W. J. Strachan) — comprises a further revision (with three additional chapters) and was given the new title of *Anamorphoses ou thaumaturgus opticus* (1984). On the title-page of this third edition, Baltrušaitis uses the designation 'Les Perspectives dépravées' as a kind of title for a series encompassing his various projects on what he terms 'dépravations optiques' (p. 2), including the work in hand (*Anamorphoses*) but also his books entitled *Aberrations: Essai sur la légende des formes* and *La Quête d'Isis: Essai sur la légende d'un mythe*. In a chapter on mirrors in *Semiotics and the Philosophy of Language*, Umberto Eco argues that such deforming or lying mirrors provide 'pleasure [...] of an aesthetic [rather than semiotic] nature' (p. 218). For more on anamorphosis, see Massey, *Picturing Space, Displacing Bodies*. See also Kodera on Della Porta's catoptric devices and amphitheatrical glasses, as well as Ulrike Feist on the design of Maignan's catoptric sundial in Rome in this volume.

[60] For more on this see, Baltrušaitis *Le Miroir*. See also the exhibition catalogue by Stafford and Terpak, *Devices of Wonder*.

[61] In *The Curious Perspective*, Gilman draws analogies between the 'visual wit' (of anamorphoses and perspective manipulation) and 'poetic wit': 'The curious perspective violates the

According to Melchior-Bonnet, the trajectory from the curved, irregular looking glass to the flat mirror free of aberrations — in effect, rendering the reflecting surface invisible[62] — also seems to coincide with a shift in perspectives, from an anagogical or allegorical view to a more 'critical and discursive' view of knowledge.[63] It is as if the transparency of the mirror as an object influenced not only the nature of representation (creating a renewed interest in mimesis and a fashion for *trompe l'œil*),[64] but it also seems to have affected its

ordered perfection of pictorial space as verbal wit violates the expository clarity of language. Both exploit the witness's uncertainty in the presence of duplicitous images but repay his puzzlement with what Tesauro would call an *imparamente veloce*, a sudden apprehension of meanings beyond the geometer's grasp' (pp. 232–33). For more on the relationship between anamorphosis and literature in medieval and early modern literature, see: Boyle, *Anamorphosis in Early Modern Literature*; Castillo *(A)wry Views*; Hallyn, 'Anamorphose et allégorie'; Margolin, 'Aspects du surréalisme au XVIe siècle' and 'Perspectivisme, relativisme et scepticisme'; as well as Frelick, 'Lacan, Courtly Love, and Anamorphosis', 'Sex, Lies, and Anamorphosis', and 'Amitié et anamorphose chez Montaigne et Holbein'.

62 As Miller puts it in the exhibition catalogue entitled *On Reflection*, 'the visibility of a surface is inversely proportional to its reflectiveness and […] the more it discloses in the way of reflection the less it reveals itself as a surface' (p. 59). In other words, the clearer the reflection, the more the mirror disappears from view, effacing itself as an instrument of mediation.

63 Melchior-Bonnet notes, in *The Mirror: A History*, that 'The technical advances that enabled the passage from the convex mirror to the plane mirror reflected humanists' new relationship with knowledge. They no longer viewed the visible and invisible as governed by the same laws. The convex mirror concentrated space and offered a global and spherical view of the world, embracing many perspectives, but its roundness distorted the image. The plane mirror, on the other hand, offered an exact but only partial image, a framed vision from a single point of view that controls what is seen like a stage director. A model of a knowledge that is no longer symbolic and analogical but rather critical and discursive, the mirror finds its place in a new philosophy of representation, responding to its own rules, and in addition to its role in organizing space, it revels in the pleasure of the spectacle' (p. 128).

64 As Stoichita explains in *The Self-Aware Image*, p. 185: 'In the culture of "resemblance," all paintings, one way or another, lend themselves as "mirrors of reality," which is why the key metaphor of the European pictorial image was, from the time of the Renaissance, the specular metaphor […] many [art theorists between the fifteenth and seventeenth century] speak of painting as the "mirror of reality." From Alberti (and Brunelleschi), through Paolo Pino and Leonardo da Vinci to Caravaggio, the mirror was selected (or used) as the metaphor (or instrument) of painting.' In 'Backgrounds of the Title *Speculum*', Bradley reminds us that this comparison goes back to Plato, who 'refers in the tenth book of the *Republic* to the artist's mirroring of the world by imitation, though he judges this representation to be of little value, since it deals only with appearances' (p. 105). The ancient preoccupation with mimesis is also reflected in the story of the contest between Zeuxis and Parrhasius in Pliny's *Natural History*. For more on Zeuxis in Erasmus, see the contribution by Cazes in this volume.

function as a sign and effaced its opacity as a signifier. There is an idea that as mirrors become flatter, clearer, more transparent (less visible as objects in themselves), signs also become more transparent, and that with the advent of what is sometimes called 'legitimate perspective', language and art also come to be regarded as more representational or mimetic, as can be seen in the paradigm summed up by Nicolas Boileau's *Art poétique*, where he writes that well-conceived ideas can be clearly and easily expressed through words: 'Ce que l'on conçoit bien s'énonce clairement, | Et les mots pour le dire arrivent aisément' — though he says later in the same first canto that verse must be polished ('Polissez-le sans cesse et le repolissez').[65] The implicit comparison suggests that poetic language is like a mirror that must be polished by a skilled craftsman in order to become clear and transparent, and to allow what is reflected therein to shine forth. Such ideas culminate in what Melchior-Bonnet terms 'the utopia of transparency' in the eighteenth century[66] — but that is beyond the scope of the present collection of essays, which ends with the seventeenth century.

In this volume on specular reflections, we are concerned with (mostly secular) mirrors in medieval and early modern culture. Contributors examine mirrors as objects, instruments, and multivalent signs in iconography and texts of the Middle Ages and early modern period, before they become commonplace and lose some of the value and symbolic charge obtained when they were rare, precious commodities.

The following chapters are arranged in roughly chronological order. Berthold Hub's contribution, 'Aristotle's "Bloody Mirror" and Natural Science in Medieval and Early Modern Europe', traces the motif of the 'bloody mirror' in the Middle Ages and early modern period, often in combination with references to the basilisk first described by Pliny in his *Natural History*. Hub shows that the image of a mirror stained by the gaze of a menstruating woman, first introduced in Aristotle's *De insomniis*, was accepted as a natural phenomenon and continued to be considered a matter of science rather than superstition throughout the premodern period. This remained the case as long as theories of extramission, relying on the idea of visual rays emitted by the eyes, continued to

[65] Boileau Despréaux, *L'Art poétique*, pp. 160–61.

[66] Melchior-Bonnet, *The Mirror: A History*, p. 153. It is interesting to note that according to Jean Starobinski, writing about Jean-Jacques Rousseau's *Confessions* in *The Living Eye*, the mirror is no longer an instrument of self-contemplation but an object used for spying on others without being seen, or (as with literary activity) for showing oneself obliquely through reflection while remaining hidden, thus serving dual tendencies to voyeurism and exhibitionism in the works of this eighteenth-century author (see especially p. 65).

hold sway, often in combination with the Aristotelian model of intromission. Indeed, the phenomenon of the 'bloody mirror', accepted as a fact in *De insomniis*, shows a fundamental contradiction in Aristotle's own theory of intromission through the introduction of evidence that relies on the notion of extramission. Hub explains the haptic nature of vision in the theory of extramission: visual rays emanating from the eye were literally seen as touching and thus having a material effect on the objects upon which the gaze fell, even at a distance — a powerful phenomenon Hub names the 'performative gaze'.

In her essay 'Specular Art and Science: Mirror Metaphor in Medieval Alchemical Texts', Anna Dysert is also concerned with scientific texts. She examines alchemical texts with mirror-titles, which are normally seen as a subset of the *speculum* genre. Going beyond the definitions set out by Herbert Grabes in his landmark study of mirror imagery in medieval and early modern titles, Dysert argues that the designation *speculum* in alchemical texts has a nuanced meaning that suggests changing understandings of technology, alchemy, and the nature of knowledge in Scholastic discourse. Dysert sees the mirror-title not only as a straightforward metaphor for the book as a reflection of a body of learning, but also as a symbol for the way such scientific and technical knowledge came to be organized and authorized in medieval Europe during a period of complex negotiations between *ars* (practice or craft-based knowledge) and *scientia* (theory or knowledge based on first principles and attained through the exercise of reason). She thus demonstrates that the intellectual heritage of the alchemical *speculum* depends on complex symbolism grounded in the idea of the mirror as an object of both 'scientific' and craft knowledge.

In his contribution, entitled 'Adjusting the Mirror: A Political Remake of Brunetto Latini's *Li Livres dou Tresor*', David Napolitano examines the translation and adaptation of Latini's political encyclopaedia from Old French to Old Italian, underscoring the importance of choosing vernacular languages over Latin in this republican 'mirror for princes'. Napolitano studies the modifications deliberately brought to the original French text by the author of a later Italian version, the Pisan notary Bondì Testario, while imprisoned in Genoa after the Battle of Meloria (1284). Focusing his analysis on the production and context of *Tesoro* manuscript L^4, Napolitano argues that instead of constituting a slavish copy or translation of the original, this particular transformation from *Tresor* to *Tesoro* served not only to depersonalize and relocate the work, but also to change its political orientation, thus allowing this 'mirror for magistrates' to cross the political divide between Guelfs and Ghibellines.

Leila Rahimi Bahmany's essay 'Mirror Metaphors in Persian Sufi Literature' focuses on three major metaphors: the mirror of the heart or soul, the mythical cup of Jamshid, and the burning mirror. According to Rahimi Bahmany, the mirror is an ambivalent (heterotopian and even utopian) space, an otherwordly threshold that can cause spatial and temporal dislocations, uniting micro- and macrocosm, and granting access to the transcendent. The catoptromantic (or divinatory) cup of Jamshid, for example, gives the viewer the power to see the whole world beyond it, while the burning mirror provides self-annihilation, purifying the mirror of the heart and clearing it of all but the divine. As Rahimi Bahmany explains, these metaphors, combined and used to relate the ineffable truths of immolating love in Persian literature, also help configure a worldview (inherited from the Hellenistic tradition via Neoplatonists like Plotinus), engaging readers in polishing both their own internal mirrors (of heart or soul), as well as the mirror of the text, in order to unveil the divine Beloved reflected therein.

In 'Language as Mirror: Semiramis and Alexander in a Late Byzantine Romance', Alison Beringer explores the anonymous *Narrative of Alexander and Semiramis* (brought to light by Ulrich Moennig's translation) that pairs the legendary Babylonian queen with Alexander the Great. Beringer examines the mirroring effects achieved through rhetorical devices, as well as in the structure and content of the narrative, which makes use of subtle references and implicit comparisons to the myth of Perseus and Medusa. As Beringer illustrates, ultimately Alexander's words prove as mighty as the mirrored shield and sword of Perseus in the face of the Gorgon-like queen of Syria, who is also Sphinx-like, as she poses riddles the Macedonian king must be able to answer in order to gaze upon her stunning and even fatal beauty, gain her hand, and rule her kingdom.

Suzanne Conklin Akbari's keynote address, entitled 'Sight Lines: The Mirror of the Mind in Medieval Poetics', explores the crucial relationship between seeing and knowing in the Middle Ages. As she puts it: 'To see is to know.' Akbari then examines way the subject is constituted through self-reflective vision in two late medieval works, demonstrating that in Christine de Pizan's *Livre de la Mutacion de Fortune*, the seeing subject discovers herself in the mirror of history, while in Geoffrey Chaucer's *Troilus and Criseyde*, the seeing subject finds herself in the self-reflective gaze of inner vision. Both works provide ekphrastic pauses and proffer mirrors for readers, inviting them to compare themselves to various historical, mythological, or religious figures portrayed therein. As Akbari explains, for both writers the subject is constructed in solitude rather than union, and comes fully into being not through erotic intimacy but through a withdrawal or abstraction from an earlier, apparently transcendent experience of the mutual gaze.

The essay by Hélène Cazes, 'Comme dans un miroir, comme dans un adage: chatoiements et reflets de la reprise et de l'amitié' (As in a Mirror, as in an

Adage: Shimmering Reflections of Repetition and Friendship), offers a close reading of selected passages from Erasmus's *Adagiorum Chiliades* which are constructed like a hall of mirrors: each adage or saying offering a reflection of itself and of the reader in a text that plays on the theme of doubling. As Cazes shows, the *Adages* borrow formulations from various traditions, recreating and commenting self-reflexively on such appropriations, even as they are used as specular reflections of the self and of the self-reflective gaze. Cazes takes readers through a dizzying labyrinth of symmetries and echo effects in the vertiginous accumulation of citations and anecdotes, as well as the apparently limitless succession of similitudes and dissimilitudes reflected and refracted in Erasmus's dialogic mirror-text. As Cazes explains, Erasmus's praise of mirrors — perhaps unexpected in the writings of a moralist, but fitting in the context of recursiveness or doubling — is part of a tribute to friendship and contributes to a larger philological reverie on the *speculum*, a tool for self-examination (or reflexivity), for knowledge of the world (speculation), and for the recognition of resemblance or similarity, as well as repetition and variations with differences.

Elizabeth Black's 'Mirror/Window, Reflection/Deflection: Regulating the Gaze Inside and Outside the House in Gilles Corrozet's *Blasons domestiques* (1539)' presents an analysis of Corrozet's images and text, comparing the domestic and the anatomical *blasons*, and revealing the complex relationship between interior and exterior, as well as public and private spaces represented in the book. Black discusses questions related to visitors' access to the home, women's confinement to the domestic sphere, and concerns surrounding the male and female gaze escaping or invading the house. She also shows the way that representations of transparent and reflective glass placed at liminal points — the mirror and windows — question the absolute nature of the boundaries set up between inside and out, and cast doubts on whether the nascent ideal of privacy can be achieved in the home. Black explains that these objects made of glass also raise questions about subjectivity and subject-hood based on visual perspective, as reader, poet-narrator, and female character exchange places and glances in a complex specular game.

In his contribution 'A Mosaic Mirror: Fracture and Fault Line of Ronsard's *Amours* of 1552', our second plenary speaker, Tom Conley, presents Ronsard's collection of poems as a protean *speculum mundi* — a fractured, scintillating mosaic mirror on which an *amorous universe* is reflected or refracted. According to Conley, Ronsard's collection, at once a cosmogony and a cosmography of love, offers up a fragmented or scattered image of the world as a loose aggregate or iteration of singularities or islands — comparable to cosmographer André Thevet's description of geographical discoveries in *Singularitez de la France antarctique* or to Bartolomeo dalla Sonetti's *Isolario* — through which readers navigate in a broad geographical and historical frame. Relating the thematic and spatial arrangement of poems — including Ronsard's famous mirror-poem, 'Je parangonne à voz yeulz ce crystal' — to reflectivity and cartography, Conley also shows that topography and typography are linked through mirror effects in an erotic geography that evokes both the poet's homeland and the female form, in a manner reminiscent of the *blasons* and (we might add) anamorphic landscape paintings of the period.

Sergius Kodera's essay 'Between Stage-Prop and Metaphor: Mirrors in Giovan Battista Della Porta and Giordano Bruno' focuses on real and imaginary mirrors in the writings of these two late Renaissance Neapolitan scholars, who use the mediating functions of optical and catoptric devices both as material tools — for practical application in science and astronomy, or for spectacular uses — and as literary and philosophical metaphors — to help render visible what cannot normally be seen or grasped by reasoned discourse. Kodera explains that the invention of Della Porta's new playful optical apparatuses, such as cabinets of mirrors or an improved *camera obscura*, served to provide a material basis for the idea of magic mirrors, leading not only to theatrical but also to metaphorical uses; these material objects were fascinating to contemporary audiences, not least because they referenced well-known discourses on reflection. In Bruno, the *speculum videns*, which functions as a mnemonic device, paradoxically representing both unity and multiplicity in the universe, is used figuratively to help visualize the functioning of human consciousness. Moreover, reflecting surfaces stand as metaphysical metaphors for his infinitist philosophy. Both scholars are also interested in experiments with distorting lenses, mirrors, and anamorphoses, which became increasingly popular in the sixteenth and seventeenth centuries.

Jean-Philippe Beaulieu's 'Portraits et miroirs dans les éloges collectifs de femmes au XVIIe siècle' (Portraits and Mirrors in Seventeenth-Century Collections in Praise of Women) explores the specular dimension of collections of exemplary women, such as Pierre Le Moyne's *La Gallerie des femmes fortes* (1647), which is set up like a metaphorical hall of mirrors, the figurative 'gallery' thus playing on the traditional link between looking glass and portrait, as well as various kinds of gazes and mirror effects, and the fashion for viewing collections of paintings or mirrors in such an architectural setting. Literary portraits of illustrious ladies from antiquity are matched with contemporary models for the benefit of female readers who are invited to identify with and emulate such paragons of piety, bravery, learning, and wisdom, as in a set of exemplary mirrors. Beaulieu also explores collections authored by women like Jacquette Guillaume (*Les Dames illustres*, 1665) and Marguerite Buffet (*Éloges des illustres sçavantes*, 1669), who adopted the genre as their own. He demonstrates that such mirror-portraits acquire new meaning when held up by women who were themselves living examples of the feminine qualities they depicted and who seem to invite their readership to relate to an actual community of exemplary women.

The final chapter by Ulrike Feist, 'The Reflection Sundial at Palazzo Spada in Rome: The Mirror as Instrument, Symbol, and Metaphor', presents an analysis of the mirror's instrumental, iconographic, and symbolic functions in a catoptric (or specular) sundial designed by French mathematician Emmanuel Maignan and painted on the vault of a gallery in the Roman palace of Cardinal Bernardino Spada in 1644. Feist examines the monumental sundial, as well as a wooden moondial (also designed by Maignan), along with Spada's collection of optical and scientific instruments, and considers the crucial roles of sun and mirror in the Latin epigrams on the walls of the palazzo's gallery. She shows that, in a literal sense, the mirror serves to capture the sun, but as the inscriptions written by Guillaume du Planté and Spada suggest, this symbolism also evokes the dawn of a new Golden Age founded by the cardinal who commissioned the sundial, a part of his self-fashioning as scientific patron and a bid for social advancement in the competitive context of the papal city.

Works Cited

Primary Sources

Boileau Despréaux, Nicolas, *Œuvres complètes*, intro. by Antoine Adam, ed. by Françoise Escal, Bibliothèque de la Pléiade (Paris: Gallimard, 1966)

The Medusa Reader, ed. by Marjorie Garber and Nancy J. Vickers, Culture Work Series (New York: Routledge, 2003)

Plato, *Republic*, ed. and trans. by Chris Emlyn-Jones and William Preddy, Loeb Classical Library, 2 vols (Cambridge, MA: Harvard University Press, 2013)

Vinci, Leonardo da, *Leonardo on Painting: An Anthology of Writings by Leonardo da Vinci with a Selection of Documents Relating to his Career as an Artist*, ed. by Martin Kemp, trans. by Martin Kemp and Margaret Walker (New Haven: Yale University Press, 1989)

Secondary Studies

Addey, Crystal, 'Mirrors and Divination: Catoptromancy, Oracles and Earth Goddesses in Antiquity', in *The Book of the Mirror: An Interdisciplinary Collection Exploring the Cultural Story of the Mirror*, ed. by Miranda Anderson (Newcastle: Cambridge Scholars, 2007), pp. 32–46

Akbari, Suzanne Conklin, *Seeing through the Veil: Optical Theory and Medieval Allegory* (Toronto: University of Toronto Press, 2004)

Anderson, Miranda, 'Early Modern Mirrors', in *The Book of the Mirror: An Interdisciplinary Collection Exploring the Cultural Story of the Mirror*, ed. by Miranda Anderson (Newcastle: Cambridge Scholars, 2007), pp. 105–20

Baltrušaitis, Jurgis, *Aberrations: Essai sur la légende des formes* (Paris: Flammarion, 1983)

——, *Anamorphic Art*, trans. by W. J. Strachan (New York: Abrams, 1976)

——, *Anamorphoses ou magie artificielle des effets merveilleux* (Paris: Perrin, 1969)

——, *Anamorphoses ou thaumaturgus opticus* (Paris: Flammarion, 1984)

——, *Le Miroir: Essai sur une légende scientifique: Révélations, science-fiction et fallacies* (Paris: Elmayan/Seuil, 1978)

——, *La Quête d'Isis: Essai sur la légende d'un mythe* (Paris: Flammarion, 1985)

Bartsch, Shadi, 'The Philosopher as Narcissus: Vision, Sexuality, and Self-Knowledge in Classical Antiquity', in *Visuality Before and Beyond the Renaissance*, ed. by Robert S. Nelson (Cambridge: Cambridge University Press, 2000), pp. 70–97

Beaujour, Michel, *Miroirs d'encre: Rhétorique de l'autoportrait* (Paris: Seuil, 1980)

Beecher, Donald, 'Windows on Contagion', in *Imagining Contagion in Early Modern Europe*, ed. by Claire L. Carlin (London: Palgrave Macmillan, 2005), pp. 32–46

Białostocki, Jan, 'Man and Mirror in Painting', in *Studies in Late Medieval and Renaissance Painting in Honor of Millard Meiss*, ed. by Irving Lavin and John Plummer, 2 vols (New York: New York University Press, 1977), I, 61–72

Boyle, Jen E. *Anamorphosis in Early Modern Literature: Mediation and Affect* (Burlington, VT: Ashgate, 2010)

Bradley, Ritamary, 'Backgrounds of the Title *Speculum* in Mediaeval Literature', *Speculum*, 29.1 (1954), 100–15

Brenkman, John, 'Narcissus in the Text', *Georgia Review*, 30.2 (1976), 293–327

Castillo, David R., *(A)wry Views: Anamorphosis, Cervantes, and the Early Picaresque* (West Lafayette, IN: Purdue University Press, 2001)

Cazes, Hélène, and others, eds, *Miroirs de la Mélancolie/Mirrors of Melancholy*, Les Collections de la République des Lettres (Paris: Hermann, 2015)

Colish, Marcia L., *The Mirror of Language: A Study in the Medieval Theory of Knowledge, Rev. edn* (Lincoln: University of Nebraska Press, 1983)

Cottrell, Robert D., 'Marguerite Porete's Heretical Discourse; Or, Deviating from the Model', *Modern Language Studies*, 21.1 (1991), 16–21

Dällenbach, Lucien, *The Mirror in the Text*, trans. by Jeremy Whiteley and Emma Hughes (Chicago: University of Chicago Press, 1989)

——, *Le Récit spéculaire: Essai sur la mise en abyme* (Paris: Seuil, 1977)

Dark, Rebecca, 'Reflections: Spenser, Elizabeth I, and Mirror Literature', *Early English Studies*, 1 (2009), 1–17

Darricau, Raymond, 'Miroirs des princes', in *Dictionnaire de spiritualité ascétique et mystique, doctrine et histoire*, ed. by Marc Viller and others, 22 vols (Paris: Beauchesne, 1937–95), x (1979–80), pp. 1303–12

Donaldson-Evans, Lance, *Love's Fatal Glance: A Study of Eye Imagery in the Poets of the 'École Lyonnaise'* (University, MS: Romance Monographs, 1980)

Dundas, Judith, '*Speculum Mentis*: Art as Mirror of the Mind', *Emblematica*, 16 (2008), 365–79

Eco, Umberto, *Semiotics and the Philosophy of Language* (Bloomington: Indiana University Press, 1986)

Edgerton, Samuel Y., *The Mirror, the Window, and the Telescope: How Renaissance Linear Perspective Changed Our Vision of the Universe* (Ithaca: Cornell University Press, 2009)

Einar Már Jónsson, *Le Miroir : Naissance d'un genre littéraire* (Paris: Belles Lettres, 1995)

Foucault, Michel, 'Technologies of the Self', in *Technologies of the Self: A Seminar with Michel Foucault*, ed. by Luther H. Martin, Huck Gutman, and Patrick H. Hutton (Amherst: University of Massachusetts Press, 1988), pp. 16–49

Frappier, Jean, 'Variations sur le thème du miroir de Bernard de Ventadour à Maurice Scève', *Cahiers de l'Association Internationale des Études Françaises*, 11 (1959), 134–58

Frelick, Nancy, 'Amitié et anamorphose chez Montaigne et Holbein', *Topiques, Études Satoriennes* 1 (2015): 14 pp < http://journals.uvic.ca/index.php/sator/article/view/11620> [accessed 25 March 2016]

——, 'Contagions of Love: Textual Transmission', in *Imagining Contagion in Early Modern Europe*, ed. by Claire L. Carlin (London: Palgrave Macmillan, 2005), pp. 47–62

——, 'Lacan, Courtly Love, and Anamorphosis', in *The Court Reconvenes: Courtly Literature across the Disciplines*, ed. by Barbara K. Altmann and Carleton W. Carroll (Woodbridge: Brewer, 2003), pp. 107–14

——, 'Lire en abyme: Les Emblèmes spéculaires de *Délie*', *textimage: revue d'étude du dialogue texte-image* (2012), <http://revue-textimage.com/conferencier/01_image_repetee/frelick1.html> [accessed 10 December 2015]

——, 'La Passion au miroir: Les Dizains spéculaires de *Délie*', in *Les Passions et leurs enjeux au seizième siècle*, ed. by Colette H. Winn (= *Renaissance and Reformation/ Renaissance et Réforme*, 38.3 (2015)), pp. 17–39
——, '*Per speculum in aenigmate*: An Introduction to *Le Miroir des melancholicques*', in *Miroirs de la Mélancolie/Mirrors of Melancholy*, ed. by Hélène Cazes and others, Les Collections de la République des lettres (Paris: Hermann, 2015), pp. 119–84
——, 'Sex, Lies, and Anamorphosis: Love as Transference in Scève's *Délie*', *Romanic Review*, 90.3 (1999), 301–16
Frontisi-Ducroux, Françoise, and Jean-Pierre Vernant, *Dans l'œil du miroir* (Paris: Odile Jacob, 1997)
Gilman, Ernest B., *The Curious Perspective: Literary and Pictorial Wit in the Seventeenth Century* (New Haven: Yale University Press, 1978)
Goldberg, Jonathan, *The Mirror and Man* (Charlottesville: University Press of Virginia, 1985)
Goldin, Frederick, *The Mirror of Narcissus in the Courtly Love Lyric* (Ithaca: Cornell University Press, 1967)
Grabes, Herbert, *The Mutable Glass: Mirror-Imagery in Titles and Texts of the Middle Ages and English Renaissance*, trans. by Gordon Collier (Cambridge: Cambridge University Press, 1982)
——, *Speculum, Mirror und Looking-Glass: Kontinuität und Originalität der Spiegelmetapher in den Buchtiteln des Mittelalters und der englischen Literatur des 13. bis 17. Jahrhunderts* (Tübingen: Niemeyer, 1973)
Greenblatt, Stephen, *Renaissance Self-Fashioning: From More to Shakespeare* (Chicago: University of Chicago Press, 1980)
Gregory, Richard L., *Mirrors in Mind* (Oxford: Freeman, 1997)
Haley, David, *Shakespeare's Courtly Mirror* (London: Associated University Press, 1993)
Hallyn, Fernand, 'Anamorphose et allégorie', *Revue de littérature comparée*, 56.3 (1982), 319–30
Hardie, Philip, *Ovid's Poetics of Illusion* (Cambridge: Cambridge University Press, 2002)
Havelange, Carl, *De l'œil et du monde: Une histoire du regard au seuil de la modernité* (Paris: Fayard, 1998)
Hugedé, Norbert, *La Métaphore du miroir dans les Épîtres de saint Paul aux Corinthiens* (Neuchâtel: Delachaux & Niestlé, 1957)
Jones, Ann Rosalind, *The Currency of Eros: Women's Love Lyric in Europe, 1540–1620* (Bloomington: Indiana University Press, 1990)
Kalas, Rayna, *Frame, Glass, Verse: The Technology of Poetic Invention in the English Renaissance* (Ithaca: Cornell University Press, 2007)
——, 'The Technology of Reflection: Renaissance Mirrors of Steel and Glass', *Journal of Medieval and Early Modern Studies*, 32.3 (2002), 519–42
Kauntze, Mark, 'Seeing through a Glass Darkly: The Interpretation of a Biblical Verse in Augustine of Hippo', in *The Book of the Mirror: An Interdisciplinary Collection Exploring the Cultural Story of the Mirror*, ed. by Miranda Anderson (Newcastle: Cambridge Scholars, 2007), pp. 60–69

Kay, Sarah, 'The *Roman de la Rose* and the Inverted Bouquet: Reflections of Love', in *Mythes à la cour, mythes pour la cour (Courtly Mythologies)*, ed. by Alain Corbellari and others, Publications romanes et françaises, 248 (Genève: Droz, 2010), pp. 295–310

Kelly, Philippa, 'Surpassing Glass: Shakespeare's Mirrors', *Early Modern Literary Studies*, 8.1 (2002), 21–32

Kodera, Sergius, *Disreputable Bodies: Magic, Medicine, and Gender in Renaissance Natural Philosophy* (Toronto: Centre for Reformation and Renaissance Studies, 2010)

——, 'Narcissus, Divine Gazes, and Bloody Mirrors: The Concept of Matter in Ficino', in *Marsilio Ficino: His Theology, his Philosophy, his Legacy*, ed. by Michael J. B. Allen and others (Leiden: Brill, 2002), pp. 285–306

Lindberg, David C., *Theories of Vision from Al-Kindi to Kepler* (Chicago: University of Chicago Press, 1976)

Margolin, Jean-Claude, 'Aspects du surréalisme au XVIe siècle: Fonction allégorique et vision anamorphotique', *Bibliothèque d'Humanisme et Renaissance*, 39 (1977), 503–30

——, 'Perspectivisme, relativisme et scepticisme: Précarité et créativité de l'Anamorphose' *Study francesi*, 138 (2002), 527–45

Massey, Lyle, *Picturing Space, Displacing Bodies: Anamorphosis in Early Modern Theories of Perspective* (University Park: Pennsylvania State University Press, 2007)

Melchior-Bonnet, Sabine, *Histoire du miroir* (Paris: Imago, 1994)

——, *The Mirror: A History*, trans. by Katharine H. Jewett, (London: Routledge, 2001)

Miller, Jonathan, *On Reflection* (London: National Gallery, 1998)

Neyrat, Yvonne, *L'Art et l'Autre: Le Miroir dans la peinture occidentale* (Paris: L'Harmattan, 1999)

Nolan, Edward Peter, *Now Through a Glass Darkly: Specular Images of Being and Knowing from Virgil to Chaucer* (Ann Arbor: University of Michigan Press, 1990)

Nouvet, Claire, 'An Allegorical Mirror: The Pool of Narcissus in Guillaume de Lorris' *Romance of the Rose*', *The Romanic Review*, 91.4 (2000), 353–74

——, 'An Impossible Response: The Disaster of Narcissus', *Yale French Studies*, 79 (1991), 103–34

Panofsky, Erwin, 'Jan Van Eyck's *Arnolfini* Portrait', *Burlington Magazine*, 64 (1934), 117–27

Pendergrast, Mark, *Mirror, Mirror: A History of the Human Love Affair with Reflection* (New York: Basic Books, 2003)

Santore, Cathy, 'The Tools of Venus', *Renaissance Studies*, 11.3 (1997), 179–207

Schmidt, Margot, 'Miroir', in *Dictionnaire de spiritualité ascétique et mystique, doctrine et histoire*, ed. by Marc Viller and others, 22 vols (Paris: Beauchesne, 1937–95), x (1979–80), 1290–1303

Schneider, Laurie, 'Mirrors in Art', *Psychoanalytic Inquiry*, 5.2 (1985), 283–324

Schwarz, Heinrich, 'The Mirror in Art', *Art Quarterly*, 15 (1952), 97–118

Shell, Marc, *Elizabeth's Glass* (Lincoln: University of Nebraska Press, 1993)

Shuger, Debora, 'The "I" of the Beholder: Renaissance Mirrors and the Reflexive Mind', in *Renaissance Culture and the Everyday*, ed. by Patricia Fumerton and Simon Hunt (Philadelphia: University of Pennsylvania Press, 1999), pp. 21–41

Sommers, Paula, '*Le Miroir de l'âme pécheresse* Revisited: Ordered Reflections on a Biblical Mirror', *Modern Language Studies*, 16.3 (1986), 101–08

Stafford, Barbara, and Frances Terpak, *Devices of Wonder: From the World in a Box to Images on a Screen* (Los Angeles: Getty Research Institute, 2001)

Starobinski, Jean, *The Living Eye*, trans. by Arthur Goldhammer (Cambridge, MA: Harvard University Press, 1989)

——, *L'Œil vivant* (Paris: Gallimard, 1968)

Stewart, Dana E., *The Arrow of Love: Optics, Gender, and Subjectivity in Medieval Love Poetry* (London: Associated University Presses, 2003)

Stoichita, Victor I., *L'Instauration du tableau: Métapeinture à l'aube des temps modernes*, 2nd rev. edn (Genève: Droz, 1999)

——, *The Self-Aware Image: An Insight into Early Modern Meta-Painting*, trans. by Anne-Marie Glasheen (Cambridge: Cambridge University Press, 1997)

——, *A Short History of the Shadow*, trans. by Anne-Marie Glasheen (London: Reaktion Books, 1997)

Summers, David, *The Judgment of Sense: Renaissance Naturalism and the Rise of Aesthetics* (Cambridge: Cambridge University Press, 1987)

——, *Vision, Reflection, and Desire in Western Painting* (Chapel Hill: University of North Carolina Press, 2007)

Taylor, Rabun, *The Moral Mirror of Roman Art* (Cambridge: Cambridge University Press, 2008)

Torti, Anna, *The Glass of Form: Mirroring Structures from Chaucer to Skelton* (Woodbridge: Brewer, 1991)

Vinge, Louise, *The Narcissus Theme in Western European Literature up to the Early 19th Century* (Lund: Gleerups, 1967)

Wack, Mary Frances, *Lovesickness in the Middle Ages: The 'Viaticum' and its Commentaries* (Philadelphia: University of Pennsylvania Press, 1990)

Wimsatt, James I., *Allegory and Mirror: Tradition and Structure in Middle English Literature* (New York: Pegasus, 1970)

Aristotle's 'Bloody Mirror' and Natural Science in Medieval and Early Modern Europe

Berthold Hub

Aristotle, De insomniis

The motif of the 'bloody mirror' — the staining of a mirror by the gaze of a menstruating woman — was first introduced by Aristotle in his *De insomniis* in the course of his explanation on the way dreams arise. The author illustrates the notion that the eye does not merely suffer passively but also exercises an active effect. Moreover, he shows that the eye or the soul perceives at lightning speed and is thus receptive even to very small differences and changes in sensory reality.[1] While the phenomena previously mentioned in the text illustrate the way strong and intense perceptions persist, the following discussion of the sensitivity of the sensory organs to small differences is intended to demonstrate the same persistence in the case of weaker sense perceptions:

[1] This is the explicit explanation given in Aristotle, *De insomniis*, 459 b 24–28 and 460 a 24–27. The question of whether the relevant passage in *De insomniis*, or even the entire text, is not actually by Aristotle need not concern us here, since our focus is on the history of its reception, during which its authenticity was never questioned even into the twentieth century. On this issue, see the commentaries listed in the following note. The contradiction to the intromission theory of vision found in Aristotle's *De anima* and *De sensu* does not constitute an objection to the authenticity of this passage, since Aristotle also presents the extramission theory in *Meteorologica*, *De caelo*, and *De generatione animalium*, the authenticity of which is undisputed. See below.

Berthold Hub (berthold.hub@univie.ac.at) is Assistant Professor in the Department of Art History at the University of Vienna and a Fellow at the Institute of Art History in Florence. His postdoctoral thesis examines the Renaissance architect Filarete and his *Libro architettonico*. Other interests include ancient aesthetics, the 'performative gaze', and Neoplatonism.

The Mirror in Medieval and Early Modern Culture: Specular Reflections, ed. by Nancy M. Frelick, CURSOR 25 (Turnhout: Brepols, 2016), pp. 31–71 10.1484/M.CURSOR-EB.5.111041
BREPOLS PUBLISHERS

> For in the case of very bright mirrors, when women during their menstrual periods look into the mirror, the surface of the mirror becomes a sort of bloodshot cloud; and if the mirror is new, it is not easy to wipe off such a stain, while if it is old it is easier. The cause is, as we said, that the eye is not only affected by the air but also has an effect upon it and moves it — as bright objects do [...]. Now it is reasonable that the eyes, like any other part whatsoever, should be affected during the menstrual period; for they are veined by nature. That is why, when the menstrual discharges occur because of a disturbance and bloody inflammation, the change in the eyes is not evident to us although it is present [...] and the air is moved by them, and has a certain effect on the air on the surface of the mirror which is continuous with it, i.e., it makes that air affected in the same way that it is itself; and the air on the mirror affects the surface of the mirror.
>
> As in the case of clothes, the purest mirrors are most quickly stained; for a pure mirror shows accurately whatever it receives, and the purest shows the smallest movements. The bronze, because it is smooth, perceives best any touch at all (one should think of the touch of the air as a sort of rubbing — like a wiping or a washing), and because it is pure the touch, however slight it may be, becomes apparent in it.[2]

Aristotle introduces the 'bloody mirror' as an empirical fact, as known and evident to everybody: one may not know how dreaming works, but everyone knows why and how the mirror becomes bloody. However, it is by no means clear how this example is intended to work. Aristotle appears to be offering what happens to the mirror as an illustration of what happens with the perceptual organ. In other words, what happens to mirrors when a menstruating woman looks at them is an illustration of what happens in the sensory organ when it receives sensory impressions. The staining of the mirror would then be comparable with the reception of the material form of the objects. The persistence of the stain on the surface of the mirror (even when the woman is no longer looking at it) would correspond to the persistence of the sensory impressions in the eyes or soul (even when the objects are no longer present). The remark about the penetration of the stain into the depth of the mirror might refer back to the effect of sensory impressions on the sensory organs.

[2] Aristotle, *De insomniis*, 459 b 24–460 a 32 (459 b 28–460 a 18); *The Complete Works of Aristotle*, ed. by Barnes, I, 73. See in particular the detailed commentary on this passage in Aristotle, *De insomniis*, ed. by van der Eijk, pp. 167–93, as well as: Preus, 'On Dreams 2'; Sprague, 'Aristotle on Red Mirrors'; Lefebvre, 'Le Miroir de l'âme'. Cf. also Rakoszy, *Böser Blick, Macht des Auges und Neid der Götter*, pp. 134–40, and Kodera, *Disreputable Bodies*, pp. 105–11. I have unfortunately not been able to consult Shankman, *Aristotle's 'De insomniis': A Commentary*.

Aristotle appears to be explaining how we receive the material of our dreams. The objects of sensory perception are often very small (or are only seen for a very short time, or lie outside of our focus), penetrating our perceptual organs so quickly that we often do not even perceive them consciously. But they are nevertheless impressed as images upon the soul, where they now provide the material from which our dreams are then formed.

Be that as it may, the only aspect that is of interest to us in this passage is what it says about the process of vision. And in this we are accurately reflecting the entire history of the reception of this passage to the end of the early modern period, during which time, without exception, the 'bloody mirror' was always torn out of its original context in the theory of dreams.[3] Due to their altered quality during menstruation, the eyes also have an effect on the air and cause movement, and the air — which is an uninterrupted medium between the eye and the mirror — transmits this movement to the surface of the mirror and affects it in the manner described. The manner in which this transmission of movement in air takes place, and more importantly the means by which this type of air-infection is capable of transporting blood particles across large distances, is not clear. But Aristotle insists that the staining of the mirror only takes place when a woman 'looks into the mirror' — assuming a strong and precise directional process. He also expresses the materiality of the process unambiguously in several ways: everyone can see the bloodstain, not only the woman who is looking at the mirror. The stain persists even when the woman is no longer looking at the mirror, and it is not easy to remove. Finally, the stain on the surface of the mirror is compared with the staining of clothes, and the residues that arise when one cleans the surface with a wet cloth. In addition, Aristotle uses the word 'touch' as many as three times in this connection (ἁφή, which always stands for physical contact).

Theories of Vision

Aristotle contradicts his own theory of vision, as expounded in *De anima* and in *De sensu*, where he maintains that the process that triggers perception consists solely of a mediation of the colours of an object *to the eye* through the

[3] This was already the case with the 'bloody mirror' in Pliny, *Naturalis Historia*, VII. xv. 64 and XXVIII. xxiii. 82 (without reference); Solinus, *Collectanea rerum memorabilium*, I. 58: *C. Ivlii Solini collectanea rerum memorabilium*, ed. by Mommsen, p. 18 (oddly enough with a reference to Democritus); and Proclus, *In Platonis Rem publicam commentarii*, XII. I. 290: Proclus, *Commentaire sur la République*, ed. by Festugière, II, 99 (with reference to Aristotle).

intervening medium.[4] In his polemical debate on the views of Democritus, Empedocles, and Plato, Aristotle rejects every materialistic conception of the perceptual process and explicitly denies that vision can be based on an emanation (of fire, light, or atoms) from the eye.[5]

However, the explanation of the 'bloody mirror' in *De insomniis* is not the only passage in which Aristotle deviates from his own theory of vision and assumes some form of extramission issuing from the eyes. He also does so repeatedly in the *Meteorologica* — for example, when explaining the phenomenon of the rainbow. Here he clearly speaks of the reflection and refraction of the *visual ray*.[6] And he again uses a peculiar example to explain this to his audience: the visual power of an old man may decrease to such an extent that his visual ray is sent forth so weakly that it is not able to push the air aside and is finally reflected by the air, so that the poor old man sees nothing but himself — another specular phenomenon with a long tradition.[7] Apparently, Aristotle turns to the extramis-

[4] Aristotle, *De Anima* II. 6–7 418 a–419 a; *De Sensu* II.

[5] Aristotle, *De Sensu* II. 437 a 19–438 b 16; cf. IV. 442 a 33–442 b 10.

[6] Aristotle, *Meteorologica*, III. 4. 373 a 35–375 b 15. A rainbow, he writes, arises due to the reflection of our visual ray from droplets of moisture and towards the sun. As a result of this 'refraction', the visual ray is 'weakened'; the various colours correspond to the degree of weakening. Cf. I. 6. 343 a 20; II. 9. 370 a 18–20; III. 2. 372 a 17–32; III. 3. 372 b 16–19; III. 6. 377 a 29–378 a 12. Cf. also *De Caelo*, II. 8. 290 a 17–24, and *De generatione animalium*, V. 2. 780 b 33–781 a 13.

[7] *Meterologica*, III. 4. 373 b 5–10. In addition to the numerous commentaries on Aristotle's meteorology — e.g., Albertus Magnus, *Meteora*, lib. III, tract. 4, cap. 13: Albertus Magnus, *Meteora*, ed. by Hossfeld, p. 189; Thomas Aquinas, *Expositio Meteorologicorum*, lib. III, lect. 5: *S. Thomae Aquinatis in Aristotelis Libros*, ed. by Spiazzi, pp. 624–27; and Nicole Oresme, *Quaestiones super quatuor libros meteorum*, lib. III, qu. 15 and 19: 'Nicole Oresme on Light, Color, and the Rainbow', ed. and trans. by McCluskey, pp. 218–21, 228–29 and 248–51. Additional, less well-known examples are cited in *Le 'Quaestiones de sensu'*, ed. by Agrimi, pp. 32–33 and 77 — see for example Seneca (*Quaestiones naturales*, I. 3. 7), Alkindi (*De aspectibus*, prop. 8), Albertus Magnus (*De homine*, Appendix ad Quaestionem 22; *De anima*, tract. 1, cap. 15), Roger Bacon (*Opus maius*, V, pars III, dist. 1, chap. 5), John Pecham (*Tractatus de Perspectiva*, cap. 4), Nicole Oresme (*De causis mirabilium*, I. 78, I. 108, and I. 179), Conrad of Megenberg (*Buch der Natur*, II. 30). Later examples that might be mentioned include Agrippa von Nettesheim, *De occulta philosophia*, I. 6; Ludwig Lavater, *De spectris, lemuribus et magnis*, I. 3, p. 16. The name of Antipheron (or Antiphon) that is often mentioned in this context can be traced back to the commentary on the *Meteorology* by Alexander of Aphrodisias, which was translated into Latin by William of Moerbeke in 1260; Alexandre d'Aphrodisias, *Commentaire sur les Météores d'Aristote*, ed. by Smet, p. 233. Cf. 'Nicole Oresme on Light, Color, and the Rainbow', ed. and trans. by McCluskey, pp. 429–30, n. 18.

sion theory of vision, with the material visual ray, when he is faced with phenomena that he is not able to explain using his own intromission theory.

The ancient extramission theory of vision included not only the visual-ray cone of mathematicians such as Euclid and Ptolemy, but also the fiery rays of the Pythagoreans, Plato's *synaugia* (radiating together) of inner and outer fire, and the visual ray consisting of a mixture of *pneuma* and air in the Stoic tradition, which was adopted by Galen's school of medicine. All of these theories assume that vision is a haptic process — that is, that vision requires physical contact between the viewer and the object being viewed. First, the visual ray that is involved, streaming out from the eyes, is of a material nature and extends spatially in straight lines until it is is reflected or deflected by an obstacle such as a mirror, or interrupted or refracted by an obstacle such as water. These are all characteristics that can only apply to a physical object. When the reflection of the visual ray is compared with the rebounding of a projectile, as it was from Heron of Alexandria to Ptolemy, then it must certainly be taken literally.[8] Second, the visual ray is an ephemeral organ that is capable of sensation, of feeling the objects in the visual field even at the greatest distance and extension. Thus, as early as Hipparchus there is an explicit comparison with the hand: the ends of the rays emitted by the eyes touch the bodies in the external world in the way that hands do, and then return to the eye.[9] Apuleius reports the assumption by numerous philosophers that 'our visual rays [...] as soon as they fall on a solid, shining, and smooth body, spring back at the same angle that they enter to the face from which they emanated, and in this way depict what they touch and see externally within the mirror'.[10] This theory of visual rays emanating from the eye, equipped with sensitivity, reaching out to objects and touching them, was the most influential theory of vision throughout antiquity.[11]

[8] Heron of Alexandria, *Catoptrics*, 2–3; Ptolemy, *Optics*, II. 20 and III. 19; also III. 22–64; cf., e.g., Pseudo-Aristotle, *Problemata*, XVI. 13.

[9] Aetius, *Placita philosophorum*, IV. 13. 9–10; Diels and Kranz, *Die Fragmente der Vorsokratiker*, 28 A 48. Cf., e.g., Pseudo-Aristotle, *Problemata*, III. 10; *Alexandri Aphrodisiensis praeter commentaria scripta minora, De anima liber cum mantissa*, ed. by Ivo Bruns, p. 130; Ptolemy, *Optics*, II. 13 (cf. in particular II. 67 and also II. 20, III. 8–12, 19, 22, 64, etc.); Galen, *De placitis Hippocratis et Platonis*, VII. 5 and 7; Plotinus, *Enneads*, II. 8 and IV. 5.

[10] Apuleius, *Apologia*, 15; Apuleius of Madauros, *Pro se de magia (Apologia)*, ed. by Hunink, I, 46–47; my translation. Cf. Ptolemy, *Optics*, III. 3; Pseudo-Aristotle, *Problemata*, XVI. 13; Seneca, *Naturales Quaestiones*, I. 13. 2 and I. 15. 7.

[11] Aristotle's theory of vision, as expounded in his *De anima* and in *De sensu*, is the only ancient theory of vision that does not presuppose physical contact for vision to occur, and this

It formed the framework for notions of sight, of the process and different ways of seeing, and of the mirror, which all survived well beyond antiquity.

Performative Gaze and Humoral Pathology

By 'different ways of seeing', I mean the wide range of what I shall call the 'performative' gaze, the powerful and effective eyebeam, which is ubiquitous in ancient, medieval, and early modern texts: the gaze of the gods, the gaze of love, the evil eye, the transmission of sickness through the eye, and so forth.[12] Of particular interest in connection with the 'bloody mirror' is the deadly gaze of the basilisk (which is killed when a mirror is placed in front of it). In antiquity, the two motifs only occur separately,[13] but in the Middle Ages and early modern period, the two motifs were to be inseparably linked.[14]

comes closest to today's interpretation of the process of vision. However, with the exception of a few of his pupils, this Aristotelian theory remained without adherents throughout antiquity and the early Middle Ages. The third type of theory of vision in antiquity, which assumes the reception of material doubles that emanate from the object, also remained restricted — with few exceptions — to a single school, that of the Atomists. On theories of vision in antiquity in general, see Lindberg, *Theories of Vision*, pp. 1–17; Simon, *Le Regard, l'être et l'apparence*; Hub, *Die Perspektive der Antike*, pp. 264–321. It was usually the passages in which Aristotle assumes the existence of a visual ray that were followed. This was already the case in Theophrastus, *De sensu et sensibilibus*, 18, and *De vertigo*, 8. The pseudo-Aristotelian *Problemata* also assumed the existence of visual rays (see in particular III. 10, XI. 58, XV. 6, XV. 7, XVI. 13, XXV. 9, XXXI. 7, 8, 15–16, 19–21, and 25). It was not least due to this ambivalence in Aristotle himself and his followers that later authors, who preferred a combination of extramission and intromission or pure extramission, were able to appeal to Aristotle with a clear conscience. In addition, a mistranslation by Michael Scot led to Aristotle appearing to defend the extramission theory even in *De generatione animalium*, V. 1. 781 a 1–2 and 781 b 2–13. See Aristotle, *De animalibus*, ed. by van Oppenraaij, bk XIX, 220–21.

12 On this point, see in particular Rakoszy, *Böser Blick, Macht des Auges und Neid der Götter*.

13 Pliny, *Naturalis Historia*, VIII. xxxiii. 78 and XXIX. xix. 66; cf. above, note 3. The basilisk's deadly and reflected gaze will only be discussed below when it appears in connection with the motif of the 'bloody mirror'. The most comprehensive discussion of the basilisk tradition is provided by Sammer, *Der Basilisk*, pp. 103–31; cf. Breiner, 'The Basilisk'. For the sixteenth and seventeenth centuries, cf. Havelange, *De l'oeil et du monde*, pp. 49–56, and also Sir Thomas Browne, *Pseudodoxia Epidemica*, ed. by Robbins, I, 181–86 and commentary in II, 807–11.

14 The earliest combination of the basilisk and the 'bloody mirror' that I have identified is in Alexander Neckham. In his natural science anthology *De naturis rerum*, written in 1190, the deadly gaze of the basilisk appears as an argument in a presentation of the Platonic and Galenic visual ray theory (cap. 153, 'De visu'; Alexander Neckam, *De naturis rerum*, ed. by Wright, pp. 234–38, 237), while — as in Pliny — the soiling of the mirror by the gaze of a

For the gaze to be conceived of as performative, however, the general, basic model of a material visual ray or eyebeam needs to be extended using two possible explanations of its special mode of action. First, the performing faculty of the eye can have a quantitative cause — that is, an excess of emotion, a surplus of pressure, so to speak. In certain situations, more or stronger rays are emitted. More, for example, when we raise the eyebrows in admiration and praise, or when we glare with anger. Stronger, for example, when the soul is excited. This intensifies a person's eyebeam, which may even damage or dry up the 'vital spirit' of the person touched by the visual rays.

Second, the performing faculty of the eye can have a qualitative cause — that is, a *dyscrasia*, an imbalance of elements and humours in the human body, which may contain either too much or too little of a particular substance and may therefore transmit its own 'corruption' or 'putrefaction' to the person seen, 'infecting' him or her. This disturbance of the humoral balance may be caused physiologically (e.g., by fever, plague, or other disease, and through aging or menstruation) or psychophysiologically (e.g., by hatred, envy, love, etc.). Ultimately, this extremely influential idea in Hippocratic and Galenic humoral pathology (consistent with the Aristotelian theory of matter) is also a quantitative argument, since the disturbed humoral balance is a quantitative imbalance among the elements and properties.

The idea that the basilisk has a deadly, poisonous gaze follows sometimes one explanation, sometimes the other, and sometimes both. The basilisk's eyes emit rays that are so strong that they first destroy a human being's eyesight, the *spiritus visibilis*, and then the rest of his or her *spiritus*, ultimately leading to death. A particularly poisonous mixture of humours may be present as well, infecting the person being gazed upon. Aristotle is familiar with both models of corruption as well, however. On one hand, his story of the old man whose visual ray is so weak that it is reflected back by the air, as if by a mirror, is an example of a quantitative alteration of the visual ray. Aristotle's story of the 'bloody mirror', on the other hand, is a clear example of a qualitative cause — to which, of course, the quantitative can always give additional performative power.

The above outlines the context in which the Aristotelian passage functions. Women are cold and moist and are therefore unable to fully digest food and drink.

menstruating woman is found in a separate chapter concerned among other things with the woman as *sola animal menstruale* (cap. 156, 'Iterum de homine', pp. 248–52, 252). By contrast, in the encyclopaedic didactic poem *De laudibus divinae sapientiae*, probably written shortly afterwards, the two phenomena are for the first time mentioned directly one after the other (dist. 9, vv. 395–96 and 397–98, p. 495).

When they are pregnant, the foetus is nourished by the resulting surplus. If they are not pregnant, the excess of humours leads to contamination of the metabolism and has to be excreted as waste. This also takes place through the eyes. As the eyes are particularly transparent and rich in veins, the menstrual blood creating contamination throughout the body can emerge through them easily.

It is important to note here that the staining of the mirror is just one instance of a process that is always taking place. When we look at objects, a material ray is emitted by the eyes that extends outward and touches them. Normally, this process is not perceptible, since the ray that is emitted is extremely fine. In the case of a particularly soiled gaze, however, as in the case of a menstruating woman, the gaze becomes so saturated with matter that it can leave a bloodstain on a particularly clean and smooth surface, such as a mirror.

Christian Superstition

Aristotle is not offering a distanced description of a popular belief; instead, he is providing a natural science explanation of a purportedly empirically proven phenomenon with which everyone is familiar. As long as the theory of the visual ray was accepted, the performativity of the gaze had to be accepted as an indisputable natural phenomenon as well. It is firmly grounded in ancient scientific theory.

The performative gaze only became a matter of 'superstition' with the rise of Christianity. The early Fathers of the Church and ecclesiastical authors who were concerned with this topic also saw the cause of the evil eye as lying in a soul corrupted by disease or hatred, envy, jealousy, or vanity; but they *partly* linked this view with the genuinely Christian doctrine that the devil was involved.[15] The person possessing the evil eye was thought to be possessed by a demon, or — in a further development of the theory — to have made a pact with the devil. Christianity did not therefore bring an end to the ancient pagan 'superstition' about the evil eye. It would have had to abolish the basis for that view in natural science — that is, the extramission of a visual ray — but that was not an option. It is presupposed in numerous passages in the Bible, and in a well-known passage in the New Testament in particular, where St Paul explicitly assumes the

[15] See in particular Rakoszy, *Böser Blick, Macht des Auges und Neid der Götter*, pp. 216–26; Limberis, 'The Eye Infected by Evil'; Leyerle, 'John Chrisostom on the Gaze'; and Dickie, 'The Fathers of the Church and the Evil Eye'. Dickie's discussion, however, is marred by the fact that he does not take seriously the theory of vision, which is the only explanation for the process, and instead assumes 'pagan superstition'.

existence of the evil eye.[16] Indeed, the natural explanation only had a supernatural one *added* to it, since the basis for the processes was still the ancient theory of the visual ray. Even demons and the devil had to use this mechanism in order to be able to bring about evil. It is only a small step from a position like this one to the later witch-hunting craze. However, as we shall see, the fact that the supernatural explanation is only added to the natural one, and in no sense replaces it, makes it possible to use the concept of the performative gaze as an argument *against* witch-hunting as well.

In any case, the extramission theory of vision remained unchallenged for the first millennium of the Christian era.[17] It was only around the year 1000 that serious competition for the theory arose among the Arabic Aristotelians. Although Al-Kindi continued to defend the visual ray theory, Alhazen, Avicenna, and others for the first time worked on the assumption that light rays are received by the *spiritus* or *pneuma* within the eye. In doing so, they combined the Aristotelian theory of reception with Galen's visual *pneuma* and with Euclid's and Ptolemy's optical and geometric concepts, thus retaining the notion of the visual pyramid but reversing the direction of the rays.[18]

Perspectivists

Towards the end of the twelfth century, and particularly in the thirteenth century, Arabic optical science also became known in the West, where it was adopted by writers such as Bartholomaeus Anglicus, Robert Grosseteste, William of Auvergne, Albertus Magnus, Roger Bacon, John Pecham, Witelo, and Theodoric (Dietrich) of Freiberg.[19] Unlike Alhazen, however, the so-called perspectivists did not dismiss the extramission theory of vision altogether. Their new paradigm of intromitted *species* did *not* lead to the disavowal of the

[16] Galatians 3.1: 'O foolish Galatians, who hath bewitched you, that you should not obey the truth, before whose eyes Jesus Christ hath been evidently set forth, crucified among you?' (King James Bible); the Vulgate version reads, 'O insensati Galatae | *quis vos fascinavit* | *ante quorum oculos* Iesus Christus proscriptus est crucifixus' (my italics). See, for example, Betz, 'Matthew vi.22f and Ancient Greek Theories of Vision'; Allison, 'The Eye as a Lamp'; Marcus, 'A Note on Markan Optics'.

[17] Schleusener-Eichholz, *Das Auge im Mittelalter*, I, 51–79; Nelson, 'To Say and to See', pp. 150–56; Akbari, *Seeing through the Veil*, pp. 21–44; Hub, 'Visual Piety and Visual Theory'.

[18] Lindberg, *Theories of Vision*, pp. 18–86. This geometrical equivalence allowed many theorists of optics after Alhazen to retain the extramission theory, or a mixture of the two.

[19] Lindberg, *Theories of Vision*, pp. 87–132.

extramission theory, and least of all to the abandonment of its implications. The perspectivists joined Alhazen in positing 'intromitted' *species* as being necessary for vision and its principal cause. Yet they were convinced that our visual sense is not only passive but also active, and this led to a mixed model or compromise between intromission and extramission.

The perspectivists' concessions to the visual ray theory were closely associated with the Neoplatonic metaphysics of light in general, and the doctrine of *species* in particular. The doctrine of *species*, which had been developed by Robert Grosseteste from the Neoplatonic conception of emanation, maintained that every natural object propagates power from itself to surrounding bodies.[20] These *corporeal* entities, known as *species*, are responsible for all efficient causality in the universe.[21] Grosseteste's view was unreservedly adopted by Roger Bacon and John Pecham and applied to Alhazen's optics, since rays of light received by the eye were only a special case of *species* — that is, visible *species*. However, if *every* natural body emits *corporeal species*, then the same must also apply to the eye itself, which also propagates some power from itself to surrounding bodies — that is, visual *species*. This was precisely the point made by all of the perspectivists, although with different degrees of emphasis.[22]

There were many reasons for the continued adherence to the existence of extramitted visual rays. The Neoplatonic basis has already been mentioned. The perspectivists were clearly making an effort to harmonize the newly available

[20] See, for example, Robert Grosseteste, *Die philosophischen Werke*, ed. by Baur, pp. 76–109; Crombi, *Robert Grosseteste and the Origins of Experimental Science*, pp. 104–16 and 128–31. Cf. Lindberg, *Theories of Vision*, pp. 94–102; and Lindberg's introduction to *Roger Bacon's Philosophy of Nature*, pp. xxxv–lxxv.

[21] The corporeality of *species* is most explicitly stated in Roger Bacon's *Opus maius*, v, pars I, dist. 6, cap. 3–4: *Roger Bacon and the Origins of 'Perspectiva'*, ed. by Lindberg, pp. 194–205; and in his *De multiplicatione specierum*, III. 2–3, in *Roger Bacon's Philosophy of Nature*, ed. by Lindberg, pp. 190–205: 'I therefore state unconditionally that the species of a corporeal thing is truly corporeal and that it has truly corporeal being' (*ideo absolute diffinio quod species rei corporalis est vere corporalis et habet esse vere corporale*), p. 190.

[22] Robert Grosseteste, *De iride*, in *Die philosophischen Werke*, ed. by Baur, pp. 72–73. Roger Bacon, *Opus maius*, V (*Perspectiva*), particularly pars I, dist. 7, cap. 2–4; *Roger Bacon and the Origins of 'Perspectiva'*, ed. by Lindberg, pp. 100–07. See also pars I, dist. 5, cap. 1, I. 9. 1, I. 10. 2, II. 1. 1, II. 1. 3, II. 2. 1, III. 1. 5. Cf. Lindberg's introduction, pp. lxxxiii–lxxxvi. Similarly, Roger Bacon, *De multiplicatione specierum*, pars I, cap. 2, in *Roger Bacon's Philosophy of Nature*, ed. by Lindberg, pp. 30–33. John Pecham, *Tractatus de Perspectiva*, ed. by Lindberg, pp. 36–42. John Pecham, *Perspectiva communis*, prop. I. 46, in *John Pecham and the Science of Optics*, ed. by Lindberg, pp. 128–31; see also the Introduction, pp. 34–37. Cf. also Batholomaeus Anglicus, *De proprietatibus rerum*, III. 17.

accounts by Alhazen, Avicenna, and Averroes with those of their own tradition of extramission theories. In addition, a need to preserve the subject as playing an active part in perception (voluntary and thus responsible before God) may also have played a role. According to this outlook, conceiving of a purely receptive eye would have implied abandoning a natural science explanation of the performative gaze and thus of the performativity of the gaze itself. All of the authors mentioned were familiar with the principle and confirm it both in general terms, and specifically through the 'bloody mirror' and the deadly gaze of the basilisk, using the two motifs as proofs of the materiality of the visual *species*, or visual ray.[23]

De oculo morali

Peter of Limoges's widely published and highly influential work *De oculo morali*, written in Paris between 1275 and 1289, provides an example of the influence of the perspectivists' compromise model and is also an example of the shift to a purely extramissive model when phenomena such as the staining of mirrors by the gaze of a menstruating woman were to be explained to a wider audience. The Franciscan scholar had studied Ptolemy, Constantinus Africanus, Alhazen, and many other 'doctores speculativae' and 'auctores perspectivae', and had evidently read Roger Bacon and Roger Pecham in order to allegorize the latest discoveries in natural science for the moral instruction of the faithful. In a chapter headed 'That a Woman's Eyes Are the Darts of Unchasteness by which Many Are Wounded', Peter of Limoges warns his listeners and readers against women and their gaze in the following words:

> The basilisk kills birds flying past simply by looking at them; so, too, the appearance of a woman sometimes drags heavenly men down to the depths of lust and thus kills them spiritually [...]. Although the basilisk is a short serpent, it kills many by its sight; thus, although something might seem to be moderate in appearance, nevertheless it wounds and kills many. Namely, it seems probable that when a woman looks at a man lustfully, the lustful vapor emanating from her heart rises

[23] Robert Grosseteste, *Expositio in Epistolam Sancti Pauli ad Galatas*, III. 3, in *Opera Roberti Grosseteste*, ed. by McEvoy and others, pp. 72–74 ('bloody mirror' and basilisk). Roger Bacon, *Opus maius* IV, in *The 'Opus Majus' of Roger Bacon*, ed. by Bridges, I, 398–99, and *The Opus Majus of Roger Bacon*, trans. by Burke, I, 413–14 ('bloody mirror'). Cf. *Epistola Fratris Rogerii Baconis de secretis operibus artis et naturae, et de nullitate magiae*, cap. 3, in *Opera quaedam hactenus inedita*, ed. by Brewer, pp. 528–32 (529: 'bloody mirror' and basilisk); cf. cap. 5, p. 535 (basilisk). John Pecham, *Tractatus de Perspectiva*, ed. by Lindberg, pp. 36–42 ('bloody mirror' and basilisk).

> to her eyes, *infecting the woman's visual rays*. Thus infected, the rays travel to the man's eyes (*assuming that we see by extramission*) and when they arrive at the man's eyes and infect them, the infection passes to the man's heart, just as it proceeded from the woman's heart first [...]. And according to Aristotle, when a menstruating woman looks in a mirror, she stains and infects it.[24]

Albertus Magnus

The extensive works of the perspectivists' slightly younger contemporary Albertus Magnus are characterized by a similarly strong ambivalence. He did not link intromission and extramission into a combination as his predecessors had, but rather developed independently of them from a clearly Platonic standpoint of extramission towards an Arabic–Aristotelian position of pure intromission,[25] according to which the eye was exclusively a receptive organ, and vision only took place through the penetration of reflected light rays.[26] At the same time, however, a need to cling to the active role of the human eye and its performative power is also evident, again leading to marked ambivalences in the presentation of the visual process in numerous places in his work. As with the feared evil eye, it was simply impossible to push aside the facts, attested to by countless authorities, that basilisks can kill with their gaze and that menstruating women can stain a mirror with blood through their gaze alone.

The 'bloody mirror', together with the deadly gaze of the basilisk, thus appears suddenly in Albert's *De anima*, which otherwise radically denies that

[24] Peter of Limoges, *De oculo morali*, VIII. 7, in Peter of Limoges, *The Moral Treatise on the Eye*, ed. by Newhauser, pp. 99–100 (my italics).

[25] Albertus Magnus, *De resurrectione*, tract. 4, qu. 1, art. 9 ('De visione per speciem'), ed. by Wilhelm Kübel, in *De Sacramentis. De incarnatione. De resurrectione*, ed. by Ohlmeyer, pp. 326–31. Cf. Anzulewicz, 'Perspektive und Raumvorstellung in den Frühwerken des Albertus Magnus', pp. 255–56.

[26] Albertus Magnus takes this radical turn in *De homine*, qu. 19–22 ('De visu'), and in the Appendix ad Quaestionem 22 ('Utrum visus sit per emissionem radiorum vel per immutationem oculi a medio et medii ab obiecto visibili?'); *De homine*, ed. by Anzulewicz and Söder, pp. 145–202. Cf. Anzulewicz, pp. 258–66. Albertus Magnus repeats the position taken in *De homine* in several passages in his other writings, particularly in *De anima* and *De sensu et sensato*, ostensibly both commentaries on Aristotle's treatises of the same title. See in particular *De anima*, lib. II, tract. 3, cap. 6–16, and *De sensu et sensato*, tract. 1, cap. 7. Theiss, *Die Wahrnehmungspsychologie und Sinnesphysiologie des Albertus Magnus*; Albertus Magnus, 'Optics in Albert the Great's "De sensu et sensato"', ed. and trans. Akdogan. Cf. Albertus Magnus, *Albert's Refutation of the Extramission Theory of Vision*, ed. and trans. by Akdogan; Theiss and Grüsser, 'Vision and Cognition in the Natural Philosophy of Albert the Great'.

anything can emanate from the eye.[27] In the *Summae de Creaturis*, by contrast, the 'bloody mirror' is used deliberately in a more detailed discussion of the question of whether the sense organs are only passive in receiving the *species* from the perceived object, or whether they are also active.[28] The fact, attested to by Aristotle, that the gaze of women stains mirrors with blood when they are menstruating is used by Albertus here as an argument to support the idea that vision is also an active process, and that human beings can therefore exert an effect through their gaze on the object they are looking at. Albertus Magnus explains that this is not always the case by introducing the interesting distinction that although the eye is passive *per se* in receiving reflected light rays, it can become active *per accidens* if it emits 'warm vapours' (*vapores calidi*) and 'subtle spirits' (*spiritus subtiles*) — normally an unimportant process that cannot be observed, but which in a special disposition or indisposition of the body can lead to a transference of that disposition to other bodies. In such cases, the objects viewed can be 'infected' (*inficiunt*); and in the special case of menstrual blood, the process can even become visibly materialized: the (distilled) menstrual blood flows through the blood vessels and reaches the eye. Since the eye is particularly rich in blood vessels, the vapours constantly emitted from the eyes are heavily infected with the blood, so that this infection can reach other bodies in the immediate vicinity — as everyone can see in the case of the mirror.

In *De animalibus*, finally, in addition to the motif of the 'bloody mirror', the deadly gaze of the basilisk is also attributed to a poisonous *spiritus visivus* that flows out from the eyes. In this case, however, *radii oculorum* are explicitly excluded as the pathway of mediation:

> It [the basilisk] also kills with its gaze, for everything on which its gaze falls dies [...]. Neither is the reason that it kills with its gaze, as some say, the fact that rays come forth from its eyes and destroy whatever they fall upon. For it is not the opinion of the natural scientists that rays come forth from the eyes. Rather, the reason for the destruction is that the visual spirit (*spiritus visivus*) is dispersed very far due to the thinness of its substance. It is this that destroys and kills everything.[29]

[27] Albertus Magnus, *De anima*, lib. II, tract. 3, cap. 1; ed. by Stroick, pp. 97–98; Cf. Theiss, *Die Wahrnehmungspsychologie und Sinnesphysiologie des Albertus Magnus*, pp. 52–53.

[28] Albertus Magnus, *Summae de Creaturis*, pars II, qu. 45, art. 4; in *Secunda Pars summae de creaturis*, ed. by Borgnet, pp. 417–18. Cf. Kovach, 'The Enduring Question of Action at a Distance', pp. 172–75.

[29] Albertus Magnus, *De animalibus*, lib. XXV, tract. 2, 13 ('Basylyscus'); Albertus Magnus, *De animalibus libri* XXVI, ed. by Stadler, II, 1561–62; quoted after Albertus Magnus, *On Animals*, trans. by Kitchell Jr and Resnick, II, 1720–21. In *De animalibus*, lib. VII, tract. 2,

In the *Quaestiones super De animalibus*, however, in which the 'bloody mirror' appears in an answer to the question of 'whether the menstrual flow causes infection in the eyes', Albertus Magnus offers gradual infection of the air as an explanatory model:

> At the time of menstruation the menses flow particularly to the eyes and infect them. And this is why if the eyes' object is clean, like a polished mirror or the eye of another person, it will immediately be infected by the eye of the one menstruating, since that one's eyes infect the air, and this air infects other air, and this infects still other air, as far as the mirror.[30]

The deadly gaze of the basilisk and the bloody gaze of a woman are thus accepted as facts of natural science that require explanation. But have they been adequately explained? Albertus Magnus explicitly rejects the assumption of *radii oculorum*, but the alternative he offers, that of the spreading *spiritus visivus* that is said to be able to penetrate and destroy 'everything', even at a great distance, is unpersuasive — whether this *spiritus visivus* spreads by itself (*Summae de creaturis* and *De animalibus*) or gradually and progressively infects the surrounding air (*Quaestiones super De animalibus*). If the damaging vapours are separated from the visual rays, how does the *spiritus visivus* obtain its range, directionality, and determination?

There are further ambiguities, however. In *De animalibus*, the assumption of the performative gaze even becomes a positive theological argument: man shares in the divine intellect, and as a result 'man is not subjugated to the world but rather placed above it as its governor' — and this to the extent that he is able to impose his ideas on the world of matter. This is seen in the example of the outstanding men who are able, using their souls, to change other bodies in such a way that one speaks of miracles. These miracles include, for example, the power of *fascinatio*. Albertus thus uses the performativity of the gaze as evidence of the fact that man, created by God, governs nature rather than being subjugated to it.[31] As a more detailed description of the process of transference

cap. 5. 132–33 (I, 553, 647), basilisk and 'bloody mirror' appear together: 'There sometimes comes a noxious spirit from the eyes of a woman who looks at something up close. It is like the bloody cloud that flows onto a mirror from the eyes of a woman who has menstruated. And this is attributed to the basilisk.'

[30] Albertus Magnus, *Quaestiones super De animalibus*, lib. IX, qu. 9 'Utrum fluxus menstrui causet infectionem in oculis', in *Liber de natura et origine animae*, ed. by Filthaut, p. 207; Albert the Great, *Questions concerning Aristotle's 'On Animals'*, trans. by Resnick and Kitchell, p. 312.

[31] Albertus Magnus, *De animalibus*, lib. XXII, tract. 1, cap. 5; ed. by Stadler, II, 1553–55;

is not given in this passage, Albertus Magnus even appears here to approve of the teaching of Avicenna (and Algazel) that the soul's power of imagination *alone* — that is, without any physical mediation — is able to have effects on foreign bodies.[32]

This is an extraordinary passage, not only in relation to the entire tradition but also within the corpus of Albertus Magnus's writings. As we have seen, the performativity of the gaze being celebrated here otherwise appears primarily in misogynous contexts, in the Dominican's work.[33]

Thomas Aquinas

Thomas Aquinas, Albertus's pupil, took the first fateful step towards the witch-hunting craze. Here again we find the motif of the 'bloody mirror' being used argumentatively several times — for example, in the 117th *Quaestio* in the first book of his *Summa Theologiae*, written *c.* 1270, in response to the question 'Can men change material things by soul-power?' Thomas initially gives an affirmative answer to this now familiar question, arising in connection with Alhazen (and Algazel), with the help of biblical quotations and tradition. Yes, the human soul can change matter by its power, but — and here Thomas, along with Aristotle, departs from the Arabic scholars and thus clarifies the ambivalence that had not been elucidated by his teacher Albertus Magnus — this can only take place by means of a physical mediator, that is, the gaze, the physical nature of which is proved once again using the example of the 'bloody mirror'. At this point, however, he interrupts the natural science explanation of the phenomenon of the performative gaze and adds the remark, fraught with consequences, that the performativity of the gaze may nevertheless also be based on a pact with the devil:

trans. by Kitchell and Resnick, II, 1445–47. Cf. Kovach, 'The Enduring Question of Action at a Distance', pp. 204–05.

[32] Although Albertus Magnus does not refer to Avicenna (or Algazel) in this passage, he does so in numerous other passages, which are no less ambiguous. See Zambelli, 'L'immaginazione e il suo potere', pp. 199–200, and Hasse, *Avicenna's 'De Anima' in the Latin West*, pp. 169–71.

[33] In any case, Albertus's writings bristle with negative statements about women in general and about menstruation in particular. See, for example, Albertus Magnus, *Quaestiones super De animalibus*, lib. XV, qu. 11 'Utrum mas habilior sit ad mores quam femina'; ed. by Filthaut, pp. 265–66; ed. by Resnick and Kitchell, p. 454. Cf. Pseudo-Albertus Magnus, *Women's Secrets*, ed. and trans. by Lemay, pp. 47–49.

> Matter is not changed with respect to form, except by some kind of agent that is composed of matter and form [...]. Thus, it was said before about the angels that they cannot change matter by their natural powers *except by making use of material agents* to bring about certain effects. Much less, therefore, can the soul change matter by its natural power *save through the instrumentality of certain material things* [...]. Avicenna attributes the phenomenon of bewitchment (*fascinatio*) to the fact that matter innately obeys non-material beings rather than opposing natural agents. So when the soul's imaginative power is strong, matter is changed by it. This, he says, is the cause of the 'evil eye'. But it has been shown above that matter does not obey non-material beings at will, but rather the Creator alone. Thus it is preferable to say that the spirits joined with the body are changed by the imagination's power. This change occurs mainly in the eyes, which the more subtle spirits reach. Now the eyes infect the air to a certain definite distance, just as new and untarnished mirrors become tarnished from the glance of a menstruating woman, as Aristotle says. Thus, when certain souls are strongly stirred by wickedness, as happens especially with old women, their looks become spiteful and poisonous, particularly to children with tender and impressionable bodies. It is also possible that by God's permission, or by virtue of some hidden arrangement, the spitefulness of the demons (with whom the witches have some kind of compact) also plays a part in this.[34]

The 1277 Condemnation

The engagement with the question of effects at a distance was exacerbated by the condemnation of the teachings of Avicenna and Algazel in the famous Condemnation in 219 articles by the bishop of Paris, Stephen Tempier, in 1277, three years after the death of Thomas Aquinas. Article no. 112 condemns as heretical the view 'that superior intelligences impress inferior ones just as one intellectual soul impresses another and even a sensitive soul. And by such impression a certain enchanter by his mere gaze cast a camel into a pit'.[35] The

[34] Thomas Aquinas, *Summa Theologiae*, pars I, qu. 117, art. 3 ('Utrum homo per virtutem animae possit corporalem materiam immutare'), ad 2; Thomas Aquinas, *Summa Theologiae*, ed. by Charlesworth, pp. 139–41 (my italics). Very similarly, and again in the context of a debate with Avicenna and the argumentative use of the 'bloody mirror': *Summa contra gentiles*, lib. III, cap. 103; *De veritate*, qu. 26, art. 3, 4 (including the basilisk); *De malo*, qu. 16, art. 9. ad 13; Thomas Aquinas, *Super Epistolam B. Pauli ad Galatas Lectura*, cap. 3, lect. 1: *Commentary on Saint Paul's Epistle to the Galatians*, trans. by Larcher, pp. 65–70 (68).

[35] 'Quod intelligentie superiores imprimunt in inferiores, sicut anima una intellectiva imprimit in aliam, et etiam in animam sensitivam; et per talem impressionem incantator aliquis prohicit camelum in foveam solo visu'; *La Condamnation parisienne de 1277*, ed. by Piché, p. 112 (my translation). Cf. Flasch, *Aufklärung im Mittelalter?*, p. 190. Cf. also theses numbers

first sentence is directed against Avicenna and his followers, all those who maintain that the soul is able to alter not only its own body but also the body (and the soul) of another person (and matter in general), through its power of imagination alone.[36] The second sentence — the example of the camel being cast into a pit by the gaze alone — goes back to Algazel.[37] From the discussion so far, we can see that the two parts of the article are not really consistent. Why does the camel have to be looked at, if the imagination alone is capable of having such effects? The condemnation thus appears not only to be denouncing the teachings of Avicenna and Algazel, but also to be undermining the theory of the performative gaze in general. The Condemnation of 1277 thus triggered a whole series of texts — ranging from an anonymous *Utrum* in a Latin manuscript in the Vatican Library,[38] to Richard of Middleton around

73 and 210. Peter of Abano, who in 1303 in his *Conciliator*, Differentia 135, still claimed that by the power of imagination 'a man can be cast into a well and a camel into a hot bath', was accused of heresy and posthumously condemned to death by burning at the stake.

[36] See in particular Avicenna, *Liber sextus naturalium*, IV. 4, in *Avicenna Latinus*, ed. by van Riet and Verbeke, pp. 54–67 (pp. 64–67). See also Zambelli, 'L'immaginazione e il suo potere', and Hasse, *Avicenna's 'De Anima' in the Latin West*, pp. 154–74. This tenet from Arabic philosophy, which diverges from the Aristotelian principle that causation is based on contact, certainly met with some agreement in the Latin West in the mid-thirteenth century. Grosseteste, for example, cites not only Avicenna and Algazel with approval, but does not even hesitate to use the example of the overturned camel. See Robert Grosseteste, *Expositio in Epistolam Sancti Pauli ad Galatas III, 3* [Galatians 3. 1], in *Opera Roberti Grosseteste*, ed. by McEvoy and others, pp. 72–74 (73). Cf. William of Auvergne, *De universo* (*c.* 1230), II. iii. 16. Further examples in Zambelli, 'L'immaginazione e il suo potere', and Hasse, *Avicenna's 'De Anima' in the Latin West*, pp. 154–74. Roger Bacon and especially John Pecham, however, already explicitly contradicted Avicenna (and Algazel) by insisting on the physicality of the transmission process: Roger Bacon, *Opus maius* IV, and *Epistola Fratris Rogerii Baconis de secretis operibus artis et naturae, et de nullitate magiae*, cap. 3; cf. above, note 23. John Pecham, *Quodlibet*, V, qu. 2 ('Utrum sola apprehensione possit homo corpora transmutare'), in John Pecham, *Quodlibet Romanum*, ed. by Delorme, pp. 77–79 (79) = Fr Ioannis Pecham, *Quodlibeta quatuor*, ed. by Etzkorn and Delorme, quodlibet IV (Romanum), qu. 30, pp. 242–45 (244–45). At approximately the same time (1270), Aegidius Romanus, in his compilation of the errors of Greek, Jewish, and Arabic philosophers mentions the teachings of Avicenna and Algazel, including the example of the camel cast into a ditch by a gaze; it is from this source that it was probably taken up into the 1277 list of condemned teachings. See Aegidius Romanus, *Errores Philosophorum*, IX. 16; Giles of Rome, *Errores Philosophorum*, ed. by Koch, pp. 44–45. Cf. on Avicenna VI, 11 (pp. 32–33) and VII, 13–14 (pp. 36–37).

[37] Algazel, *Metaphysica*, II. v. 9, in *Algazel's Metaphysics*, ed. by Muckle, p. 194.

[38] Città del Vaticano, Biblioteca Apostolica Vaticana, MS Lat. 1121. See Thorndike, 'Imagination and Magic', pp. 353–56.

1290,[39] to Gentile da Cingoli and Engelbert of Admont around the turn of the century,[40] to the famous Nicole Oresme and Conrad of Megenberg in the fourteenth century — all of whom attempted to exempt the performative gaze from the condemnation. Either explicitly or implicitly, but unmistakably, they all refer to the condemned article, and they all draw on the 'bloody mirror' in order to prove the physical nature of the process. Thomas Aquinas had provided the pattern for the response. They agree on the condemnation of the first sentence, that the soul can transmute the body of another person by imagination alone. But they disagree on the second sentence by divorcing it from the first. They certainly do not maintain that a camel can be cast into a ditch by the gaze alone, but they insist on the performativity of the gaze. Its performativity is not caused by imagination alone, however, but by the materiality of the gaze, the quantity and quality of which, of course, do indeed follow the imagination. And the phenomenon of the mirror being stained by the gaze of the menstruating woman is the perfect argument at this point. It not only proves the fact of the performativity of the gaze but also its materiality, or the materiality of the process of transference. This alone makes it possible to save the performativity of the gaze from the Condemnation. And in the process, the assumption of *radii oculorum*, rejected by Albertus Magnus and silently passed over by Thomas Aquinas, is reintroduced once again.

Nicole Oresme

Nicole Oresme turns to the issue in several of his writings, in the greatest detail in Chapter 4 of his essay *De causis mirabilium* and in Book II of his *Tractatus*

[39] Richardus de Mediavilla, *Quodlibet*, III, qu. 12; quoted in Ubl, *Engelbert von Admont*, pp. 51–52, n. 150.

[40] Occasionally surviving *Quaestio* 'Utrum species sensibilis vel intelligibilis habeat virtutem alterandi corpus ad caliditatem vel frigiditatem?' by Gentile da Cingoli (Gentilis de Cingulo), professor of philosophy in the Faculty of Arts at the University of Bologna, Biblioteca Apostolica Vaticana, MS Lat. 772, printed in Grabmann: 'Gentile da Cingoli, ein italienischer Aristoteleserklärer aus der Zeit Dantes'. The text of the *Quaestio* appears on pp. 68–88 (pp. 78–81). We may dispense here with a presentation of the remarks by Gentile da Cingoli — particularly interesting in the present context — which in examining Avicenna defend in detail the evil eye, the deadly gaze of the basilisk, and the staining gaze of the menstruating woman, since reference may be made to the excellent analysis by Siraisi, *Taddeo Alderotti and his Pupils*, pp. 204–08; pp. 206–07 and 218 refer to the tradition of the 'bloody mirror' in an unpublished commentary, surviving only in manuscript form, by his teacher the physician Taddeo Alderotti, on the *Isagoge* by Johannitius (Hunayn ibn Ishaq).

de configurationibus qualitatum et motuum. The following quotation is from the latter, written *c.* 1350:

> Sometimes the soul alters its own subject [...]. But also it can be the cause of the transmutation of an alien body in two ways: (1) In one way, locally, by the application of members, as by moving something with the hand. (2) In the other way, without any such exterior application, but (rather) by a certain latent action. And it is in this way that the soul sometimes acts on an alien body by means of a forceful thought or imagination, as happens in the case of bewitchment (*fascinatio*). [...] Moreover, such action of the soul can be imagined as taking place in two ways: (1) in one way by the command of the soul alone; and, if the soul is noble, *yle*, or the matter of the world, obeys it (immediately) without another, intermediate alteration. And it is by this method that Avicenna in his Sixth Book of Natural Things and Algazel in the fifth of his Physics [Metaphysics], held that bewitchments (*fascinatio*) take place [...]. But that such action arises from the soul by thought alone is excessively absurd and irrational [...]. Furthermore, it is far removed from philosophy and out of harmony with our faith. Accordingly an article at Paris has condemned this. Therefore, there is (2) a second, more rational way as follows: It is said that imagination or affection can be so increased intensively and its difformity figured in such a way that it acts to change significantly the body or some member of the person who is so imagining: i.e. towards health or sickness, or towards some other special disposition [...]. Then the body or member of the person so altered sometimes can alter the medium and a body near to it at a certain distance, and most of all if that body is disposed to receive easily such an alteration or impression, (e.g.) in the way that the eye of a menstruating woman infects the air and by means of the air infects a clean mirror. And this kind of alteration principally arises *where visual rays are directed* (at someone) (*ubi diriguntur radii visuales*).[41]

[41] *Tractatus de configurationibus qualitatum et motuum*, pars II, cap. xxxviii, in *Nicole Oresme and the Medieval Geometry of Qualities and Motions*, ed. by Clagett, pp. 381–85 (my italics). Cf. I. xxxiv ('Bloody Mirror'), II. xxviii, and II. xxxv (further refutation of Avicenna). Similarly, *De causis mirabilium*, IV. 14. 529–45 (538) and IV. 19. 920–1105 (1050–59), and the attached *Tabula problematum*, qu. 160, in 'Nicole Oresme and the Marvels of Nature', ed. and trans. by Hansen, pp. 312–15 (314–15) and 344–59 (354–55), as well as p. 379. The text, which represents a reworking of responses to more than two hundred *Quaestiones* and whose overarching topic is the evidence that supposedly marvellous phenomena do in fact have natural causes, used to be given the title 'Quodlibeta', as in the earlier edition by the same author: 'Nicole Oresme and the Marvels of Nature', ed. and trans. by Hansen. Cf. *Questiones super de sensu et sensato*, qu. 3 ('bloody mirror' and basilisk) and *Questiones super septem libros Physicorum*, lib. II, qu. 14 (Algazel, Pliny, evil eye), quoted in *Oresme and the Medieval Geometry of Qualities and Motions*, ed. by Clagett, pp. 458 and 490, and 481. Cf. also *Le 'Quaestiones de sensu'*, ed. by Agrimi, pp. 75–84 (questio 4, 'Queritur utrum visus fit extra-mictendo vel intus-suscipiendo': 'bloody mirror' and basilisk); pp. 32–33 and p. 84 refers to two unpublished

This argument is based on the theory of the infection of the air proposed by Thomas (and by Aristotle in *De insomniis*). But Oresme, honestly enough, explicitly adds to it the assumption that there are rays emanating from the eyes. Without them, the process would remain incomprehensible.

Engelbert of Admont

Engelbert of Admont's text *De fascinatione*, written *c.* 1295 — the first to bear this title, so far as I am aware — also represents an analysis of the teachings of Avicenna and Algazel, attempting to save the phenomena associated with the performative gaze from the Condemnation of 1277.[42] The example of the camel being cast into the ditch by the soul's power of imagination already appears in the second chapter of the first book.[43] Engelbert considers this quite possible (!), but regards the explanation as inadequate. It was impossible for an imagination to effect this alone, according to him, since the soul and its ideas (imaginations) were of an immaterial nature (*pure spiritualis*). The action of the soul on foreign bodies, he claimed, was only possible through physical mediation. This is achieved by the gaze, which is partly immaterial and partly material (*partim spirituale et partim corporale*).[44] Later, in the eighth and ninth chapters, he defines the visual process in a way that initially confirms the assumption of 'perspective science' — that vision takes place through the reception of external light rays — as being correct; however, it ends with a reversion to the Platonic–Galenic variant of the visual ray theory that goes beyond the perspectivists' compromise model: vision requires light rays that emanate (or are reflected or refracted) from the object, but also a radiation of *spiritus visibilis* from the eye that 'unites' with the external light ray.[45] It is only this union of the two rays that

texts by Johannes Buridanus, surviving only in manuscript, which also argue on the basis of the 'bloody mirror'. On Oresme in this context, see also Thorndike, *A History of Magic and Experimental Science*, pp. 424–39.

[42] Engelbert of Admont, *De fascinatione*, I. 8–13, in 'Engelberti Admontensis Tractatus de fascinatione', ed. by Fowler, pp. 187–231, reproducing the text of *De fascinatione* on pp. 195–231 (201–11). Cf. Thorndike, *A History of Magic and Experimental Science*, pp. 433–37, and in particular Ubl, *Engelbert von Admont*, pp. 49–57.

[43] *De fascinatione*, I. 2; ed. by Fowler, p. 197 (here incorrectly attributed to Algazel and Avicenna). Constant reference is made in the work to Avicenna; Algazel is also mentioned in I. 12 and I. 13 (pp. 209 and 210).

[44] *De fascinatione*, I. 7; ed. by Fowler, p. 202.

[45] *De fascinatione*, I. 8; ed. by Fowler, pp. 202–04.

makes the 'diaphanous medium' (elsewhere called the 'diaphanous body'), that is, the air, into a medium of transference, which takes place in both directions. In this way, excitations of the soul by means of the blood, which combines with the *spiritus visibilis*, could be transferred, via the channel formed by the two rays within the visual cone, to the object seen and could cause a corresponding change in the object.[46] Once again, the physical quality of this process is proved through a reference to Aristotle and his 'bloody mirror' (as well as the deadly gaze of the basilisk):

> All these things are seen to be confirmed from what the Philosopher says [...], that the gaze not only receives something from its object, but also acts upon it, namely *by means of the visual spirit that is commingled with the eye's rays* and stains the air all the way to the object of the gaze, upon which the stain is impressed by the mediating air. And he demonstrates this using women's mirrors. When they are looked at by women when they are menstruating, it makes the surface of the mirror like a blood-stained cloud. And if it is an old mirror, the stain can be more easily wiped off; but if it is fresh and new, it is more difficult to wipe off.[47]

Thomas of Cantimpré and Conrad of Megenberg

A radicalization of the visual process and its mechanism of effect can be seen once again during the dissemination of these ideas. Thomas of Cantimpré, a student of Albertus Magnus, in his *Liber de natura rerum*, written between 1230 and 1243 and widely published, initially explains the evil eye of women and the staining of the mirror by the gaze of a menstruating woman through direct infection resulting from the bad humours in the gaze.[48] In the case of the basilisk, he simply speaks of poisonous *radii oculorum*, which are responsible for the death of the person gazed at.[49] Rays emanate from the basilisk's eyes,

[46] *De fascinatione*, I. 9; ed. by Fowler, pp. 204–05.

[47] *De fascinatione*, I. 11; ed. by Fowler, pp. 207–08 (my italics). Cf. I. 10; ed. by Fowler, pp. 205–07 (206–07), where Engelbert adds another mirror example as further empirical evidence for the materiality of the gaze. He describes a man who made such a smooth and clean mirror that whenever a criminal looked into it, it was obscured by a black cloud that was not easy to remove, as the consequence of the evil in the criminal's humours and spirits — a kind of medieval lie-detector.

[48] Thomas of Cantimpré, *De natura rerum*, I. 6: 'De oculis', in Thomas Cantimpratensis, *Liber de natura rerum*, ed. by Boese, pp. 18–21 (19).

[49] Thomas of Cantimpré, *De natura rerum*, VIII. 4, 'De basilisco'; ed. by Boese, pp. 178–79.

which first destroy the vision, the *spiritus visibilis* of the person affected, and then the remaining *spiritus*, leading to death.

Conrad of Megenberg's *Buch der Natur* (*c.* 1350), which represents a German version of the Latin text by Thomas of Cantimpré, takes the same position. The book was the first natural history written in German and met with overwhelming success that continued for nearly two hundred years; it even became one of the most widely distributed and most-read German books in general. It reproduces the passage more or less verbatim, but then adds, 'Avicenna therefore also says that a woman can cast a camel into a pit by her gaze.'[50] Notable here — apart from the confusion of Avicenna and Algazel — is first the fact that this is no longer a justification in principle of the phenomenon of the performative gaze, but rather a dissemination of the story of the camel overturned by the evil eye itself, as an empirical fact; and second, that the woman here takes the place of the gender-neutral *incantator* in the Paris Condemnation.

De secretis mulierum

The topic of the performative gaze grew in importance within and alongside the growth of a misogynous view of women, which spread wider and wider starting in the mid-thirteenth century, particularly in educated circles. The central catalyst for this misogyny was the taboo on menstrual blood, which is as old as the human race, but only in the thirteenth century began to spread with fateful consequences along with the recovery and assimilation of Aristotelian writings.[51]

We have already noted this tendency above in the writings of Albertus Magnus and his pupil, Thomas Aquinas. But its high point is certainly represented by the *De secretis mulierum*. Numerous copies of this text ascribe it to Albertus Magnus, but it was certainly written by one of his pupils or successors in the late thirteenth or early fourteenth century.[52] The text, which belongs

[50] Conrad von Megenberg, *Buch der Natur*, I. 5: 'Von den Augen', in Konrad von Megenberg, *Das 'Buch der Natur'*, ed. by Luff and Steer, p. 33; cited after *Das Buch der Natur von Conrad von Megenberg*, ed. and trans. by Schulz, p. 6. Cf. III. E. 3 (pp. 291–92; 222–23) on the basilisk (my translation). A crude theory of visual rays is also used to explain the phenomenon of the rainbow (II. 30: pp. 123–25; 79–81). Cf. Gottschall, *Konrad von Megenbergs Buch von den natürlichen Dingen*, pp. 273–79 and 321–39 (333–34).

[51] See, for example, Allen, *The Concept of Woman*, and the excellent introduction by Lemay in Pseudo-Albertus Magnus, *Women's Secrets*, ed. and trans. by Lemay.

[52] Pseudo-Albertus Magnus, *Women's Secrets*, ed. and trans. by Lemay.

to the literary genre of *Quaestiones* or *Problemata*, was probably written for a monastic context, but it soon spread well beyond that. Its importance is also indicated by the fact that it is often — both in manuscript and in print — accompanied by anonymous commentaries, which further radicalize the text's misogyny. The two most important of these are included in the critical edition by Helen Rodnite Lemay.[53] Remarks that are still scattered asides in Albertus himself and in Thomas Aquinas are here made into the central and sole topic of the text.

In the tenth chapter, Pseudo-Albertus warns us above all against two types of women. First, 'those who have their menstrual periods, since the menstrual flow can poison children lying in their cradles by infecting the eyes of women, and transmitting this poison through the air to the child'.[54] Although the author refers in this connection to Aristotle's *De somno et vigilia* (*De insomniis*), which the contemporary reader would automatically have taken as a reference to the staining of the mirror by the gaze of a menstruating woman, it is only the commentaries that explicitly introduce the 'bloody mirror'. The following quotation is from the introduction to the most widely disseminated 'Commentary A' on the overall text:

> The moving cause [for writing the book] was a certain priest who asked Albert if he would write for him a book on the secrets of women. The reason for this is that women are so full of venom in the time of their menstruation that they poison animals by their glance; they infect children in the cradle; they spot the cleanest mirror; and whenever men have sexual intercourse with them they are made leprous and sometimes cancerous.[55]

But this is not all; even more dangerous are old women, whose stock of evil humours is even greater, as they are no longer able to excrete their physical waste via the menstrual blood; and the situation is even worse when they are poor, for in this case they also consume coarse food, which further contributes to the poisonous matter.[56]

[53] Pseudo-Albertus Magnus, *Women's Secrets*, ed. and trans. by Lemay, pp. 59–150. For the tradition of the text in Germany, cf. 'Pseudo-Albertus Magnus: *Secreta mulierum*', ed. by Schleissner.

[54] Pseudo-Albertus Magnus, *Women's Secrets*, X; ed. and trans. by Lemay, p. 129.

[55] Pseudo-Albertus Magnus, *Women's Secrets*, Commentary A to the Preface; ed. and trans. by Lemay, p. 60. Cf. Chapter X, Commentary B; ed. and trans. by Lemay, p. 131. Cf. 'Pseudo-Albertus Magnus: *Secreta mulierum*', ed. by Schleissner, pp. 188–89.

[56] Pseudo-Albertus Magnus, *Women's Secrets*, X; ed. and trans. by Lemay, p. 129. Com-

Malleus Maleficarum

From there it was only a small step to the notorious *Malleus Maleficarum* of 1487. The second subchapter of the first part deals in particular with the evil eye, and the authors discuss whether witches and demons are also able to achieve their evil purposes independently of each other.[57] The positions taken by Avicenna and Algazel are initially discussed and quoted with distortion in relation to the topic of the evil eye: 'Avicenna even thinks (his position here is not accepted) that the virtue of the imagination can change external bodies even without vision.'[58] With the gaze, however, it certainly was possible. For the imaginations of the soul alter one's own body and through this physical change in one's own body, the soul can also have effects on foreign bodies, particularly by means of visual rays, and especially in menstruating women and old women, a fact to which the 'bloody mirror' (as well as the deadly gaze of the basilisk) can attest. Although the Thomist model of infection of the air, from the *Summa Theologiae*, is then quoted in detail and paraphrased again in greater detail, the prerequisite of a physical mediator established in the response to Avicenna and Algazel, without which an effect at a distance is not possible, requires that the authors of the *Malleus Maleficarum* also conceive of the infection of the air as being caused and directed by physical

mentary A adds (p. 130) that 'the eye of a soothsayer once forced a camel into a ditch, for the evil within her generated evil humors, which exited through the eyes, and caused the camel to fall down in an attempt to flee'. In the same way, Commentator A continues, a basilisk is killed when a mirror is placed in front of it, for the poisonous humours that are emitted from its eyes are reflected back to it. A very similar passage is found in a text known as the *Problemata varia anatomica*, a collection of *Quaestiones* from the same period (late thirteenth or early fourteenth century; known also by the first words of its *incipit*, 'Omnes homines'), which was similarly widely circulated and continued to be printed until 1686. See *Problemata varia anatomica*, ed. by Lind, p. 20 ('bloody mirror' and basilisk). Cf. Schleissner, 'Sexuality and Reproduction in the Late Medieval *Problemata Aristotelis*'; also Jacquart and Thomasset, *Sexuality and Medicine in the Middle Ages*. In Germany, the *Problemata varia anatomica* was sometimes bound along with *De secretis mulierum*; see 'Pseudo-Albertus Magnus: *Secreta mulierum*', ed. by Schleissner, pp. 51–59, 61–62, 74–75.

[57] *Malleus Maleficarum*, I. 2; Henricus Institoris and Jacobus Sprenger, *Malleus Maleficarum*, ed. and trans. by Mackay, I: *The Latin Text and Introduction*, pp. 229–43 (pp. 236–39); II: *The English Translation*, pp. 57–73 (pp. 56–68).

[58] *Malleus Maleficarum*, I. 2; ed. and trans. by Mackay, I, 231 (II, 59). Cf. I, 236–37 (II, 65–66), where the views of Avicenna and Algazel are not only rejected once again, but explicit reference is also made to the Condemnation of 1277. Cf. also *Malleus Maleficarum*, I. 5; ed. and trans. by Mackay, I, 271–72 (II, 101–02).

rays.[59] In any case, a natural explanation for the performative gaze is involved, which admittedly is immediately undermined again, since — as the two authors both hasten to admit — the performativity of the gaze may also be the work of a demon or a pact with the devil; one can never be quite certain, so the possibility can never be excluded. The authors also warn judges and everyone involved in a trial that they should never allow themselves to be touched or looked at by the witch. For safety reasons, she should only be brought before the court walking backwards.[60]

Helen Rodnite Lemay has rightly stressed the direct line of influence from Albertus Magnus to Pseudo-Albertus's *De secretis mulierum*, to the infamous fifteenth-century *Malleus Maleficarum*:

> The authors of the *Malleus* used the *Secrets* and the tradition they represent as an ideological basis for concluding that women are prone to witchcraft, for which crime they deserve death [...]. The *De secretis mulierum* sets down the scientific justification for this persecution, which will be incorporated into the manual of witchcraft for inquisitors almost two centuries later.[61]

This is only partly correct, however. It is correct in relation to the *Malleus Maleficarum*, but not in relation to manuals of witchcraft and inquisitorial documents in general. What is more remarkable in the latter is the absence of any reference to theories and concepts of vision in general and to the 'bloody mirror' in particular, despite all the 'evil eyes'.[62] The two authors of the *Malleus*

[59] *Malleus Maleficarum*, I. 2; ed. and trans. by Mackay, I, 236 (II, 65) and I, 237–38 (II, 66–67).

[60] *Malleus Maleficarum*, II. 1. 12 und III. 2. 15; ed. and trans. by Mackay, I, 465 (II, 318) and p. 621 (pp. 503–04). Cf. 'verhüllen', in Bächtold-Stäubli and Hoffmann-Krayer, *Handwörterbuch des Deutschen Aberglaubens*, VIII, cols 591–1609 (1601).

[61] See the introduction to Pseudo-Albertus Magnus, *Women's Secrets*, by Lemay, pp. 49–58 (50 and 54). Lemay argues for direct borrowing of the arguments from *De secretis mulierum*. This is probably correct, but it should be noted that nothing is found in the text that is not also in Thomas, who is also quoted in detail. And her argument that the incorrect reference to *De somno et vigilia* instead of *De insomniis* in both authors is evidence of this is incorrect, since — as we have seen above — almost all of the authors mentioned (including Thomas: *De veritate*, qu. 26, art. 3, 4, and *De malo*, qu. 16, art. 9. ad 13) who discuss Aristotle's 'bloody mirror' refer to *De somno et vigilia*, and correctly, since *De somno et vigilia*, *De insomniis*, and *De divinatione per somnum* were passed down during the Middle Ages under the common title *De somno et vigilia* (sometimes also *De sompno et vigilia*).

[62] A few blindly repeated examples of the passages in Thomas and the *Malleus Maleficarum* represent exceptions to this. See, for example, *Tractatus contra demonum invocatores, compilatus per sacre theologie professorem fatrem Johannem Vineti, ordinis Predicatorum, inquisitorem apostolicum Carcassone*, *c.* 1450, quoted in Hansen, *Quellen und Untersuchungen zur Geschichte*

Maleficarum, Heinrich Kramer (Institoris) and Jakob Sprenger, mention the natural explanation for the evil eye not because they are making any effort to be scientific — this is the only natural science argument in the entire text — but rather because of their effort to provide the inquisitors with every conceivable argument for condemning witches. In the fervour of their argumentation, they go beyond the mark, since a natural science explanation of the performative gaze of menstruating women or old women might undermine the supposition of the woman's guilt and the assumption of a pact with the devil. If the corruption of their visual rays is a natural phenomenon, it can hardly always be due to their evil intentions or to a pact with the devil.

Medical Enlightenment

This is confirmed by the fact that precisely the same arguments are used to exonerate women in theological texts, and particularly in medical treatises, which oppose any form of superstition in general and witch-hunting in particular. This genre developed in the early sixteenth century, emphasizing the natural ability of women (and of men) to have an evil eye or 'performative' gaze.

Among the early, still rather ambivalent examples is the *Tradado de las supersticiones y hechizerias* by Fray Martin de Castañega of 1529. The text, probably written in the vernacular at the request of the bishop of Calahorra, is aimed at the clergy in his diocese, who are to receive a handbook to help them cope with the accumulation of alleged cases of witchcraft in Vizcaya and Navarra. Despite this, however, the treatise clearly differs in style and content from the notorious *Malleus Maleficarum* used by the German inquisitors. Although the Franciscan 'theologian, philosopher and preacher of the Holy Office' also believes in the existence of the devil, as he does in the influence of the stars and the efficacy of amulets, and even considers that witches are able to fly during the night, he nevertheless proves in many points in his discussion to be a sceptic and materialist who is making an effort to trace what seems miraculous back to natural processes whenever possible. In a chapter entitled 'To Give the Evil Eye is a Natural Act and Not Witchcraft', basing his argument on the assumption that the body excretes everything that is not useful for its own substance and

des Hexenwahns, pp. 124–30 (p. 129: 'bloody mirror'); Martín de Andosilla, *Tractatus de Superstitionibus*, in Goñi Gaztambide, 'El tradato "De Superstitionibus" de Martín de Andosilla', p. 287, and in *Príncipe de Viana*, p. 1041, 'bloody mirror'; Arnaldus Albertini, *De agnoscendis assertionibus catholicis et haereticis*, c. 1540, quoted in Hansen, *Quellen und Untersuchungen zur Geschichte des Hexenwahns*, pp. 344–54 (p. 349: 'bloody mirror').

nourishment, Castañega — referring to Pliny's *Natural History*, Aristotle's *De insomniis*, and Thomas Aquinas's *Summa Theologia* — writes:

> The natural expulsive power, which is one of the powers of the nutritive faculty, expels and throws from the body all the impurities that can't be used for sustenance and maintenance of the body. What is thickest is expelled by the nether parts that nature provided for that purpose; and what is not so thick comes out of the natural windows, like the mouth, the nose, and the ears; and what is more subtle through the pores and spongy parts as sweat; and what is the most subtle is expelled by the glassy parts of the eyes. Thus the most subtle impurities of the body come out through the eyes like rays, and the more subtle they are the more they penetrate and are most infectious. This is how a menstruating woman can stain a new and clean mirror *with the rays that come from her eyes*, because the body's natural powers at such a time throw off *corporeal* impurities, and the most subtle ones come from the eyes and thereby dirty the mirror. If at such a time the woman should stare closely at a tender and delicate child she would imprint on him those *poisonous rays* and distemper his body in such a way that he would be unable to open his eyes or to hold his head up [...]. Nor should mothers suppose that the sickness comes only from witches and is always part of witchcraft, because it is a natural thing that can come from any indisposed person; although it is true, as I said, that it can be increased and made more intense by malice in the heart of she who looks at the child, and thus it may be done by some malicious person or witch or sorceress that wishes to do harm to innocent creatures to serve her lord the devil.[63]

As a protective measure, Fray Martin de Castañega recommends that small children should have pieces of mirror attached to their hair in such a way that they hang over the forehead, where they will be able to deflect (and reflect) the gaze. In case of infection, one should resort not to some miracle-worker or other but to tasty herbs and fragrant incense and similar natural remedies.

The physicians in the following generation go even further, taking the defence of witches into their own hands and providing natural explanations not only for the evil eye, but also for other natural phenomena, using the tools provided by Hippocrates and Galen. Their theory of the four humours allowed almost every disease to be traced back to disturbances of the physical elements, *humores* (and *spiritus*) and their balance. On this basis, not only the performa-

[63] Fray Martín de Castañega, *Tratado de las supersticiones y hechizerias [...]*, cap. 14, in Fray Martín de Castañega, *Tratado de las supersticiones y hechizerias*, ed. by Campagne, pp. 115–17; quoted after Darst, 'Witchcraft in Spain', pp. 309–10 (my italics). Similarly, Geiler von Kaysersberg, *Die Emeis*, fol. 47, sermon 27: 'Ob man die kind verzauberen mög mit dem gesicht — De Fastinatione', printed as no. 11 in *Zur Geschichte des Volks-Aberglaubens im Anfange*, ed. by Stöber, pp. 44–48 (p. 46: 'bloody mirror').

tivity of the gaze but even the maddest ideas (such as the ability of witches to fly), as well as convulsions and visions that were considered to be symptoms of possession, could be interpreted as the effects of a severe disturbance of the body's humours, and the alleged offenders could be declared victims. Here again, it is in no sense claimed that the performative gaze does not exist; on the contrary, its existence is confirmed but regarded as a natural process.[64]

The most outstanding and influential representative of this approach was certainly Johannes Weier, personal physician to Duke Wilhelm V of Kleve-Jülich-Berg. His treatise *De praestigiis daemonum*, written in 1563 explicitly in opposition to the *Malleus Maleficarum*, was devoted to a confrontation with witch-hunting. In a short chapter about the evil eye, he cites the arguments mentioned above in order to show that it is a natural process and to deny that 'witches' are responsible.[65] The decisive aspect for Weier is that the feature of possessing the evil eye must either be congenital or due to a transient form of corruption, and was therefore not due in any way to a voluntary choice (or even a pact with the devil) by the person affected, who had no control over such processes. Weier, who significantly gave his son the name Galenus, is attempting from a medical point of view to regard 'witches' as melancholy women who

[64] Admittedly, not all physicians shared these enlightened views. Paracelsus, for example, instead gives the screw of misogyny yet another turn. In *De natura rerum libri*, we read that the basilisk 'is a monster above all monsters. For none of them is more to be feared, since it can suddenly kill any person with its face and sight. For it is a poison beyond all poisons that none in the world can match. It carries this poison hidden in its eyes, it is an imagined poison not unlike that of a woman who is in her monthly period and who also has hidden poison in her eyes. This may be seen from the fact that she sees strips or spots in a mirror [caused by looking into it] and soils it and stains it solely with her face'. This is not merely a case paralleling the natural process, however; instead, there is a kind of kindred nature between the basilisk and the woman, since — as Paracelsus continues — the monster basilisk owes its congenitally deadly gaze to the female *menstruo* and can be artificially engendered by taking menstrual blood, 'the greatest impurity of women', filling it into a glass bottle and letting it rot in the belly of a horse. See Paracelsus (Theophrast von Hohenheim), *De natura rerum libri* IX (Strasbourg, 1584), lib. I: 'De generatione rerum naturalium'; quoted after Paracelsus, *Sämtliche Werke*, ed. by Aschner, III, 221–29 (225); my translation. In *Von Ursprung und Heilung der natürlichen Pestilenz*, tract. 1, Paracelsus considers that infection with the plague is (also) transferable through the visual rays, and he uses the 'bloody mirror' twice to illustrate the process; I, 682–804 (pp. 701 and 705). On the basilisk, cf. Harrer and Harrer, 'Paracelsus und der Basilisk'.

[65] Johannes Weier, *De praestigiis daemonum: Von Teuffelsgespenst, Zauberern und Gifftbereytern*, III. 34, pp. 226–27; Johannes Wier, '*De praestigiis daemonum*', ed. by Mora, pp. 265–67. Similarly, for example, John Webster, *The Displaying of Supposed Witchcraft*, chap. 2, p. 23, and chap. 8, pp. 180–83.

deserve medical assistance and loving care, rather than persecution.[66] Weier himself does not mention the 'bloody mirror', but it is found among several other representatives of the medical profession. The clouding of the mirror by a menstruating woman is adduced as evidence from experience for the natural process involved in *fascinatio*.[67]

Natural Magic

These representatives of the medical profession are in agreement with the proponents of natural magic and also draw on the latter to support their hypotheses. Weier, for example, cites among others Cardano, Pomponazzi, and Della Porta, to whom Reginald Scot also refers in this context.[68] Among the representatives of natural magic, the evil eye is also presented as an involuntary,

[66] The development and spread of numerous other diseases were also explained using this assumption of a material visual ray, as in the case of the plague. This is the case, for example, in Paracelsus (see above, note 64) or Athanasius Kircher, *Scrutinium*, II. iv. 6, in *Athanasii Kircheri e Soc. Iesu Scrutinium physico-medicum Contagiosae Luis*, pp. 152–53 (my translation): 'Infection with the plague through the face can only result by means of fascination: for just as a menstruating woman infects a mirror and those with watery eyes stain others through the poison and impure humours that they radiate from their eyes, I have no doubt that a person who has contracted the plague can only infect another by firmly staring at him.' Cf. Campbell, *The Black Death and Men of Learning*, pp. 48–56, listing earlier examples of this hypothesis of gaze transference (but without mention of the 'bloody mirror'). Love was also often regarded as a sickness caught through the eyes. I need only refer here to the most influential (Neoplatonic) text that is relevant, which in this context resorts not only to the theory of the visual ray dating from antiquity, but also uses the Aristotelian 'bloody mirror' as empirical evidence for the fact that infection through the gaze involves a physical process: Marsilio Ficino, *De amore*, VII. 4, in Marsilio Ficino, *Commentary on Plato's 'Symposium' on Love*, trans. by Jayne, pp. 159–61 (160). Cf. Hub, 'Material Gazes and Flying Images', pp. 102–06; Kodera, 'Narcissus, Divine Gazes and Bloody Mirrors', pp. 303–04. Incidentally, it was from here that the 'bloody mirror' as evidence for the theory of physical visual rays even found its way into a text on the theory of art: Gregorio Comanini, 'Il Figino overo del fine della pittura (1591)', pp. 300–01; cf. Weststeijn, 'Seeing and the Transfer of Spirits in Early Modern Art Theory', pp. 149–50.

[67] See Diego Alvarez Chanca, *Tractatus de fascinatione* (Seville, 1499), Antonio de Cartagena, *Libellus de fascinatione* (Alcalà de Henares, 1530), Alonso López de Corella, *Trezientas preguntas de cosas naturales* (Valladolid, 1546); all three cited and contextualized in Salmón and Cabré, 'Fascinating Women', pp. 57, 59, and 62–64.

[68] Reginald Scot, *The Discovery of Witchcraft*, ed. by Summers, bk XVI, chap. 9, 'Of inchanting or bewitching eies', pp. 281–82 ('bloody mirror' as evidence of the naturalness of the process of the evil eye); cf. p. 281, chap. 8, 'Of natural witchcraft or fascination'; and pp. 282–83, chap. 10, 'Of natural witchcraft for love, etc.'.

natural process belonging to, or at least possible, for all human beings. Their purpose here was a different one, however. They were concerned not so much with protecting women but rather with establishing (natural or white) magic as a science for the 'modern age'.[69]

The 'bloody mirror' (as well as the basilisk's deadly gaze) is found as an argument for the naturalness of the performativity of the gaze in Girolamo Cardano's treatise *De subtilitate*, written *c.* 1550.[70] And in Giovan Battista Della Porta's *Magia naturalis* of 1558, we read:

> It is manifest, that all women which have two pupils in one eye, can bewitch with it [...] and that they kill with their looks, making the persons languish and consume away as in a consumption [...] so great is the power and witchcraft of their eyes: for though the mischief be often caught in copulation with them, yet it is the eyes that work; for they send forth spirits, which are presently conveyed to the heart of the bewitched, and so infect him. Thus it cometh to pass, that a young man, being full of thin, clear, hot, and sweet blood, sendeth forth spirits of the same nature; for they are made of the pure blood, by the heat of the heart: and being light, get into the uppermost parts of the body, and flye out by the eyes, and wound those who are most porous, which are fair persons, and the most soft bodies. With the spirits there is sent over also a certain fiery quality, as red and blear eyes do, who make those that look on them, fall into the same disease: I suffered by such an accident my self: for the eye infecteth the air; which being infected, infecteth another: carrying along with it self the vapours of the corrupted blood, by the contagion of which, the eyes of the beholders are overcast with the like redness. So [...] the Cockatrice [basilisk] killeth, who poysoneth with looking on, and giveth venimous wounds *with the beam of his eyes*: which being reflexed upon himself, by a looking-glass, kill the Author of them. So a bright Mirror dreadeth the eyes of an unclean woman, saith Aristotle, and groweth cloudy and dull, when she looketh on it: by reason that the sanguine vapour is contracted by the smoothness of the glass into one place; so that it is spotted with a kind of little mist, which is plainly seen; and if it be newly gathered there, will be hardly wip'd off [...]. In like manner almost, if you breath upon a clear glass, it will wax moist as it were with a sprinkling of spettle, which condensing will drop down: *so this efflux of beams out of the eyes, being the*

[69] We are following traditional terminology here, although the distinction between natural (white) and supernatural (black) magic appears extremely questionable, at least for the sixteenth century, since this contrast is hardly found anywhere in the writings of the so-called natural magicians, who were concerned rather with subjecting *all* forms of magic to a natural explanation.

[70] Girolamo Cardano, *De subtilitate*, XII, in *Opera omnia*, III. 2, lib. IX 'De animalibus, quae ex putredine generantur', p. 557; cf. *De venenis*, lib. I, cap. 17: 'De Veneficio, fascino, praestigiis', in *Opera Omnia*, VII. 9, pp. 297–99.

> *conveyers of spirits*, strike through the eyes of those they meet, and flye to the heart, their proper region, from whence they rise; and there being condensed into blood, infect all his inward parts [...] there are two kinds of Fascinations mentioned by Authors: One of Love, the other of Envy or Malice.[71]

Tommaso Campanella's treatise *De sensu rerum*, written in the 1590s but only printed for the first time in 1620, leads this tradition into the seventeenth century,[72] which is marked by a flood of treatises on 'fascination', although these are beyond the scope of the present study.[73] It must suffice here to note the astonishing fact that the hypothesis of a visual ray emanating from the eye, and thus the 'bloody mirror', by no means disappeared despite the ever more widespread familiarity with the principles of the *camera obscura* and the discovery of the retinogram by Johannes Kepler around 1600.[74]

[71] Giovan Battista Della Porta, *Magia naturalis* (1558), lib. VIII, cap. 14 'De fascinatione, & fascinationis amuleta'; quoted after Giovan Battista Della Porta, *Natural Magick in Twenty Books*, pp. 229–32 (230) (my italics). Cf. Agrippa von Nettesheim, *De occulta philosophia libri tres* (Graphaeus edn), in Agrippa von Nettesheim, *De occulta philosophia libri tres*, ed. by Compagni, lib. I, cap. 42 (pp. 161–65: 'bloody mirror' and basilisk), lib. I, cap. 65 (p. 125: basilisk), lib. I, cap. 65 (pp. 225–27: evil eye, basilisk, discussion of Avicenna and Algazel). Cf. Agrippa's disciple Caspar Schott, *Magia universalis* (1657), I. II, praeclusio 4, prop. 2 ('bloody mirror' and basilisk). Cf. Unverzagt, 'Philosophia, Historia, Technica', pp. 48–64. The only author in this group who explicitly disputes the staining of the mirror by the gaze of a menstruating woman (and *fascinatio* by means of an emitted *spiritus* in general) and notes that this hypothesis can easily be disproved by experiments, is Thomas Erastus, *Disputationum De Medicina nova*, p. 91.

[72] Tommaso Campanella, *De sensu rerum*, lib. IV, cap. 14: 'Ex oculis manare vim immutativam, & per consensum activam' ('bloody mirror' and basilisk), in Tommaso Campanella, *Del senso delle cose e della magia*, ed. by Bruers, pp. 284–85. Cf. Tommaso Campanella, *De homine inediti theologicorum Liber* IV, ed. by Amerio, IV. 4, pp. 145–67, where Campanella pursues the old question of whether the soul is able to move or change its own and other bodies solely through the power of imagination.

[73] See for example: Leonardo Vairo, *De fascino libri tres*, lib. I, cap. 3, p. 14 ('bloody mirror' and basilisk), lib. I, cap. 10, p. 44 (basilisk) and lib. II, cap. 9, pp. 114–17 ('bloody mirror' and basilisk); Sebastian de Covarrubias, *Tesoro de la Lengua Castellana*, Aojar' on fols 76v–r ('bloody mirror'); Rodericus à Castro (Estêvão Rodrigues de Castro), *Medicus-politicus*, lib. IV, cap. 1: 'De fascinatione' — *Roderici à Castro Lusitani Medicus-politicus*, pp. 205–15 ('bloody mirror' and basilisk, in the context of a discussion of Avicenna); Johann Christian Frommann, *Tractatus De fascinatione novus et singularis. In quo fascinatio vulgaris profligatur, naturalis confirmatur, & magica examinatur[...]*, lib. I, pars I, sectio I, cap. 3, pp. 17–20 ('bloody mirror' & basilisk), lib. I, pars II, sectio I, cap. 8, pp. 108–14 ('bloody mirror'), cap. 9: pp. 115–24 (basilisk); Etienne (Stephanus) Chauvin, *Lexicon philosophicum* (1692), 'Fascinatio' ('bloody mirror' and basilisk); repr. of the second edition, Leeuwarden: Halma, 1713, p. 239.

[74] For example, the pseudo-Albertian text *De secretis mulierum* continued to be widely

Conclusion

The aim of the above discussion, tracing the tradition of the Aristotelian motif of the 'bloody mirror' through the Middle Ages to the early modern period, was to show that the claim that mirrors can be stained by the gaze of a menstruating woman — which may at first seem very strange — by no means represents a popular superstition. Instead, as we have seen, it had very specific argumentative functions in discourses which, by the standards of their time, could be regarded as belonging to the field of natural science. A second purpose of the discussion was to show that the astonishing continuity of the tradition of the 'bloody mirror' was due to the perpetuation of the underlying natural science theory that there is a material emanation from the eye, which extends to the objects gazed at, and physically touches them. The continuity of this theory was never interrupted, either by the introduction of intromissive aspects of the visual process by the so-called perspectivists, or by the ultimately unpersuasive model of a gradual infection of the air by humoral vapours or spirits that was juxtaposed to a purely intromission theory of vision. It was only after 1600 that the theory began to lose its general persuasiveness.

The 'bloody mirror' and the theory of the visual ray became mutually explanatory. Sometimes it was the staining of the mirror by the gaze of a menstruating woman that was the phenomenon in the foreground, which was 'rescued' by reference to the natural explanation of a material visual ray flowing from the eye. Conversely, on other occasions, the emphasis was on the theory of the material visual ray and the associated performativity of the gaze, the existence of which was proved by reference to the empirical phenomenon of the 'bloody mirror', familiar to everyone and certainly attested to by countless authorities. Usually, both aspects were involved simultaneously.

In addition to this general finding, there were two main groups of topics that prompted arguments based on the 'bloody mirror'. On the one hand, the misogynous view of women became increasingly widespread from the mid-thirteenth century onwards, even in educated circles, with the recovery and assimilation of Aristotelian writings. The catalyst for such views was the taboo on menstrual blood. This misogyny is to be expected of all of the authors cited, but as was seen above, it is particularly pronounced in Albertus Magnus,

popular. A French variant, including the foreword of Commentary A, which includes the 'bloody mirror', was republished at least twenty-seven times (!) during the eighteenth and nineteenth centuries. See *Le Grand et le Petit Albert*, ed. by Husson, p. 59 ('bloody mirror'); cf. pp. 92–93.

Thomas Aquinas, the pseudo-Albertian text *De secretis mulierum*, the *Malleus Maleficarum*, and successors who sought to use the 'bloody mirror' to prove the inferiority of women on the basis of natural science. This first group of topics also includes the sixteenth-century theologians, physicians, and 'natural magicians' who resorted to the same 'bloody mirror' argument in their debates with these misogynous texts, stressing the naturalness of the performative effects of the gaze in order to place such effects beyond the control of human beings and of the devil.

The second context that prompted usage of the 'bloody mirror' argument was the engagement with the teachings of Avicenna and Algazel on the capacity to move and alter a foreign body by the imaginative power of the soul alone. This engagement was already encountered among the so-called perspectivists, Albertus Magnus and Thomas Aquinas, but it was considerably intensified by the condemnation of the teaching in 1277. It led not only the philosophers of the late thirteenth and fourteenth centuries, but also numerous writers in the fifteenth and sixteenth centuries, to resort to the 'bloody mirror' argument as evidence of the materiality of the gaze in order to defend the performativity of the gaze in the face of doctrinal condemnation.

In 1739, the following experiment was attempted at the Academy in Paris: an old woman was placed in front of a mirror and was asked to stare into it intently; a vapour did in fact condense on it, which although it was not blood-red proved to be 'toxic'.[75] Paradoxically, it was this first documented 'successful' experiment carried out to obtain empirical evidence of the staining of the mirror by the gaze of a woman that led to the motif of the 'bloody mirror' being consigned to the realm of superstition after a reception history extending over two thousand years. The intention in carrying out the experiment was to prove something that could not be proven according to the natural science assumptions of the time, since the hypothesis that there is a material emanation from the eye during the visual process was by then no longer regarded as valid. As long as the material visual ray had been regarded as a valid assumption in natural science, there had been no reason to question the phenomenon of the 'bloody mirror' that had been attested by numerous witnesses.

[75] Gifford, 'The Evil Eye in Medical History', p. 239. This was enthusiastically taken up by the Neapolitan jurist and historian Nicola Valletta as outstanding evidence for the existence of the evil eye: Nicola Valletta, *Cicalata sul fascino volgarmente detto jettatura*, p. 55; *Fascino volgarmente detto jettatura*, ed. by Attardi, p. 43.

Works Cited

Primary Sources

Agrippa von Nettesheim, Heinrich Cornelius, *De occulta philosophia libri tres* (Antwerp: Johannes Graphaeus, 1533)

——, *De occulta philosophia libri tres*, ed. by V. Perrone Compagni (Leiden: Brill, 1992)

Albertus Magnus, *Albert's Refutation of the Extramission Theory of Vision and his Defence of the Intromission Theory: English Translation and Analysis of his Muslim Sources*, ed. and trans. by Cemil Akdogan (Kuala Lumpur: International Institute of Islamic Thought and Civilization, 1998)

——, *De anima*, ed. by Clemens Stroick, Opera Omnia, 8 (Münster: Aschendorff, 1968)

——, *De animalibus libri* XXVI, *nach der Kölner Urschrift*, ed. by Hermann Stadler, 2 vols (Münster: Aschendorff, 1916–20)

——, *De homine*, ed. by Henryk Anzulewicz and Joachim R. Söder, Opera Omnia, 27. 2 (Münster: Aschendorff, 2008)

——, *De Sacramentis. De incarnatione. De resurrectione*, ed. by Albert Ohlmeyer, Opera Omnia, 26 (Münster: Aschendorff, 1958)

——, *Le Grand et le Petit Albert*, ed. by Bernard Husson (Paris: Belfond, 1974)

——, *Liber de natura et origine animae. Liber de principiis motus processivi. Quaestiones super De animalibus*, ed. by Ephrem Filthaut, Opera Omnia, 12 (Münster: Aschendorff, 1955)

——, *Meteora*, ed. by Paulus Hossfeld, Opera Omnia, 6.1 (Münster: Aschendorff, 2003)

——, *On Animals: A Medieval 'Summa Zoologica'*, trans. and annotated by Kenneth F. Kitchell, Jr and Irven Michael Resnick, 2 vols (Baltimore: Johns Hopkins University Press, 1999)

——, 'Optics in Albert the Great's *De sensu et sensato*: An Edition, English Translation, and Analysis', ed. and trans. by Cemil Akdogan (unpublished doctoral dissertation, University of Wisconsin-Madison, 1978)

—— [Albert the Great], *Questions concerning Aristotle's 'On Animals'*, trans. by Irven M. Resnick and Kenneth F. Kitchell, Jr (Washington, DC: Catholic University of America Press, 2008)

——, *Secunda Pars summae de creaturis*, ed. by Auguste Borgnet, Opera Omnia, 35 (Paris: Vivès, 1896)

Alexander of Aphrodisias, *Alexandri Aphrodisiensis praeter commentaria scripta minora, De anima liber cum mantissa*, ed. by Ivo Bruns (Berlin: Reimer, 1887)

Alexandre d'Aphrodisias [Alexander of Aphrodisias], *Commentaire sur les Météores d'Aristote*, ed. by A. J. Smet (Paris: Béatrice-Nauwelaerts, 1968)

Algazel [Al-Ghazzali], *Algazel's Metaphysics: A Mediaeval Translation*, ed. by J. T. Muckle (Toronto: St Michael's College, 1933)

Apuleius of Madauros, *Pro se de magia (Apologia)*, ed. by Vincent Hunink, 2 vols (Amsterdam: Gieben, 1997)

Aquinas, Thomas, *S. Thomae Aquinatis in Aristotelis Libros De caelo et mundo, De generatione et corruptione, Meteorologicorum expositio*, ed. by Raymundi M. Spiazzi (Turin: Marietti, 1952)

——, *'Summa Theologiae': Latin Text and English Translation, Introductions, Notes, Appendices, and Glossaries*, XV: *The World Order (1a 110–19)*, ed. by M. J. Charlesworth (Cambridge: Cambridge University Press, 1970; repr. 2006)

——, *Super Epistolam B. Pauli ad Galatas Lectura: Commentary on Saint Paul's Epistle to the Galatians*, trans. by F. R. Larcher (Albany, NY: Magi Books, 1966)

Aristotle, *The Complete Works of Aristotle*, ed. by Jonathan Barnes, 2 vols (Princeton: Princeton University Press, 1984)

——, *De animalibus: Michael Scot's Arabic Latin Translation*, Part 3: Books XV/XIX: *Generation of Animals*, ed. by Aafke M. I. van Oppenraaij (Leiden: Brill, 1992)

——, *De insomniis, De divinatione per somnum*, ed. by Philip J. van der Eijk (Berlin: Akademie, 1994)

——, *Problemata varia anatomica: The University of Bologna MS 1165*, ed. by J. R. Lind (Lawrence: University of Kansas Publications, 1968)

Avicenna, *Avicenna Latinus: Liber de Anima seu Sextus de naturalibus* IV–V, ed. by S. van Riet and G. Verbeke (Leiden: Brill, 1968)

Bacon, Roger, *Opera quaedam hactenus inedita*, I: *Opus tertium, Opus Minus, Compendium Philosophiae*, ed. by J. S. Brewer (London: Longman, Green, and Roberts, 1859; repr. Nendeln: Kraus, 1965)

——, *The 'Opus Majus' of Roger Bacon*, ed. by John Henry Bridges, 2 vols (Oxford: Clarendon Press, 1897–1900; repr. Frankfurt am Main: Minerva, 1964)

——, *The Opus Majus of Roger Bacon*, trans. by Robert Belle Burke, 2 vols (Philadelphia: University of Pennsylvania Press, 1928; repr. Bristol: Thoemmes, 2000)

——, *Roger Bacon and the Origins of 'Perspectiva' in the Middle Ages: A Critical Edition and English Translation of Bacon's 'Perspectiva' with Introduction and Notes*, ed. and trans. by David C. Lindberg (Oxford: Clarendon Press, 1996; repr. 2006)

——, *Roger Bacon's Philosophy of Nature: A Critical Edition, with English Translation, Introduction, and Notes, of 'De multiplicatione specierum' and 'De speculis comburentibus'*, ed. and trans. by David C. Lindberg (Oxford: Clarendon Press, 1983)

Browne, Sir Thomas, *Pseudodoxia Epidemica*, ed. by Robin Robbins, 2 vols (Oxford: Clarendon Press, 1981)

Campanella, Tommaso, *De homine inediti theologicorum liber* IV, ed. by Romano Amerio (Rome: Centro Internazionale di Studi Umanistici, 1960)

——, *Del senso delle cose e della magia: Testo inedito Italiano con le varianti dei codici e delle edizioni Latine*, ed. by Antonio Bruers (Bari: Laterza, 1925)

——, *De sensu rerum* (Frankfurt: Tobia Adami, 1620)

Cardano, Girolamo, *Opera omnia*, 10 vols (Lyons: Huguetan & Ravaud, 1663); <http://www.cardano.unimi.it/testi/opera.html> [accessed 1 May 2013]

Castañega, Fray Martín de, *Tratado de las supersticiones y hechizerias y de la possibilidad y remedio dellas* (1529), ed. by Fabián Alejandro Campagne (Buenos Aires: UBA, Facultad de Filosofía y Letras, 1997)

Chauvin, Etienne (Stephanus), *Lexicon Philosophicum, 2nd edn* (Düsseldorf: Stern-Verlag Janssen, 1967 [1692])

La Condamnation parisienne de 1277, ed. by David Piché (Paris: Vrin, 1999)

Conrad of Megenberg, *Das 'Buch der Natur'*, vol. II: *Kritischer Text nach den Handschriften*, ed. by Robert Luff and Georg Steer (Tübingen: Niemeyer, 2003)

——, *Das Buch der Natur von Conrad von Megenberg: Die erste Naturgeschichte in deutscher Sprache: In Neu-Hochdeutscher Sprache bearbeitet und mit Anmerkungen versehen*, ed. and trans. by Hugo Schulz (Greifswald: Julius Abel, 1897)

Covarrubias, Sebastian de, *Tesoro de la Lengua Castellana, o Española* (Madrid: Luis Sanchez, 1611; repr. Barcelona: Alta Fulla, 1993)

Della Porta, Giovan Battista [John Baptista Porta], *Natural Magick in Twenty Books [...] wherein are set forth all the riches and delights of the natural sciences* (London: Young & Speed, 1658); <http://hdl.loc.gov/loc.rbc/pre1801.23451.1> [accessed 1 May 2013]

Diels, Hermann, and Walther Kranz, *Die Fragmente der Vorsokratiker* (Berlin: Weidmann, 1903)

Engelbert von Admont, 'Engelberti Admontensis Tractatus de fascinatione', ed. by George B. Fowler, *Recherches de Théologie ancienne et médiévale*, 37 (1970), 187–231

Erastus, Thomas, *Disputationum De Medicina nova Philippi Paracelsi Pars Prima [...]* (Basel: Peter Perna, *c.* 1572)

Ficino, Marsilio, *Commentary on Plato's 'Symposium' on Love*, trans. by Sears Jayne (Dallas: Spring Publications, 1985)

Geiler von Kaysersberg, Johann, *Die Emeis: Dis ist das buoch von der Omeissen: Vnnd auch herr der künnig ich diente gern: Vnd sagt von Eigenschaft der Omeissen: Vnd gibt vnderweisung von den Vnholden oder Hexen, vnd von gespenst der geist [...]* (Strasbourg: Johann Grienninger, 1517)

——, *Zur Geschichte des Volks-Aberglaubens im Anfange des* XVI. *Jahrhunderts: Aus Dr. Joh. Geilers von Kaisersberg Emeis*, ed. by August Stöber (Basel: Schweighauser'sche Verlagsbuchhandlung, 1856)

Giles of Rome, *Errores Philosophorum*, ed. by Joseph Koch, trans. by John O. Riedl (Milwaukee: Marquette University Press, 1944)

Grosseteste, Robert, *Die philosophischen Werke des Robert Grosseteste, Bischofs von Lincoln*, ed. by Ludwig Baur (Münster: Aschendorff, 1912)

——, *Opera Roberti Grosseteste Lincolniensis*, ed. by James McEvoy and others (Turnhout: Brepols, 1995)

Institoris [Kramer], Henricus, and Jacobus Sprenger, *Malleus Maleficarum*, ed. and trans. by Christopher S. Mackay, 2 vols (Cambridge: Cambridge University Press, 2006)

Kircher, Athanasius, *Athanasii Kircheri e Soc. Iesu, Scrutinium physico-medicum Contagiosae Luis, quae Pestis dicitur: Quo origo, causa, signa [...] una cum appropriatis remediorum Antidotis nova doctrina in lucem eruuntur* (Rome: Typis Mascardi, 1658)

Lavater, Ludwig, *De spectris, lemuribus et magnis atque insolitis fragoribus [...] liber unus* (Zurich: Johannes Crispinus, 1570)

Neckam, Alexander, *De naturis rerum: De laudibus divinae sapientiae*, ed. by Thomas Wright (London: Longman, 1863; repr. Lavis: La Finestra, 2003)

Oresme, Nicole, 'Nicole Oresme and the Marvels of Nature: A Critical Edition of his *Quodlibeta* with English Translation and Commentary', ed. and trans. by Bert Hansen (unpublished doctoral dissertation, Princeton University, 1973)
——, *Nicole Oresme and the Marvels of Nature: A Study of his 'De causis mirabilium' with Critical Edition*, trans. by Bert Hansen (Toronto: Pontifical Institute of Mediaeval Studies, 1985)
——, *Nicole Oresme and the Medieval Geometry of Qualities and Motions: A Treatise on the Uniformity and Difformity of Intensities known as 'Tractatus de configurationibus qualitatum et motuum'*, ed. by Marshall Clagett (Madison: University of Wisconsin Press, 1968)
——, 'Nicole Oresme on Light, Color, and the Rainbow: An Edition and Translation, with Introduction and Critical Notes, of Part of Book Three of his *Questiones Super Quatuor Libros Meteorum*', ed. and trans. by Stephen Clement McCluskey (unpublished doctoral dissertation, University of Wisconsin, 1974)
——, *Le 'Quaestiones de sensu' attribuite a Oresme e Alberto di Sassonia*, ed. by Jole Agrimi (Florence: La Nuova Italia, 1983)
Paracelsus, *Sämtliche Werke: Nach der 10bändigen Huserschen Gesamtausgabe (1589–1591): Zum erstenmal in neuzeitliches Deutsch übersetzt*, ed. by Bernhard Aschner, 4 vols (Jena: Fischer, 1926–32)
Pecham, John, *John Pecham and the Science of Optics — 'Perspectiva communis'*, ed. by David C. Lindberg (Madison: University of Wisconsin Press, 1970)
——, *Quodlibet Romanum*, ed. by Ferdinand Delorme (Rome: Cuggiani, 1938)
——, *Tractatus de Perspectiva*, ed. by David C. Lindberg (St Bonaventure, NY: Franciscan Institute, 1972)
——— [Pecham Fr Ioannis], *Quodlibeta quatuor*, ed. by Girard J. Etzkorn and Ferdinand Delorme (Grottaferrata: Collegio S. Bonaventura, Padri Editori di Quaracchi, 1989)
Peter of Limoges, *The Moral Treatise on the Eye*, ed. by Richard Newhauser (Toronto: Pontifical Institute of Mediaeval Studies, 2012)
Petrus de Abano, *Conciliator differentiarum philosophorum et medicorum* (Venice: Johannes Herbort, 1483)
Proclus, *Commentaire sur la République*, ed. by A. J. Festugière, 3 vols (Paris: Vrin, 1970)
Pseudo-Albertus Magnus, 'Pseudo-Albertus Magnus: *Secreta mulierum cum commento*, Deutsch: Critical Text and Commentary', ed. by Margaret Schleissner (unpublished doctoral dissertation, Princeton University, 1987)
——, *Women's Secrets: A Translation of Pseudo-Albertus Magnus's 'De Secretis Mulierum' with Commentaries*, ed. and trans. by Helen Rodnite Lemay (Albany: State University of New York Press, 1994)
Rodericus à Castro (Estêvão Rodrigues de Castro), *Roderici à Castro Lusitani Medicus-politicus: sive de officiis medico-politicis tractatus, quatuor distinctus Libris [...]* (Hamburg: Georg Ludwig Frobenius, 1614)
Scot, Reginald, *The Discovery of Witchcraft* (1584), ed. by Montague Summers (New York: Dover, 1972)
Solinus, *C. Ivlii Solini collectanea rerum memorabilium*, ed. by T. Mommsen (Berlin: Friedrich Nicolai, 1864)

Thomas Cantimpratensis, *Liber de natura rerum: Editio princeps secundum codices manuscriptos*, ed. by Helmut Boese (Berlin: de Gruyter, 1973)

Trattati d'arte del cinquecento, fra manierismo e controriforma, ed. by Paola Barocchi, 3 vols (Bari: Laterza, 1960–62)

Vairo, Leonardo, *De fascino libri tres, in quibus omnes fascini species et causae optima methodo describuntur, & ex philosophorum ac theologorum sententiis scitè & eleganter explicantur [...]* (Paris: Nicolaus Chesneau, 1583)

Valletta, Nicola, *Cicalata sul fascino volgarmente detto jettatura* (Naples: M. Morelli, 1787)

——, *Fascino volgarmente detto jettatura*, ed. by Umberto Attardi (Naples: Colonnese Editore, 1988)

Webster, John, *The Displaying of Supposed Witchcraft...* (London: J. M., 1677); <http://astronomie-rara.ethbib.ethz.ch/cgj/alch/content/titleinfo/2581467> [accessed 1 May 2013]

Wier [Weier], Johannes, *'De praestigiis daemonum': Von Teuffelsgespenst, Zauberern und Gifftbereytern, Schwartzkünstlern, Hexen und Unholden, darzu irer Straff, auch von den Bezauberten, und wie ihnen zuhelffen sey [...]* (Frankfurt am Main: Basseus, 1586; repr. Darmstadt: Bläschke, 1969)

——, *'De praestigiis daemonum': Witches, Devils, and Doctors in the Renaissance*, ed. by George Mora (Binghamton: Center for Medieval and Early Renaissance Studies, 1991)

Secondary Studies

Akbari, Suzanne Conklin, *Seeing through the Veil: Optical Theory and Medieval Allegory* (Toronto: University of Toronto Press, 2004)

Allen, Prudence, *The Concept of Woman: The Aristotelian Revolution 750 BC–AD 1250* (Grand Rapids, MI: Eerdmans, 2002)

Allison, Dale C., 'The Eye as a Lamp: Finding the Senses', in *The Jesus Tradition in Q* (Harrisburg, PA: Trinity Press, 1997), pp. 133–67

Anzulewicz, Henryk, 'Perspektive und Raumvorstellung in den Frühwerken des Albertus Magnus', in *Raum und Raumvorstellung im Mittelalter*, ed. by Jan A. Aertsen and Andreas Speer (Berlin: de Gruyter, 1998), pp. 249–86

Bächtold-Stäubli, Hanns, and Eduard Hoffmann-Krayer, eds, *Handwörterbuch des Deutschen Aberglaubens*, 10 vols (Berlin: de Gruyter, 1927–42)

Betz, Hans Dieter, 'Matthew vi.22f and Ancient Greek Theories of Vision', in *Text and Interpretation: Studies in the New Testament Presented to Matthew Black*, ed. by Ernest Best and Robert McLachlan Wilson (Cambridge: Cambridge University Press, 1979), pp. 43–56

Breiner, Laurence A., 'The Basilisk', in *Mythical and Fabulous Creatures: A Source Book and Research Guide*, ed. by Malcolm South (Westport, CT: Greenwood, 1987), pp. 112–22

Campbell, Anna M., *The Black Death and Men of Learning* (New York: Columbia University Press, 1931)

Crombi, Alistair C., *Robert Grosseteste and the Origins of Experimental Science, 1100–1700* (Oxford: Clarendon Press, 1953)
Darst, David H., 'Witchcraft in Spain: The Testimony of Martín de Castañega's Treatise on Superstition and Witchcraft (1529)', *Proceedings of the American Philosophical Society*, 123 (1979), 298–322
Dickie, Matthew, 'The Fathers of the Church and the Evil Eye', in *Byzantine Magic*, ed. by Henry Maguire (Washington: Dumbarton Oaks, 1995), pp. 9–34
Flasch, Kurt, *Aufklärung im Mittelalter? Die Verurteilung von 1277: Das Dokument des Bischofs von Paris, eingeleitet, übersetzt und erklärt* (Mainz: Dieterich, 1989)
Gifford, E. S., 'The Evil Eye in Medical History', *American Journal of Ophthalmology*, 44 (1957), 237–42
Goñi Gaztambide, José, 'El tratado "De Superstitionibus" de Martín de Andosilla', *Cuadernos de Etnología y Etnografía de Navarra*, 9 (1971), 249–322; *Príncipe de Viana* 71, 251 (2010), 1011–69
Gottschall, Dagmar, *Konrad von Megenbergs Buch von den natürlichen Dingen: ein Dokument der deutschsprachigen Albertus Magnus-Rezeption im 14. Jahrhundert* (Leiden: Brill, 2004)
Grabmann, Martin, 'Gentile da Cingoli, ein italienischer Aristoteleserklärer aus der Zeit Dantes', in *Sitzungsberichte der Bayerischen Akademie der Wissenschaften, Philosophisch-Historische Abteilung* 1940, 9 (Munich: Verlag der Bayerischen Akademie der Wissenschaften, 1941), pp. 1–88
Hansen, Joseph, *Quellen und Untersuchungen zur Geschichte des Hexenwahns und der Hexenverfolgung im Mittelalter* (Hildesheim: Georg Olms, 1963)
Harrer, Michael, and Gerhart Harrer, 'Paracelsus und der Basilisk — Eine Reise in das Reich der Fabeltiere und Monstren', in *Paracelsus und die Tiere, 21. Paracelsustag 2002* (Vienna: Österr. Kunst- und Kulturverlag, 2003), pp. 43–57
Hasse, Dag Nikolaus, *Avicenna's 'De Anima' in the Latin West: The Formation of a Peripatetic Philosophy of the Soul 1160–1300* (London: Warburg Institute, 2000)
Havelange, Carl, *De l'oeil et du monde: Une histoire du regard au seuil de la modernité* (Paris: Fayard, 1998)
Hub, Berthold, 'Material Gazes and Flying Images in Marsilio Ficino and Michelangelo', in *Spirits Unseen: The Representation of Subtle Bodies in Early Modern European Culture*, ed. by Christine Göttler (Leiden: Brill, 2008), pp. 93–120
——, *Die Perspektive der Antike: Archäologie einer symbolischen Form* (Frankfurt am Main: Lang, 2008), pp. 264–321
——, 'Visual Piety and Visual Theory: Was There a Paradigm Shift?', in *Semantics of Vision: Art Production and Visual Cultures in the Middle Ages*, ed. by Raphaèle Preisinger (Turnhout: Brepols, forthcoming)
Jacquart, Danielle, and Claude Thomasset, *Sexuality and Medicine in the Middle Ages* (Cambridge: Polity Press, 1988)
Kodera, Sergius, *Disreputable Bodies: Magic, Medicine, and Gender in Renaissance Natural Philosophy* (Toronto: Centre for Reformation and Renaissance Studies, 2010)

——, 'Narcissus, Divine Gazes and Bloody Mirrors: The Concept of Matter in Ficino', in *Marsilio Ficino: His Theology, his Philosophy, his Legacy*, ed. by Michael J. B. Allen and others (Leiden: Brill, 2002), pp. 285–306

Kovach, Francis J., 'The Enduring Question of Action at a Distance in Saint Albert the Great', in *Albert the Great: Commemorative Essays*, ed. by Francis J. Kovach and Robert W. Shahan (Norman: University of Oklahoma Press, 1980), pp. 161–235

Lefebvre, R., 'Le Miroir de l'âme (Aristote, *De insomniis* 459b 25 — 460a 26)', *Les Études Philosophiques*, 63 (1988), 195–206

Leyerle, Blake, 'John Chrisostom on the Gaze', in *Journal of Early Christian Studies*, 1 (1993), 159–74

Limberis, Vasiliki, 'The Eye Infected by Evil: Basil of Caesarea's Homily on Envy', *Harvard Theological Review*, 84 (1991), 163–84

Lindberg, David C., *Theories of Vision from Al-Kindi to Kepler* (Chicago: University of Chicago Press, 1976)

Marcus, Joel, 'A Note on Markan Optics', *New Testament Studies*, 45 (1999), 250–56

Nelson, Robert S., 'To Say and to See: Ekphrasis and Vision in Byzantium', in *Visuality before and beyond the Renaissance: Seeing as Others Saw*, ed. by Robert S. Nelson (Cambridge: Cambridge University Press, 2000), pp. 143–68

Preus, Anthony, 'On Dreams 2, 459b24–460a33, and Aristotle's οψις', *Phronesis*, 13 (1968), 175–82

Rakoszy, Thomas, *Böser Blick, Macht des Auges und Neid der Götter: Eine Untersuchung zur Kraft des Blickes in der griechischen Literatur* (Tübingen: Narr, 1996)

Salmón, Fernando, and Montserrat Cabré, 'Fascinating Women: The Evil Eye in Medical Scholasticism', in *Medicine from the Black Death to the French Disease*, ed. by Roger French and others (Aldershot: Ashgate, 1998), pp. 53–84

Sammer, Marianne, *Der Basilisk: Zur Natur- und Bedeutungsgeschichte eines Fabeltieres im Abendland* (Munich: Institut für Bayerische Literaturgeschichte, 1998)

Schleissner, Margaret, 'Sexuality and Reproduction in the Late Medieval *Problemata Aristotelis*', in *Licht der Natur: Medizin in Fachliteratur und Dichtung: Festschrift für Gundolf Keil zum 60. Geburtstag*, ed. by Joseph Domes (Göppingen: Kümmerle, 1994), pp. 383–98

Schleusener-Eichholz, Gudrun, *Das Auge im Mittelalter*, 2 vols (Munich: Fink, 1985)

Shankman, A., *Aristotle's 'De insomniis': A Commentary* (Leiden: Brill, 1992)

Simon, Gérard, *Le Regard, l'être et l'apparence dans l'optique de l'antiquité* (Paris: Seuil, 1988)

Siraisi, Nancy G., *Taddeo Alderotti and his Pupils: Two Generations of Italian Medical Learning* (Princeton: Princeton University Press, 1981)

Sprague, Rosamond Kent, 'Aristotle on Red Mirrors (*On Dreams* II 459b24 — 460a23)', *Phronesis*, 30 (1985), 323–26

Theiss, Peter, *Die Wahrnehmungspsychologie und Sinnesphysiologie des Albertus Magnus — ein Modell der Sinnes — und Hirnfunktion aus der Zeit des Mittelalters: Mit einer Übersetzung aus 'De anima'* (Frankfurt am Main: Lang, 1997)

Theiss, Peter, and Grüsser, Otto-Joachim, 'Vision and Cognition in the Natural Philosophy of Albert the Great', *Documenta Ophthalmologica: The Journal of Clinical Electrophysiology of Vision*, 86.2 (1994), 123–51

Thorndike, Lynn, *A History of Magic and Experimental Science*, III: *Fourteenth and Fifteenth Centuries* (New York: Columbia University Press, 1934; repr. 1980)
——, 'Imagination and Magic: The Force of Imagination on the Human Body and of Magic on the Human Mind', in *Mélanges Eugène Tisserant*, vol. 7: Bibliothèque Vaticane (Città del Vaticano: Biblioteca Apostolica Vaticana, 1964), pp. 353–58
Ubl, Karl, *Engelbert von Admont: Ein Gelehrter im Spannungsfeld von Aristotelismus und christlicher Überlieferung* (Vienna: Oldenbourg, 2000)
Unverzagt, Dietrich, 'Philosophia, Historia, Technica — Caspar Schotts *Magia universalis*' (unpublished doctoral dissertation, Technische Universität Berlin, 2000)
Weststeijn, Thijs, 'Seeing and the Transfer of Spirits in Early Modern Art Theory', in *Renaissance Theories of Vision*, ed. by John Shannon Hendrix and Charles H. Carman (Farnham: Ashgate, 2010), pp. 149–69
Zambelli, Paola, 'L'immaginazione e il suo potere: Da al-Kindi, al-Farabi e Avicenna al Medioevo latino e al Rinascimento', in *Orientalische Kultur und Europäisches Mittelalter*, ed. by Alberti Zimmermann and Ingrid Craemer-Ruegenberg (Berlin: de Gruyter, 1985), pp. 188–206

Specular Art and Science: Mirror Metaphor in Medieval Alchemical Texts

Anna Dysert

Scientific texts form a small but intriguing subset of mirror-title texts which became popular during what Herbert Grabes identifies as the flowering period of the *speculum* genre, the thirteenth and fourteenth centuries. The appearance of these 'scientific' mirrors overlaps tellingly, this article will argue, with the peak of Scholastic study in Western Europe. The integration of the rational approach and theoretical framework of Arabic natural philosophy, adapted and adopted from the Greek tradition, transformed European learned culture through the vehicle of twelfth-century Arabic-Latin translations and culminated in the epistemological methodology known as Scholasticism. This method of teaching and learning aided in establishing new textual genres, ways of using books, and avenues of book production and exposition. Scholasticism, with its Aristotelian framing, shaped not only the ways in which knowledge was parsed, transmitted, and elaborated, but the objects of study themselves — namely, what was considered to fall under the rubric of 'scientific' knowledge. As William Wallace notes, in centres of learning such as Paris and Oxford around the year 1200, scholars were faced with 'vast branches of new learning' coming from the Jewish and Arab world which 'had to be assimilated

Anna Dysert (anna.dysert@mcgill.ca) is a PhD candidate in history at McGill University and Assistant Librarian at the Osler Library of the History of Medicine. Her dissertation centres on the diffusion of medical knowledge in Europe during the twelfth century through the study of manuscripts, especially the Isaac Israeli *Diets*.

The Mirror in Medieval and Early Modern Culture: Specular Reflections, ed. by Nancy M. Frelick, CURSOR 25 (Turnhout: Brepols, 2016), pp. 73–87 10.1484/M.CURSOR-EB.5.111042
BREPOLS PUBLISHERS

into their existing syntheses',[1] most notably in the realm of natural philosophy. Thus, the appearance of newly composed mirror-texts such as the *Specula alchemiae* attributed respectively to Roger Bacon and Arnold of Villanova and Vincent of Beauvais's *Speculum maius* occurs during this period of negotiation of knowledge categories. More precisely, the texts' appearance coincides with a philosophical discussion over the relationship of the medieval concepts of *ars*, craft-based knowledge, to *scientia*, the Aristotelian definition of knowledge based on first principles, particularly in the types of new learning that are being derived from non-Latin traditions. Fittingly, the intellectual heritage of alchemical *speculum* texts is dependent upon a complex mirror-symbolism that takes into account the mirror as an object of both scientific (in the Aristotelian sense) and technological investigation.

There have been various attempts to explain the metaphor at work in the medieval mirror title: Grabes defines four principal types of mirror texts (encyclopaedic, exemplary, prognostic, and fantastic) with various subdivisions. In the literary sphere, critics have looked for classical literary antecedents for mirror symbolisms — finding them in the work of such writers as Seneca and the Neoplatonists — and defined the mirror function as both a metaphor for self-reflection and examination as well as a vessel for seeing indirectly that which cannot usually be seen (a function which automatically seems to make the category particularly apt for theological and moral works).[2] In Grabes's classification, alchemical works fall under the category of 'factually informative mirrors', those titles for which the reflective function of the metaphorical mirror is straightforward, laying out a body of learning as a mirror reflects a human face.[3] As we shall see, the reflection is less clear when we take into consideration the intellectual context surrounding both mirror technology and alchemical theory and practice. In the relationship between the subject matter of an alchemical text and the designation *speculum*, the mirror serves not only as a metaphor for the book as a reflection of a body of learning, but also as a symbol for how knowledge, specifically scientific knowledge, had come to be organized and

[1] Wallace, 'Philosophical Setting of Medieval Science', p. 95.

[2] An example is Einar Már Jónsson, *Le Miroir*. Another early piece of scholarship on the mirror-title is Bradley, 'Backgrounds of the Title *Speculum*'.

[3] Grabes, *The Mutable Glass*, trans. by Collier. He describes the function of the titular mirror in the 'factually informative mirror' of a 'specific branch of knowledge' as follows: 'the value placed on the exposition of a particular kind of subject-matter, on the recording and conveying of one area of knowledge, is such that there can be no question of the mirror's reflecting function being outweighed by other intentions' (pp. 45–46).

understood after the intellectual expansion of the twelfth-century Renaissance. The medieval scientific understanding of the mirror as a material technology built into a theoretical structure provides a parallel with what fields such as alchemy and medicine were attempting to do to redeem their practical orientation for wider acceptance and academic legitimization — creating a rational explanatory framework.

During the Renaissance of the twelfth century, the movement of ideas and a growing interest in the rediscovery and absorption of earlier forms of written knowledge and technical know-how changed learned Latin European society intellectually and commercial society economically. Since the first half of the twentieth century, scholars have described a thirteenth-century industrial revolution and attempted to parse its implications for modernity and civilizational progress.[4] More recently, scholars such as Elspeth Whitney have been interested in a more nuanced 'medieval philosophy of technology', that is, the 'metaphysical and ontological status accorded to craftsmanship and the process of invention in medieval thought'.[5] Among the craft technologies that Western Europe relearned during the period in question was glass mirror technology. Polished brass mirrors were slowly replaced by glass ones as glass manufacturers began producing flattened glass in the twelfth century and colourless glass in small European workshops around the beginning of the thirteenth century.[6] Glass technologies and instruments, which had been a part of Arabic scientific usage, such as mirrors and lenses, were reintroduced and adopted.[7] Concurrently, Arabic works of natural philosophy, including reworkings, abridgements, and compilations of ancient Greek scientific learning, were entering Europe in Latin translation and bringing with them more developed theories of vision and optics, as well as the new or expanded fields of astrology, medicine, mathematics, and alchemy.[8] The impact of these intellectual imports and rediscover-

[4] The desire to link medieval technological developments and accompanying intellectual and socio-cultural shifts to modern institutions and modes of thinking can be found notably in the work of Lynn White and Pierre Duhem. See, e.g., Duhem, *Le Système du monde*; White, 'Technology and Invention in the Middle Ages' and *Medieval Technology and Social Change*. An overview of the study of the Middle Ages related to narratives of progress can be found in Ovitt, *The Restoration of Perfection*.

[5] Whitney, 'Paradise Restored', p. 1.

[6] Glick, Livesey, and Wallis, *Medieval Science, Technology, and Medicine*, p. 200.

[7] MacFarlane and Martin, *Glass: A World History*, pp. 27–50.

[8] Some examples in the field of optics are the works of Al-Kindi (*c.* 810–73), whose work on optics in Latin translation was particularly influential for Roger Bacon, and Ibn al-Haytham

ies was one of cultural revitalization, as was originally explored in the work of Charles Homer Haskins and updated by many medievalists since.[9]

With these developments came various attempts to redress the inherited Aristotelian hierarchy of knowledge. While various theories about the nature of knowledge were postulated during the twelfth century, such as Hugh of St Victor's concept of salvific knowledge,[10] the integration of new texts and commentaries of Aristotle by the thirteenth century pushed the philosopher's categories to the forefront. Aristotelian tradition set up a particular knowledge taxonomy that puts 'practice' in opposition to 'theory', *ars*, as in technological, mechanical, or craft knowledge, in opposition to *scientia*, or knowledge based on first principles and attained through the exercise of reason. The distinction between theory and practice was retained by the Arabic rational science that came into the Latin West, but the relationship between the two is not quite the same as, say, in the stricter interpretation of Thomas Aquinas. Rather, as a relationship was posited between the arts and the sciences, the practical arts were thus recast and reclassified. Although medieval philosophy is not monolithic, there is distinguishable movement, as Whitney writes, to develop 'positive frameworks — both religious and scientific — for integrating technology as a category of knowledge into their thought'.[11] One notable case is with the Islamic import of mathematics. Mathematics was described as 'speculative science', that is, a *scientia* in Aristotelian terms, but was further divided into rational mathematics, a branch practiced through reason alone, and sensible mathematics, the investigation of which lends itself to an understanding of practical crafts. This is the hierarchy constructed by al-Farabi, whose tenth-century taxonomy of the arts and sciences was influential in its Latin translation, created in Toledo during the mid-twelfth century.[12] Another well-explored case is with medicine, as university physicians and medical writers redubbed their

(Alhazen) (965–1040), author of the influential *Kitab al Manazir* (Book of Optics). An overview is available in Lindberg, *Theories of Vision*.

[9] Although Haskins did not originally include technology in his identification of the twelfth-century Renaissance, as Whitney notes ('Paradise Restored', p. 6), it seems that the idea has since then been absorbed wholeheartedly into his vision of a revitalized cultural landscape after the twelfth century. See, e.g., Benson, Constable, and Lanham, *Renaissance and Renewal in the Twelfth Century*, and Collette and Garrett-Goodyear, *The Later Middle Ages*, which includes the section '*Scientia*: Knowledge Practical, Theoretical, and Historical'.

[10] Whitney, 'Paradise Restored', pp. 89–90.

[11] That is, Aristotelian 'knowledge'; Whitney, 'Paradise Restored', p. 8.

[12] Franklin-Brown, *Reading the World*, p. 78.

field *physica*, aligning it in name with natural philosophy and in text by elaborating it with the new expositional methods of Scholasticism, as we will explore further.[13] In the Aristotelian framework as modified by the injection of Arabic philosophical adaptions and promoted by practitioners, craft knowledge was regarded less pejoratively and given a place in the organization of knowledge by virtue of its connection to rational knowledge.

Vincent of Beauvais's treatment of the art and science of mirrors gives us an insight into the Scholastic mirror-text.[14] His *Speculum maius* was written in the 1240s and revised during the 1250s after the full, formal acceptance of the Aristotelian curriculum into the university of Paris. It aimed to represent the whole range of Scholastic learning and was notably 'the only Latin encyclopedia of the century to include, in addition to natural history, [...] the liberal and mechanical arts'.[15] In the first part, called the *Speculum naturale*, the section of the work dedicated to the exposition of natural philosophy, Vincent presents a discussion of mirrors literally framed within a scientifically oriented section explaining the properties and theories of metals. He discusses the nature of silver followed by considerations of its practical applications for alchemy, medicine, and making mirrors.[16] Through this association, mirror making is seen to occupy the same medial location between art and science. Earlier in Book II of the *Speculum naturale*, Vincent explains how mirrors function in a handful of chapters tucked within a section devoted to the nature of light.[17] In the *Speculum doctrinale*, Vincent follows his explanation of the science of *perspectiva* within his section on mathematics with an exposition of mirrors. In the 1494 printed edition, the ordering of sections reflects this intellectual arrangement: *De scientia perspectiva* (On the Science of Perspective) immediately precedes *De radiis visualibus et de speculis* (On the Roots of Vision and of Mirrors).[18] In this organizational schema, he is effectively aligning the *practica* of optics with the *theorica* of optics, placing everything side by side under the same heading of

[13] Faith Wallis provides an overview of this concept in her volume of source texts, *Medieval Medicine*, pp. 129–30.

[14] The *Speculum maius* as a Scholastic example of the *speculum* genre is also discussed by Einar Már Jónsson in 'Le Sens du titre *Speculum*'; in this case, it is situated within an analysis based upon earlier monastic and theological mirrors.

[15] Franklin-Brown, *Reading the World*, p. 96.

[16] Vincent of Beauvais, *Speculum naturale*, bk VII, chap. 72.

[17] Vincent of Beauvais, *Speculum naturale*, bk II, pp. 72–76.

[18] Vincent of Beauvais, *Speculum doctrinale*, bk XVIII, chaps 43–44, fol. 247r of the 1494 edn.

speculum. The craft of mirror making is shown through this hierarchic decision to be a derivative of speculative science and is even considered a 'way of supplying information' to the theoretical science in a way that seems more rooted in an Arabic hierarchy of the sciences than in a stricter Aristotelian scheme.[19] Indeed, Vincent of Beauvais's treatment of mirrors is believed to have been based on a lost treatise on the division of the sciences by Michael Scot, who, following Aristotle, divides philosophy into theory and practice, but adapts this division to allow the practical and mechanical arts to be subsumed under science. Practical types of knowledge are classified according to the speculative sciences to which they are related: natural philosophy, for example, allows for the practical applications, most notably, of medicine, alchemy, and mirrors.[20] In this way, craft technologies with strong theoretical roots and relations are subject to the same taxonomical difficulty as disciplines perhaps better known for their division into practical and theoretical parts, such as medicine.

Medicine is the most immediate and perhaps best-studied example of the classificatory dilemma at hand: the new textual medicine being translated and studied during the twelfth century had at its base the intellectual method and framework of natural philosophy. Its proponents and elaborators styled it a *physica*, or natural philosophy, and developed a body of canonical texts and glosses, created programs of study, and designed books that reflected this new epistemological status. New textual models inspired by the Scholastic *mise-en-page* developed in theology and canon law texts such the *Glossa ordinaria* and the *Decretals* were exploited for the teaching and exposition of medical texts.[21] There remained a separate (and according to a strictly Aristotelian perspective, unrelated) genre of so-called medical *practica*, but many writers, compilers, and commenters were attempting to impose rational trappings on areas of *practica*, thereby absorbing it into a rational, theory-based framework.

Alchemy also faced a classification problem, although it was not as successful in integrating its craft elements into a philosophical order to give it some intellectual staying power or academic credibility, as William Newman discusses.[22] Another twelfth-century import of the Arab world, alchemy had during this early period many points in favour of its being recognized as a 'legitimate'

[19] As described in Whitney, 'Paradise Restored', pp. 129–45.

[20] Whitney, 'Paradise Restored', p. 135.

[21] See, e.g., Parkes, 'Influence of the Concepts of *Ordinatio* and *Compilatio*', and Rouse and Rouse, '*Statim invenire*'.

[22] Newman, 'Technology and Alchemical Debate in the Late Middle Ages'.

scientia — and indeed, there were certainly practitioners and intellectuals with an interest in seeing it legitimized. As in all areas of medieval knowledge, intellectual authority was derived from real or imagined roots in antiquity. Medieval alchemists actively extolled their field's illustrious heritage: Adam, Noah, Moses, Aristotle, and John the Evangelist were among its claimed practitioners, as we can see in one fourteenth-century example of the pseudo-Avicennan 'On the Soul in Alchemy'.[23]

The creator of this witness, as well as many other people such as educated clerics, physicians, university masters, and students, would most likely have been interested in seeing the study of alchemy 'canonized' or intellectually sanctioned. One example is found in Taddeo Alderotti, a professor of medicine and logic at Bologna during the second half of the thirteenth century, who was an advocate of absorbing the new Latin translations of Galen and Avicenna into the university medical canon, as well as an adept of alchemical techniques. He was interested particularly in techniques of alcohol distillation, an alchemical operation of course.[24] The refining operations employed were imagined to correspond to a process of spiritual purification, but it was also literally a search to refine the elements to 'improv[e] the nature of things', as the author of the pseudo-Avicennan alchemical treatise writes, or as the author of an anonymous fourteenth-century *Speculum alchemiae* (at one time attributed to Arnold of Villanova) indicates, 'ut ad perfectam scientam pervenire possimus' (in order that we are able to arrive at the perfection of science),[25] and thus presented opportunities for both improved pharmacological and laboratory techniques. In the thirteenth and fourteenth centuries, the legitimization of alchemy as a science, as in the case of medicine, was a question of rehabilitating practice and aligning it with theory.

[23] The list of 'artis magistri' given in the *Speculum doctrinale*, bk XI, chap. 107 and in the *Speculum naturale*, bk VII, chap. 87, appears indebted to the longer listing given in *De anima*. Pseudo-Avicenna, *De anima in arte alchemiae*, Montreal, Osler Library of the History of Medicine, B.O. 480, fols 334–33v. See Moreau, 'Les Sources alchimiques de Vincent de Beauvais', p. 64, for comparison of the passages.

[24] Siraisi, *Taddeo Alderotti and his Pupils*. For a discussion of the development of the medical curriculum at Bologna, see Wallis, *Medieval Medicine*, pp. 197–204. For the crossover between medicine and alchemy during the Middle Ages, see Crisciani and Bagliani, *Alchimia e medicina nel Medioevo*, particularly McVaugh's chapter, 'Alchemy in the *Chirurgia* of Teodorico Borgognoni'.

[25] Moreau, 'Some considerations concerning the alchemy of the *De anima in arte alchemiae* of Pseudo-Avicenna', p. 50; Pseudo-Arnold of Villanova, *Speculum alchymiae*, p. 515.

Thus, considering the context for the medieval understanding of the mirror as a scientific and material object in conjunction with medieval alchemy's intellectual challenges, what does the designation of *speculum* signify in the case of alchemical works? It would appear that a scientific understanding of the mirror does influence the use of the moniker *speculum* for this particular branch of scientific mirror-books. The *Speculum maius* brings this out in its defense of the mirror title and provides a comparison to the intellectual and generic efforts undertaken in two early fourteenth-century alchemical texts, the *Speculum alchemiae* attributed to Roger Bacon and that attributed to Arnold of Villanova.

Vincent of Beauvais, a Dominican, spent some time at the University of Paris in the early thirteenth century and was attached to the Scholastic method of the schools and the intellectual culture of his order. His encyclopaedia work was originally written in two parts (*bifaria*), also referred to as *specula*: the *naturale* and the *historiale*. Grabes lists manuscript witnesses of these dating to 1244. A final section, the *Speculum doctrinale*, joined the first two in the 1250s to form what is known as the *trifaria* edition. The *Speculum maius* was intended to summarize all aspects of human knowledge: *naturale* describes the natural world according to a scheme based on the creation, *doctrinale* organizes and explains various disciplines, in both their practical and theoretical aspects. By absorbing technology, crafts, and the mechanical arts into this compendium of universal *knowledge*, Vincent of Beauvais is choosing between competing visions of the organization of knowledge.

We can glimpse Vincent's reasoning in the choice of title that he made; he states it explicitly in the preface to the work:

> Speculum maius appellari decreui: Speculum quidem, eo quod quicquid fere speculatione id est admiratione uel imitatione dignum ex hiis que in mundo uisibili uel inuisibili, ab initio usque ad finem, facta uel dicta sunt, siue etiam adhuc futura sunt, ex innumerabilibus fere libris colligere potui, in hoc uno breuiter continentur; Maius, autem, ad differentiam parui libelli iamdudum editi, cuius titulus est Speculum uel Ymago mundi, in quo scilicet huius mundi sensibilis dispositio et ornatus paucis uerbis describitur.[26]

> (I decided to call it 'speculum'. Indeed, [it is] a 'speculum' in that everything in it is generally worthy of consideration, that is, admiration and even imitation, out of everything that has been said or done or even will be in the future in both the vis-

[26] *Speculum naturale, Prologus*, cap. 2, edited in Lusignan, *Préface au 'Speculum maius' de Vincent de Beauvais*, p. 117. All translations are my own.

> ible world and the invisible from its beginning to its end, is contained in this one brief volume, which I was able to assemble from a nearly innumerable number of books. [It is called] 'Maius' in order to differentiate it from a small book published some time ago now, whose title is the Mirror or Image of the world, in which is briefly described the disposition and accoutrements of the sensible world.)

Even though he cites a previous encyclopaedic 'speculum', Vincent renounces generic and conceptual similarities: his own *speculum* is both practical and philosophical, touching the 'visible and invisible world'; it is timeless and contains unchanging knowledge — the definition of true science, according to Aristotle. The *Imago mundi* describes 'the decoration' of the world, that is, the particulars of the world such as its geography, something that is material and thus inconstant and uninteresting to an Aristotelian view of science.

Although he is deliberately differentiating his work from the *Imago mundi* for seemingly Aristotelian reasons, his choice of title also differentiates his text from works of traditional Aristotelian science. He chooses not to use the other seemingly appropriate generic term, *summa*, which also serves the purpose of a manual of 'necessary knowledge for the use of those unlikely or unready to pursue advanced knowledge themselves'.[27] Instead, he is using this new generic term and effectively inaugurating a type of mirror-title that reflects the more recent epistemological view of the mirror. Rather than designing his work as an overview of science in the strictly Aristotelian sense, he is interested in absorbing everything that is *ars* as well into the structure, realigning practice and theory.

Vincent's treatment of alchemy describes its practical origins and aims, but also includes numerous philosophical trappings: 'Alchimia proprie est ars transmutandi corpora mineralia a propriis speciebus ad alias [...]. Haec descendit ab illa parte naturalis philosophiae que est de mineralis, sicut agricultura ab illa parte, quae est vegetabilibus' ([Alchemy] rightly said is the art of transmuting mineral bodies from their own types into others [...]. It comes from the part of natural philosophy which treats minerals, as agriculture comes from that part which treats vegetation).[28] Interestingly, alchemy is identified as descending intellectually from a branch of *natural philosophy*, rather than natural history, as Albertus Magnus categorizes the vegetable world in his *De*

[27] Franklin-Brown, *Reading the World*, p. 69. In the case of the most famous *summa*, that of Thomas Aquinas, there is a problem in the view of knowledge being espoused, which Vincent does not embrace. Aquinas is against the inclusion of any of the inferior, practical crafts as in any way related to speculative philosophy. Alchemy, e.g., cannot be 'part' of physics because 'physics in itself and in its parts is speculative': Whitney, p. 139.

[28] *Speculum doctrinale* (1494), bk XI, chap. 105.

vegetabilibus. Although alchemy is listed under the mechanical arts, the discussion yet places it, as Newman notes, within the realm of speculative science, giving 'long theoretical descriptions of the generation of minerals from sulfur and mercury within the earth'.[29] There is a work of synthesis underway within the organizational structure and disciplinary definitions that resembles a dressing up and rehabilitation of *practica*: enveloping it within a rational, philosophical structure, and outlining specifically its connections to natural philosophy. This branch of *speculum* texts exemplifies this synthetic genre, addressing both *practica* and *theorica*, as does the unifying metaphor.

The two notable *specula* devoted to alchemy produced during this period thus rely on the mirror categorization that Vincent is espousing and explaining in the *Speculum maius*. First is the *Speculum alchemiae* generally ascribed to Roger Bacon, although the author is given sometimes as a certain Simon of Cologne. Bacon himself does write at length about alchemy in the *Opus tertium* (1267), in which he describes two levels of alchemy. He calls the first *alkimia operativa et practica* and explains that it 'docet facere metalla nobilia, et colores, et alia multa melius et copiosius per artificium quam per naturam fiant' (teaches how to make noble metals and colours and [make] other things better and in greater abundance through artifice than what is done by nature), and the second *alkimia speculativa*, 'quae speculatur de omnibus inanimatis et tota generatione rerum ab elementis' (which reflects on all inanimate things and on the whole generation of things from elements).[30] Bacon's work here reflects the belief in a basic division into theory and practice in the field of alchemy, which is echoed in the pseudoepigraphical *speculum*:

> Cap. I: In pluribus antiquorum codicibus plures inveniuntur istius artis diffinitiones [...]. Alius quidam dicit: Alchemia est scientia docens transformare omne genus metalli in alterum et hoc per medicinam propriam, sicut patet in multis philosophorum libris. Quare Alchemia est scientia, docens facere, & genere quandam medicinam.[31]
>
> (In many volumes of the ancients, many definitions of this art have been found [...] A certain one says: Alchemy is the science that teaches how to transform any kind of metal into another through the use of a special medicine, as is shown in many

[29] Newman, 'Technology and Alchemical Debate in the Late Middle Ages', p. 431.

[30] Roger Bacon, *Opus tertium*, ed. by Brewer.

[31] Pseudo-Roger Bacon, *Speculum alchemiae*, printed in *Alchemiae Gebri Arabis philosophi solertissimi libri [...]* as *Doctissimi viri Rogerii Bachonis de alchemia libellus cui titulum fecit 'Speculum Alchemiae'*, pp. 208–09.

philosophers' books. Therefore, alchemy is the science that instructs us how to make and generate this particular medicine.)

The description of the use of alchemical procedures concentrates on practical applications — the creation of an elixir that has the power to transform imperfect metals and substances and presumably prolong life — but *ars* and *scientia* are used interchangeably. The *Speculum alchemiae* work is divided into seven chapters, likely representing the seven stages of alchemical operations. From this organization, it can immediately be seen as being oriented towards practice. Yet its provided definition of alchemy situates the rest of the work philosophically: after outlining the goal of the 'science', the author promises to make an exposition of the metals which form the basis of alchemical operations, concentrating on a theoretical discussion of their 'natural principles', or properties, before continuing on to discuss the more practical concerns of the alchemical laboratory: the vase, the furnace, and instructions for their use.

Arnold of Villanova also has a fourteenth-century *Speculum alchemiae* attributed to him.[32] Although attributions in the manuscript witnesses disagree (and cite variously other authors including Nicholas Comes, 'Saint Astrob', and a disciple of St Francis named Brother Elie),[33] the misattribution to Arnold was fixed in printing in the *Theatrum chemicum*. The text itself appears to date from the early part of the fourteenth century and exists in eighteen fourteenth- and fifteenth-century manuscripts.[34] Pseudo-Arnold's *Speculum alchemiae* is again based on a seven-book division, although there is significant variation in its manuscript tradition. Its philosophical trappings are indicated in its format — the form of a dialogue between a master and a student, whom he calls

[32] Arnold is, interestingly enough, the actual author of a *Speculum medicine*, a compendium with a textual structure reminiscent of the *Isagoge* of Johannitius, one of the earliest and more prominent theoretical medical works coming from the Arab-speaking world and translated into Latin by Constantine the African at the end of the eleventh century. A premier medical textbook for centuries, the *Isagoge* formed the core of the academic medical canon in Latin Europe. Arnold's *Speculum medicine*, alternately dated to either 1300–01 or around 1308, goes into great detail discussing remedies and their medicinal qualities. He demonstrates familiarity with the use of alchemical techniques in pharmacy, including the medical usage of two alchemical operations, *distillatio* (distillation) and *sublimatio* (boiling and drawing off steam through the use of an alembic). McVaugh, 'Chemical Medicine in the Medical Writings of Arnau of Vilanova'; see esp. p. 252 for discussion of Arnold's description of chemical medicine in the *Speculum medicine*, and pp. 242–43 for a summary of the debated date of the text.

[33] Calvet, *Les Œuvres alchimiques*, p. 63.

[34] Based on the handlist found in Calvet, *Les Œuvres alchimiques*, pp. 64–65.

'dearest son' — the dialogue, being, of course, an important genre convention of scientific and philosophical writing. The incipit aligns the text with speculative scientific goals: '*ut ad perfectam scientiam pervenire possimus primum oportet scire quod tres lapides et tres sales sunt*' ('that we might arrive at perfect knowledge, it is necessary first to know which are the three stones and the three salts'). The author relies not only on generic signals of authority such as the dialectical format, but also writes a prologue that addresses a discipline:

> Scias igitur carissime fili quod ita scientia nihil aliud est quam perfecta inspiratio dei quia totum magisterium ex una sola reconsistit & ita tibi ostendemus per dicta philosophorum. Et similiter secundum quod nos vidimus & tetigimus cum magnis laboribus & cum magna industria illam solam rem cognovimus perfectam esse.[35]
>
> (Know therefore, dearest son, that knowledge is nothing other than the perfect inspiration of God, because all power comes out of one. This we will demonstrate through the sayings of the philosophers. Likewise, based on what we have seen and touched through great labour and industry, we have found this one thing to be perfect.)

Aside from these trappings, what follows is of a practical character, with the master sharing mostly technical advice with the student and outlining the process of transmutation.

Both of these alchemical mirror-texts derive authority and publicity from their identification with medieval learned masters, that is, through their misattributions to Roger Bacon and Arnold of Villanova (or others). Yet the lack of classical authority makes this new generic designation appropriate. There were indeed many alchemical texts falsely attributed to classical figures and early Arabic authorities, such as the *De anima in arte alchemiae* attributed to Avicenna or the *Hermetica* of Hermes Trismegistus. The pseudoepigraphical tradition in alchemy was present before the medieval period and was certainly due in part to alchemy's attractive associations with antiquity and semi-mystical origins,[36] but attribution also had consequences for the intellectual fate and acceptance of a work. Even though false attribution to ancient authorities and claims of antique authority were rampant in alchemy, contemporary writers used the titular designation of these *speculum* texts as a way of classifying new works. Indeed, Vincent in the *Speculum maius* had to describe and defend his adoption and usage of the term. These new, homegrown Latin texts of alchemy

[35] Pseudo-Arnold of Villanova, *Speculum alchymiae*, p. 515.

[36] Calvet, *Les Œuvres alchimiques*, p. 1.

used the *speculum* moniker as part of their repackaging of practically oriented new texts, those texts whose traditional intellectual credentials are weak. These new texts are both taking part in a traditional method of establishing authority — that is, through attribution to an existing *auctoritas* — as well as participating in a form of referential intellectual legitimization based on adopting the *speculum* designation from other contemporary texts.

The designation *speculum*, when given to medieval alchemical texts, has nuanced meaning that is related to a post-twelfth-century Renaissance understanding of the mirror, the organization of knowledge in the scholastic tradition, and debates about the relative value of different categories of knowledge. Just as new mirror technology is diffused and integrated into a rational scientific framework, alchemy is as well: a way of integrating art into science, or *practica* into *theorica*, and legitimizing *practica*. The choice of the new generic term *speculum* is a way of situating a work on one side of an academic debate about the relationship of practical science to speculative science. The history of medieval attitudes to technology are naturally more complicated than described here, but the period is marked by a tendency towards an integration of types of craft knowledge, such as glass-making, or practical knowledge, such as medical examination, into inherited knowledge schemas. Thus, the *speculum* genre is made more complicated when we consider mirror symbolism within the context of the influx of new knowledge during the twelfth century and the subsequent debates. The interpretation of the mirror functions on several levels: at a spiritual level, it reflects beliefs about the mirror's connection to the soul and vices such as vanity; it has a scientific dimension through its inclusion in studies of optics and vision; and it has a technological, material reality — the last two of these functions changed radically during the twelfth century, and their importance was hotly debated throughout the thirteenth. Thus it is fruitful to consider the mirror in relation to mirror-texts whose subject matter was also an intellectual product of the twelfth century and subject to the same epistemological and classificatory problems.

Works Cited

Manuscripts

Pseudo-Avicenna, *De anima in arte alchemiae*, Montreal, Osler Library of the History of Medicine, B.O. 480

Primary Sources

Bacon, Roger, *Opus tertium*, in *Opera quaedam hactenus inedita*, I: *Opus tertium, Opus Minus, Compendium Philosophiae*, ed. by J. S. Brewer (London: Longman, Green, and Roberts, 1859), pp. 3–310

Pseudo-Arnold of Villanova, *Speculum alchymiae*, in *Theatrum Chemicum*, IV (Strasbourg: Lazarus Zetzner, 1613), pp. 515–42

Pseudo-Roger Bacon, *Speculum alchymiae*, in *Alchemiae Gebri Arabis philosophi solertissimi libri* (Nuremberg: Johannes Petreius, 1545), pp. 208–18

Vincent of Beauvais, *Speculum doctrinale* (Venice: Hermann, Liechtenstein, 1494)

——, *Speculum natural*e (Venice: Hermann Liechtenstein, 1494)

Secondary Studies

Benson, Robert L., Giles Constable, and Carol D. Lanham, eds, *Renaissance and Renewal in the Twelfth Century* (Cambridge, MA: Harvard University Press, 1982)

Bradley, Ritamary, 'Backgrounds of the Title *Speculum* in Mediaeval Literature', *Speculum*, 29.1 (1954), 100–15

Calvet, Antoine, *Les Œuvres alchimiques attribuées à Arnaud de Villeneuve: Grand œuvre, médecine et prophétie au Moyen-Âge* (Paris: Société d'Étude de l'Histoire de l'Alchimie; Milan: Archè, 2011)

Collette, Carolyn P., and Harold Garrett-Goodyear, *The Later Middle Ages: A Sourcebook* (New York: Palgrave Macmillan, 2011)

Crisciani, Chiara, and Agostino Paravicini Bagliani, eds, *Alchimia e medicina nel Medioevo* (Florence: SISMEL, Edizioni del Galluzzo, 2003)

Duhem, Pierre, *Le Système du monde: Histoire des doctrines cosmologiques de Platon à Copernic* (Paris: Hermann, 1954)

Einar Már Jónsson, *Le Miroir: Naissance d'un genre littéraire* (Paris: Belles Lettres, 1995)

——, 'Le Sens du titre *Speculum* aux XIIe et XIIIe siècles et son utilisation par Vincent de Beauvais', in *Vincent de Beauvais: Intentions et réceptions d'une œuvre encyclopédique au Moyen Âge*, ed. by Monique Paulmier-Foucart, Serge Lusignan, and Alain Nadeau (Montréal: Bellarmin, 1990), pp. 11–32

Franklin-Brown, Mary, *Reading the World: Encyclopedic Writing in the Scholastic Age* (Chicago: University of Chicago Press, 2012)

Glick, Thomas F., Steven J. Livesey, and Faith Wallis, eds, *Medieval Science, Technology, and Medicine: An Encyclopedia* (New York: Routledge, 2005)

Grabes, Herbert, *The Mutable Glass: Mirror-Imagery in Titles and Texts of the Middle Ages and English Renaissance*, trans. by Gordon Collier (Cambridge: Cambridge University Press, 1982)

Lindberg, David C., *Theories of Vision from Al-Kindi to Kepler* (Chicago: University of Chicago Press, 1976)

Lusignan, Serge, *Préface au 'Speculum maius' de Vincent de Beauvais: Réfraction et diffraction* (Montréal: Bellarmin; Paris: Vrin, 1979)

MacFarlane, Alan, and Gerry Martin, *Glass: A World History* (Chicago: University of Chicago Press, 2002)

McVaugh, Michael, 'Alchemy in the *Chirurgia* of Teodorico Borgognoni', in *Alchimia e medicina nel Medioevo*, ed. by Chiara Crisciani and Agostino Paravicini Bagliani (Florence: SISMEL, Edizioni del Galluzzo, 2003), pp. 55–76

——, 'Chemical Medicine in the Medical Writings of Arnau of Vilanova', in *Actes de la II Trobado Internacional d'Estudis sobre Arnau de Vilanova*, ed. by Josep Perarnau (Barcelona: Institut d'Estudis Catalans, 2005), pp. 239–67

Moreau, Sébastien, 'Some Considerations Concerning the Alchemy of the *De anima in arte alchemiae* of Pseudo-Avicenna', *Ambix*, 56.1 (2009), 49–56

——, 'Les Sources alchimiques de Vincent de Beauvais', *Spicae, Cahiers de l'Atelier Vincent de Beauvais*, n.s., 2 (2012), 5–118

Newman, William, 'Technology and Alchemical Debate in the Late Middle Ages', *Isis*, 80.3 (1989), 423–45

Ovitt, George, Jr, *The Restoration of Perfection: Labor and Technology in Medieval Culture* (New Brunswick, NJ: Rutgers University Press, 1987)

Parkes, M. B., 'Influence of the Concepts of *Ordinatio* and *Compilatio* on the Development of the Book', in *Medieval Learning and Literature: Essays Presented to Richard William Hunt*, ed. by J. J. G. Alexander and M. T. Gibson (Oxford: Clarendon Press, 1976), pp. 115–41

Rouse, Mary A., and Richard H. Rouse, '*Statim invenire*: Schools, Preachers, and New Attitudes to the Page', in *Authentic Witnesses: Approaches to Medieval Texts and Manuscripts* (Notre Dame, IN: University of Notre Dame Press, 1991), pp. 191–220

Siraisi, Nancy G., *Taddeo Alderotti and his Pupils: Two Generations of Italian Medical Learning* (Princeton, NJ: Princeton University Press, 1981)

Wallace, William A., 'Philosophical Setting of Medieval Science', in *Science in the Middle Ages*, ed. by David C. Lindberg (Chicago: University of Chicago Press, 1978), pp. 91–119

Wallis, Faith, *Medieval Medicine: A Reader* (Toronto: University of Toronto Press, 2010)

White, Lynn, *Medieval Technology and Social Change* (Oxford: Oxford University Press, 1962)

——, 'Technology and Invention in the Middle Ages', *Speculum*, 15.2 (1940), 141–59

Whitney, Elspeth, 'Paradise Restored: The Mechanical Arts from Antiquity through the Thirteenth Century', *Transactions of the American Philosophical Society*, 80.1 (Philadelphia: American Philosophical Society, 1990), pp. 1–169

ADJUSTING THE MIRROR: A POLITICAL REMAKE OF BRUNETTO LATINI'S *LI LIVRES DOU TRESOR*

David Napolitano

Introduction

The effective transmission of knowledge presupposes constant adjustment to changing circumstances. This truism applied to thirteenth-century encyclopaedias. The core of this genre consisted of encyclopaedias compiled in a clerical context and written in Latin, such as Thomas of Cantimpré's *De naturis rerum* (1237–40), Bartholomaeus Anglicus's *De proprietatibus rerum* (*c.* 1250), and Vincent of Beauvais's *Speculum maius* (1244–59).[1] However, in the second half of the thirteenth century, this encyclopaedic model also started to be used by secular authors writing in the vernacular. Brunetto Latini's *Li Livres dou Tresor* is a prime example of this new trend, and it is the starting point for this case study.[2] More specifically, I will examine the translation of this encyclopaedia from one vernacular (Old French) into another (Old Italian), and I will argue that, in the specific case of manuscript L^4, this transition from *Tresor* to *Tesoro*

[1] For an introduction to these encyclopaedias: Roux, *Mondes en miniatures*, pp. 32–38. See also Anna Dysert's contribution in this volume.

[2] On this trend: Meier, 'Vom Homo Coelestis zum Homo Faber', p. 163. Specifically on the *Tresor*: Meier, 'Cosmos Politicus'.

David Napolitano (dphn2@cam.ac.uk) is a British Academy Postdoctoral Fellow at the Faculty of History of the University of Cambridge, working on Western-European medieval cities and city magistrates, including the writing of a monograph dealing with the topic of his PhD thesis, entitled *The Profile and Code of Conduct of the Professional City Magistrate in Thirteenth-Century Italy*. He is also completing a new critical edition of the *Oculus pastoralis*.

The Mirror in Medieval and Early Modern Culture: Specular Reflections, ed. by Nancy M. Frelick, CURSOR 25 (Turnhout: Brepols, 2016), pp. 89–111 10.1484/M.CURSOR-EB.5.111043
BREPOLS PUBLISHERS

did not constitute a slavish copy of the original but resulted in the depersonalization, relocation, and political reorientation of its contents. In other words, I will demonstrate that the original French composition (1260–66/7), which fits into a tradition of political mirrors, was adjusted in a triple manner to reflect the different historical circumstances under which this particular Italian copy (1284–99) was produced.

Brunetto Latini

Brunetto Latini (*c.* 1220–94) was born into a privileged family of notaries,[3] and he played a significant role in the political machinery of Florence under the *Primo Popolo* regime (1250–60), functioning, amongst other duties, as *scriba ancianorum*, a notary-scribe attached to the highest executive institution of this regime, the *Anziani*, the Elders.[4] Archival evidence particularly highlights Latini's involvement in diplomatic affairs during this period, such as his involvement in the strategic purchase of a series of castles in order to create a defensive ring around Florence or his role in the redaction of treaties between Florence and its rivals, Pisa and Siena, in 1254 and 1255. Prior to the Battle of Montaperti (1260), Brunetto Latini was instructed to go on an embassy to Alfonso X of Castile in order to gain his support in the upcoming hostilities. This embassy could, however, not prevent the devastating defeat of Florence by its Ghibelline, pro-imperial opponents.[5] The city barely escaped a razing. Following this turn of events, Brunetto Latini had to join the ranks of the *fuoriusciti*, the Florentine exiles. Unlike his father, he did not go into

[3] The classic work on the life and work of Brunetto Latini remains Sundby, *Della vita e delle opere di Brunetto Latini*, trans. by Renier. The standard work is Ceva, *Brunetto Latini*. For other discussions of this topic: d'Addario, 'Brunetto Latini'; Bolton Holloway, *Twice-Told Tales*; Carmody and Fery-Hue, 'Brunetto Latini'; De Vincentiis, 'Le parole di ser Brunetto'; Inglese, 'Brunetto Latini'; Mazzoni, 'Brunetto Latini'.

[4] The classic work on the history of Florence remains Davidsohn, *Storia di Firenze*, trans. by Klein and rev. by Palmarocchi. Italian historians have recently produced a series of monographs covering different stages of Florentine history. See, in chronological order: Faini, *Firenze nell'età romanica (1000–1211)*; De Rosa, *Alle origini della repubblica fiorentina*; Diacciati, *Popolani e magnati*; Gualtieri, *Il comune di Firenze tra Due e Trecento*. See also: Raveggi and others, *Ghibellini, guelfi e popolo grasso*; Zorzi, *La trasformazione di un quadro político*. For an introduction in English: Abulafia, *Italy in the Central Middle Ages*; Green, 'Florence'; Najemy, *A History of Florence, 1200–1575*; Tanzini, 'Tuscan States: Florence and Siena'.

[5] On the Guelf-Ghibelline dichotomy: Canaccini, 'Restano i termini, mutano i significati'; Dessì, 'Guelfi e Ghibellini'; Gentile, 'Factions and Parties: Problems and Perspectives'.

exile in an Italian city supportive of the Guelf cause but moved on to France.[6] Traces in the archives show him travelling to and from Arras, Paris, and Bar-sur-Aube.[7] He continued to act as a trusted *notaio* for the exiled Florentine elite, and he also played a part in the formalization of the papal-Angevin alliance in 1263.[8] As a result of this collaboration the tables of fortune turned at the Battle of Benevento (1266). A successful military expedition, led by Charles of Anjou (1226/27–85),[9] promoted by the pope, and backed by Florentine bankers, signalled not only the beginning of the end for the imperial House of Hohenstaufen in Italy but resulted also in a fundamental break in the leadership over Florence.[10] Under the protectorate of Charles of Anjou, elected as *podestà* of Florence in April 1267, a party regime, led by the *Parte Guelfa*, was eventually put into place (1267–80).[11] This new power constellation caused an exodus of Florentine Ghibellines, while loyal supporters of the regime, such as Brunetto Latini, took up leading offices. He acted, for instance, as *protonotarius* (1269–70), head of the chancery of the general-vicar of Charles of Anjou, Jean Britaud de Nangis (d. 1278), and as consul of the guild of judges and notaries (1275) during the Angevin protectorate.

Brunetto Latini's Tresor

Brunetto Latini wrote the *Tresor*, the centrepiece of his collected works, during his exile in France (1260–66/67). His language choice is said to reflect this deplorable situation. The *Tresor* was not written in Latin, the language of learning at the time and a language that Latini, as a notary, was perfectly capable of writing, but in the vernacular. Interestingly, this vernacular was not Latini's native tongue but Old French, identified in the work as *francès* or *roman selonc*

[6] On this choice: Cella, 'Gli atti rogati da Brunetto Latini in Francia', p. 385.

[7] For a detailed discussion: Cella, 'Gli atti rogati da Brunetto Latini in Francia'. See also: Maffia Scariati, 'Gli atti rogati da Brunetto Latini in Francia'.

[8] On the negotiation process leading up to and the principles underlying this alliance: Dunbabin, *Charles I of Anjou*, pp. 131–32; Herde, *Karl I von Anjou*, pp. 40–46.

[9] For an introduction to this figure: Abulafia, 'The State of Research'; Dunbabin, *Charles I of Anjou*; Herde, *Karl I von Anjou*.

[10] In 1268, Conradin lost the Battle of Tagliacozzo and was decapitated in Naples. After twenty-three years of imprisonment, Enzo died in Bologna in 1272.

[11] Initially, he was elected for six years, but this appointment was prolonged until 1279.

le patois de France.[12] In an often cited passage of the prologue, Latini lists three motives in support of this linguistic preference: his presence in France, the attractiveness of Old French, and its widespread use.[13] However, the *Tresor* not only reflects the day-to-day realities of political adversity, but it also constructs and legitimates a way forward. Scholars have qualified it as a skilfully designed political encyclopaedia, in which Latini worked his source material as malleable wax.[14] Its innovative tripartite structure culminates in an instruction manual for the professional magistrate of an Italian city, the *podestà*.[15] It is the fine gold that completes the treasure metaphor in which coins symbolize the first book dealing with theoretical knowledge, precious stones represent the topic of ethics treated in the second book, and fine gold stands for rhetoric and, above all, politics discussed in the third book.[16] In combination with the other components of the *Tresor* this political apotheosis, entitled 'des governemenz des citez' (III.73–105), is intended to provide its recipient with a strong knowledge base, an effective moral compass, and the necessary rhetorical skills to efficiently govern a city in accordance with Italian customs.[17] The patron of this hard-won treasure of functional knowledge is not explicitly named in the *Tresor*, but is

[12] For other Italians who made a similar language choice (such as Aldobrandino da Siena, Martino da Canal, Niccolò da Verona, Rustichello da Pisa): Cigni, 'I testi della prosa letteraria', pp. 175 and 179–80.

[13] *Tresor*, I.1.7: 'Et se aucun demandoit por quoi ceste livre est escrit en roman selonc le patois de France, puis nos [so]mes ytaliens, je diroie que ce est par.ii. raisons: l'une que nos [so]mes en France, l'autre por ce que la parleure est plus delitable et plus comune a touz languaiges.' Citations are based upon the most recent edition of the *Tresor* (with Italian translation): Brunetto Latini, *Tresor*, ed. and trans. by Beltrami and others. The Roman number refers to the relevant book, while the Arabic numerals refer to the relevant chapter and paragraph respectively. For an English translation: Brunetto Latini, *The Book of the Treasure*, trans. by Barrette and Baldwin.

[14] For Meier's typology of encyclopaedias: Meier, 'Enzyklopädischer Ordo und sozialer Gebrauchsraum', pp. 520–32. On the image of malleable wax: *Tresor*, III.12.2.

[15] Exceptionally, extant manuscripts consist of four or five books as a result of a subdivision of the second and/or third book: Roux, *Mondes en miniatures*, p. 49, n. 25. On the innovative character of this structure: Beltrami, 'Appunti su vicende del *Tresor*', pp. 313 and 316. For an interpretation of this innovation as a shift from a theological to an anthropocentric perspective: Meier, 'Enzyklopädischer Ordo und sozialer Gebrauchsraum', p. 517; Meier, 'Cosmos Politicus', pp. 340–41.

[16] For an interpretation of this symbolism as a case of 'choix des termes, choix de l'ordre': Miller, 'Le Poète dans la Cité', p. 62.

[17] For the reference to Italian customs: *Tresor*, I.1.4.

merely referred to in the prologue as a *biau douz amis*, a beloved friend — a reference that is echoed in the introduction to the political section.[18] The exact identity of Latini's patron will probably remain forever elusive, but it is generally accepted that there was an actual commission,[19] and the intended audience is usually identified as French, either an exiled Florentine or the royal court — possibly even Charles of Anjou or someone in his entourage.[20]

A Tradition of Mirrors

The political section of the *Tresor* did not emerge out of a vacuum. Rather than falling back upon his personal experience under the *Primo Popolo* regime, Brunetto Latini opted to build upon the existing *podestà* literature. Apart from the *Tresor*, this literary tradition consists of the anonymous *Oculus pastoralis* (1220s),[21] Giovanni da Viterbo's *Liber de regimine civitatum* (1228–64),[22] and

[18] On the intentional character of this anonymity: Cella, 'Gli atti rogati da Brunetto Latini in Francia', p. 408; Luff, *Wissensvermittlung im europäischen Mittelalter*, p. 284. For other explanations: Van Welie-Vink, 'Was Charles d'Anjou Brunetto Latini's *biaus dous amis*?', p. 321.

[19] *Contra*: Roux, *Mondes en miniatures*, pp. 50–51 (speaking of a generic dedication to a fictive person). See also: Beltrami, 'Introduzione', pp. viii and xix–xx (presenting a general category of public officeholders, and in particular the *podestà*, as its intended audience).

[20] For an introduction to this debate: Plisson, 'Science et littérature au XIIIème siècle', pp. 135–36. For the hypothesis of an exiled Florentine: Carmody, 'Introduction', p. xviii (referring to Davizzo della Tosa); Maffia Scariati, *Dal Tresor al Tesoretto*, pp. 26–27, n. 2; Swiggers, *Le 'Tresor' de Brunetto Latini*, p. 9. *Contra*: Cella, 'Gli atti rogati da Brunetto Latini in Francia', p. 403; Roux, *Mondes en miniatures*, p. 50, n. 31; Van Welie-Vink, 'Was Charles d'Anjou Brunetto Latini's *biaus dous amis*?', pp. 319–23. For a link to Charles of Anjou: Bolton Holloway, *Twice-Told Tales*, pp. 60–61; Van Welie-Vink, 'Was Charles d'Anjou Brunetto Latini's *biaus dous amis*?', pp. 321–22 and 332. *Contra*: Roux, *Mondes en miniatures*, p. 50.

[21] For an edition of this work: *Oculus pastoralis*, ed. by Muratorio; *Oculus pastoralis*, ed. by Franceschi; 'Oculus pastoralis', ed. by Tunberg; *Speeches from the 'Oculus pastoralis'*, ed. by Tunberg. These editions are based upon the only then known copy of this text (Cleveland, Public Library, MS Wq 789.0921 M–C 37). Following up on a lead in a footnote in an article by Diego Quaglioni and information generously shared by Gérard Giordanengo, I have however confirmed the existence of a second copy of this text (Paris, Bibliothèque nationale de France, MS Latin 4686). I discuss this newly found copy, and its implications for our understanding of this text, in Napolitano, 'Verborgen schatten'. I am currently working on a new critical edition of this text.

[22] For an edition of this work: Giovanni da Viterbo, *Liber de regimine civitatum*, ed. by Salvemini.

Orfino da Lodi's *De regimine et sapientia potestatis* (mid-1240s).[23] Some scholars even extend this canon to include Fra Paolino minorita's *De regimine rectoris* (1313–15).[24] These treatises on city government were contrived to hold up a mirror of expected behaviour to Italian city magistrates, a *specchio del podestà* — that is an Italian, republican variant of the better-known mirrors for princes.[25] Gustav Hanauer and Wilhelm Berges speak of a *Podestatenspiegel*, while Philip Jones refers to these works as 'mirrors of magistrates'.[26] Hans Hubert Anton, a specialist on the mirror-for-princes genre, calls it 'eine neue Form der Spiegel'.[27] As underlined by Nancy M. Frelick in her introduction to this collection of essays, such *specula* functioned as an aid to self-examination and self-improvement, providing the city magistrates in question with both a model and a challenge.[28]

Prior to the discovery of the *De regimine civitatum* in the late nineteenth century, the political section of Latini's *Tresor* was deemed largely original. In 1869 Mussafia was the first to draw attention to the fact that Brunetto Latini must have known the *Oculus pastoralis* — or at least its first two *divisiones*.[29] However, once the *De regimine civitatum* was discovered, a second, and more fundamental link became apparent. While Francesco Novati already referred to the existence of 'numerous close links' between both works in 1888,[30] Gaetano Salvemini deserves the credit for developing this intuition. He speaks of Latini's political section as 'a true and proper abbreviated translation' of the *De*

[23] For an edition of this work: *Orfini Laudensis poema De regimine et sapientia potestatis*, ed. by Ceruti; Orfino da Lodi, *De regimine et sapientia potestatis*, ed. by Castelnuovo; Orfino da Lodi, *De regimine et sapientia potestatis*, ed. by Pozzi.

[24] Hertter put this canon on the map; see Hertter, *Die Podestàliteratur Italiens im 12. und 13. Jahrhundert*. In the past these works have been a favoured topic of research, but the latest overview studies of the corpus date back to the mid-twentieth century. See: Franchini, 'Trattati "De regimine civitatum" (sec. XIII–XIV)'; Sorbelli. 'I teorici del reggimento comunale'. For an introduction to the scholarly debate surrounding this literary tradition: Napolitano, 'Pleidooi voor een hernieuwde kennismaking'.

[25] For this Italian term: Fenzi, 'Brunetto Latini', p. 359.

[26] Berges, *Die Fürstenspiegel*, p. 106; Hanauer, 'Das Berufspodestat im dreizehnten Jahrhundert', p. 411; Jones, *The Italian City-State*, pp. 486 and 530.

[27] Anton, *Fürstenspiegel*, p. 26.

[28] The term 'mirror-for-princes' has been popularized by the seminal work of Berges, *Die Fürstenspiegel des hohen und späten Mittelalters*. On this genre: Eberle, 'Mirror of Princes'; Bratu, 'Mirrors for Princes (Western)'. See also: Anton, *Fürstenspiegel*, pp. 3–37.

[29] Mussafia, 'Sul testo del Tesoro di Brunetto Latini', pp. 371–74.

[30] Novati, *La giovinezza di Coluccio Salutati*, pp. 82–83, n. 3.

regimine civitatum.[31] Despite a sharply formulated challenge by Albano Sorbelli to this hypothesis of direct borrowing,[32] this view is now generally accepted.[33]

At the same time, the incorporation of this political section into a larger encyclopaedic project by Brunetto Latini deliberately transcends the borders of the Italian city for which the *podestà* literature was originally constructed. Characterized by a careful balancing of French and Italian elements throughout the encyclopaedia, the *Tresor* looks beyond the Florentine experiment of the *Primo Popolo* and revives — and legitimates — an established, elastic institutional structure, the *podestà*.[34] Through the promotion of this institute, Brunetto Latini is able to accommodate a drastically changed political context and to pave the way for a temporary overlap of interests between Charles of Anjou, the papacy, and the Florentine Guelfs.[35]

From Tresor *to* Tesoro*: Manuscript L⁴*

Having briefly put the original composition of the *Tresor* in context, I will now turn my attention to its Italian adaptation, the less-studied *Tesoro*. In fact, the success of the *Tresor* quickly extended back to the Italian peninsula in the form of a translation into Old Italian.[36] Although the connections between these

[31] Salvemini, 'Il "Liber de regimine civitatum" di Giovanni da Viterbo', pp. 293–94.

[32] Sorbelli, 'I teorici del reggimento comunale', pp. 77–79, 99–100, and 106–14.

[33] For a discussion of the different positions in this debate: Napolitano, 'The Profile and Code of Conduct of the Professional City Magistrate', pp. 114–17; on the type of modifications needed to permit Giovanni da Viterbo's Ghibelline manual to be incorporated into a Guelf encyclopaedia, see pp. 308–09.

[34] For an introduction to the *podestà* office: Artifoni, 'I podestà professionali e la fondazione retorica'. On its institutional functioning: Bognetti, *Storia del diritto italiano pubblico*; Franchini, *Saggio di ricerche sull'instituto del podestà*; Hanauer, 'Das Berufspodestat im dreizehnten Jahrhundert'. On the staffing of this office: Maire Vigueur, *I podestà dell'Italia comunale*. On the elastic character of this institute: Ottokar, *Studi comunali e fiorentini*, pp. 27–28.

[35] On this balancing act: Napolitano, 'The Profile and Code of Conduct of the Professional City Magistrate', pp. 309–17.

[36] For a general introduction to this manuscript tradition: Marshall, 'The Manuscript Tradition of Brunetto Latini's *Tresor*'. See also the overview lists prepared by Squillacioti, 'La tradizione manoscritta delle opera di Brunetto Latini'. These lists have been updated by Divizia, 'Integrazioni al censimento dei codici italiani di Brunetto Latini'; Divizia, 'Aggiunte (e una sottrazione) al censimento'. For a specific introduction to the Italian manuscript tradition: Giola, 'Sul volgarizzamento italiano del "Tresor" di Brunetto Latini'. See also: Giola, *La*

French and Italian textual traditions are likely to be multiple according to the latest scholarship,[37] I will focus on one specific instance in this contribution, embodied by manuscript L^4 currently preserved at the Biblioteca Medicea Laurenziana in Florence.[38]

The explicit of this luxury manuscript reveals the identity of the scribe and sheds light on its production context:

> Bondì Pisano mi scrisse Dio lo be | nedisse Testario sopranome Dio | lo chavi di Gienova di prigione e | a llui e a li autri che vi sono, e da | Dio abiano benisione. Amen, amen.
>
> (Bondì, the Pisan, wrote me. God blessed him. Testario was his surname. May God free him from prison in Genoa and may he and the others who are there receive God's blessing. Amen, amen.)[39]

According to this explicit, the manuscript is the product of a Pisan copyist, Bondì Testario, who was imprisoned in Genoa. Based upon this fact, scholars have advanced the hypothesis that this manuscript was produced after the sea battle between Pisa and Genoa at Meloria (9 August 1284), resulting in such imprisonment, and prior to the peace treaty between both cities (31 July 1299), accompanied by the release of the last prisoners.[40]

tradizione dei volgarizzamenti toscani del "Tresor" di Brunetto Latini. On the print history of the *Tesoro*, see Napolitano, 'Brunetto Latini's *Tesoro* in Print'.

[37] Giola, 'Sul volgarizzamento italiano del "Tresor" di Brunetto Latini', pp. 30, 37–38 and 296; Marshall, 'The Manuscript Tradition of Brunetto Latini's *Tresor*', p. 130. For a long time this translation was erroneously attributed to Bono Giamboni, a Florentine Ghibelline notary. This attribution was based upon the preamble of two manuscripts (Br and M), which had been uncritically reproduced in the early print editions. Recently, this attribution has been refuted and the hypothesis of a multiplicity of translators has been advanced. For this debate: Marshall, 'The Manuscript Tradition of Brunetto Latini's *Tresor*', pp. 155–79. See also: Artifoni, 'I podestà professionali e la fondazione retorica', p. 712, n. 26; Roux, *Mondes en miniatures*, p. 56.

[38] Florence, Biblioteca Medicea Laurenziana, MS Pluteo 42.23 (which can be freely consulted via the library's website <http://www.bml.firenze.sbn.it/it/fondi.htm>).

[39] Transcription and translation mine. On the luxury character of this manuscript: Bertelli, 'Tipologie librarie e scritture nei più antichi codici fiorentini di ser Brunetto', pp. 222, 231–32 and 252 (with details on its material aspects and a reprint of a folio). On the reliability of the explicit in the present case: Cigni, 'Copisti prigionieri (Genova, fine sec. XIII)', pp. 429–30. Italian scholars sometimes refer to a partial student edition of manuscript L^4 by Diego Dotto, but he had run out of copies when I contacted him. I produced my own transcription for this contribution.

[40] Bertelli, 'Tipologie librarie e scritture nei più antichi codici fiorentini di ser Brunetto',

Manuscript production in prison was not exceptional in medieval Italy. For a prisoner this type of labour was one of the ways to cover his living expenses,[41] especially if he was not locked up in jail but living under a prison regime in the home of a citizen.[42] Roberto Benedetti speaks bluntly of 'scrivere per sopravvivere'.[43] The presence of an important contingent of Pisan notaries, living and working under such conditions in Genoa, is well documented.[44] Fabio Zinelli even calls the Genovese prisons, with their imprisoned Pisans after the Battle of Meloria, 'le prigioni più letterarie del medioevo italiano'.[45]

In the second half of the thirteenth century, Pisa was an established centre for the translation of French texts.[46] At the same time Genoa was known as a lively city where, amongst others, French and Italian cultural spheres intersected.[47] Not surprisingly, therefore, scholarship has identified this bustling port as an important link between the *Tresor* and *Tesoro* traditions. After an initial attribution to southern Italy, scholars now ascribe two of the oldest *Tresor* manuscripts (R and V) to a Genovese hand.[48] Brigitte Roux even argues that Bondì Testario used one of these manuscripts as his exemplar based upon similarities in the illumination of the manuscripts.[49]

p. 222; Cigni, 'I testi della prosa letteraria', p. 161; Cigni, 'Copisti prigionieri (Genova, fine sec. XIII)', p. 426; Cigni, 'Genova e una versione toscana', p. 60.

[41] Ceccarelli Lemut, 'I Pisani prigionieri a Genova', p. 81; Signorini, 'Il copista di testi volgari', p. 143.

[42] Fabbri, 'Romanzi cortesi', p. 15; Giannini, 'Comptes rendus: A scuola con ser Brunetto', pp. 237–38.

[43] Benedetti, 'Qua fa' un santo e un cavaliere', p. 41.

[44] Cigni, 'Manuscrits en français, italien et latin', p. 197; Giannini, 'Comptes rendus: A scuola con ser Brunetto', p. 238.

[45] Zinelli, 'Tradizione "mediterranea" e tradizione italiana del "Livre dou Tresor"', p. 74.

[46] Fabbri, 'Romanzi cortesi', p. 14; Giannini, 'Comptes rendus: Pisa crocevia di uomini, lingue e culture', p. 514.

[47] Cigni, 'La ricezione medievale della letteratura francese'; Fabbri, 'Romanzi cortesi', p. 14. The designation of Genoa as European Capital of Culture in 2004 has resulted in conference proceedings which focused on the prominent role of this city as an intersection of cultures during the Middle Ages: *Poeti e poesia a Genova (e dintorni) nell'età medievale*.

[48] On this debate: Fabbri, 'Romanzi cortesi'; Giannini, 'Comptes rendus: A scuola con ser Brunetto', p. 239; Maffia Scariati, *Dal Tresor al Tesoretto*, p. 166, n. 82; Roux, *Mondes en miniatures*, pp. 100–01; Zinelli, 'Tradizione "mediterranea"', pp. 61–64 and 72.

[49] Roux, *Mondes en miniatures*, pp. 102–03, especially n. 92.

Not much is known about the figure of Bondì Testario, except for the fact that he belonged to a well-known family of notaries in Pisa.[50] Nor do we have any concrete, non-speculative information about the initial recipients of this manuscript.

Bondì Testario's Tesoro: *Erasure of Identifying Characteristics*

A confrontation of the incipit, the prologue, and the historical section of the *Tresor* with their corresponding parts in the *Tesoro* shows that the amount of scribal intervention in manuscript L[4] — and its coherence in direction — is highly revealing.[51]

From the very start, the fact that the original composition was written in Old French is deleted in the incipit and prologue of L[4]. Only in the explicit is this information, as well as the translation into Old Italian, mentioned:

> Explicit libro lo quale fue conposto per lo maestro Brunetto Latino di Fiorensa e poi traslectato di franciescho in latino.
>
> (Here ends the book which was composed by master Brunetto Latini of Florence and then translated from French into Italian.)[52]

Cigni stresses that '*latino*' in this context should not be misconstrued as Latin, but as Old Italian:

> la locuzione 'in latino' ha indotto molti, compreso chi scrive [...], nell'errore di (far) credere all'esistenza di una traduzione latina del testo di Brunetto, mentre qui è ovviamente da intendere 'in italiano', secondo una consuetudine diffusa, e di cui dà conto Zinelli.[53]
>
> (the phrase 'in Latin' has induced to many, including this writer [...], the error of believing in the existence of a Latin tradition of the text of Brunetto, while it

[50] Bertelli, 'Tipologie librarie e scritture nei più antichi codici fiorentini di ser Brunetto', pp. 221–22; Castellani, 'La Toscana dialettale d'epoca antica', p. 8; Castellani, *Saggi di linguistica e filologia italiana e romanza*, III, 351–52, n. 87; Cigni, 'Manuscrits en français, italien et latin', p. 194, n. 30; Cigni, 'Copisti prigionieri (Genova, fine sec. XIII)', pp. 434–36; Fabbri, 'Romanzi cortesi', p. 15; Giola, 'Sul volgarizzamento italiano del "Tresor" di Brunetto Latini', p. 11; Marshall, 'The Manuscript Tradition of Brunetto Latini's *Tresor*', p. 308; Roux, *Mondes en miniatures*, p. 102; Squillacioti, 'Ricerche sulla tradizione del "Tesoro" toscano', p. 563, n. 43.

[51] *Tresor*, incipit, I.1, and I.19–98.

[52] Transcription and translation mine.

[53] Cigni, 'Copisti prigionieri (Genova, fine sec. XIII)', p. 435, n. 29; Zinelli, '"Donde noi metremo"', p. 161, n. 1.

should be clearly understood as 'in Italian' here, in accordance with a wide-spread custom, as stated by Zinelli.)[54]

To be precise, Bondì Testario used the linguistic variety of his hometown, Pisa.[55] Medieval scribes often put the copied work into their own dialect.[56]

Moreover, Brunetto Latini is only briefly acknowledged as the original compiler of the encyclopaedia in the incipit.[57] It simply states that Latini began (*cominciò*) this work. This wording implies that Latini did not have the final word and that the encyclopaedia was open to completion. Explicit references to the personal experiences or exiled status of Latini have also been systematically deleted. Along the same lines, the affectionate and individualized dedication to a beloved friend in the prologue is modified into a more neutral and impersonal *amico*.

Illustrative of L^4's transfer to a new local setting is the mentioned use of Pisan, while explicit references to Florence and its virulent factionalism have been removed, and a string of references to Pisan characters and events (discussed below) has been added.

In other words, through a series of small but carefully interacting scribal interventions, the work has been thoroughly depersonalized and relocated, thus reflecting the different historical circumstances under which this particular Italian copy of Latini's *specchio del podestà* was made.

Bondì Testario's Tesoro*: Political Reorientation*

This erasure of essential identifying characteristics is only a preparatory step for an even more profound manipulation, that of its political orientation, as illustrated by a fundamental remake of the historical section of the encyclopaedia.

To fully grasp the direction and intensity of this revision, it is important to keep in mind that the historical section of the *Tresor* consists of three major components. The first part deals with ancient biblical history, from the creation of Adam until the birth of Christ, covering five of the six ages, a traditional medieval chronology.[58] This basic framework is interrupted by

[54] Translation mine.

[55] On Pisan: Castellani, 'Pisano e lucchese'; Manni, *Storia della lingua italiana*, pp. 41–46.

[56] Beltrami, 'Per il testo del Tresor', p. 1008.

[57] See also the cited explicit.

[58] *Tresor*, I.19.3. On this chronology: Ribémont, 'Le Temps de l'histoire', p. 238; Roux, *Mondes en miniatures*, p. 193.

a lengthy secular digression,[59] which deals extensively with the Trojan diaspora, the Roman republic, and various royal dynasties up to the Carolingian period (Charlemagne).[60] The second component consists of a set of biographical sketches of central figures of the Old and New Testament, while the third part deals with the sixth age, starting with the reign of Constantine and ending in the year 1260 (in a first redaction — the start of Brunetto Latini's exile) or 1268 (in a second redaction — the year of Charles of Anjou's victory over Conradin at Tagliacozzo, ending the Hohenstaufen rule in Italy). Scholars have stressed the pro-Guelf and pro-Angevin orientation of this historical section, which occupies the central part of the first book of the *Tresor* and covers almost half of its two hundred chapters.[61]

Out of these three components, the second one has, by far, known the least degree of intervention. Its main modifications are limited to the addition of a qualifier of holiness to the names of saints, that of beatitude to Maria and that of prophet in certain rubrics. This observation is consistent with the attitude of the rubricator of L^4. He even adds a new rubric specifically intended to visually assist the reader in skipping this part of the historical section and picking up where the third component starts. The rubric reads: 'Come ritorna a ssua materia. lxxxvj.' (How it returns to its subject-matter. lxxxvj.).[62]

In the first and third component of the historical section, the combination of French and Italian elements is, first of all, replaced by an unequivocal stress on the Italian situation. This shift is discernible in some minor, but not less telling, modifications. Contrary to the *Tresor*, the *Tesoro* contains precise place names when these places are familiar to an Italian audience,[63] while clarification of geographical information generally known to an Italian audience is deleted.[64] This change in focus can also be illustrated by the more extensive discussion of Aeneas's journey to Italy or the more enlivened treatment of the story

[59] Ribémont, 'Le Temps de l'histoire', p. 239.

[60] On the Trojan legend as a foundation myth for the French monarchy: Ribémont, 'Le Temps de l'histoire', p. 239. On the link between the Roman Republic and communal Italy: Fenzi, 'Brunetto Latini', pp. 361–64.

[61] Ceva, *Brunetto Latini*, p. 133; Mussafia, 'Sul testo del Tesoro di Brunetto Latini', pp. 344 and 358–59; Plisson, 'Science et littérature au XIIIème siècle', p. 288.

[62] Transcription and translation mine.

[63] For instance, replacing a reference to the middle of Italy by a mention of Spoleto, or the specification that Aeneas landed at the estuary of the Tiber.

[64] Such as deleting the specification that Fiesole is located in Tuscany, or removing the clarifications that Venice was founded in the sea or that Treviso is situated near to Venice.

of Romulus and Remus, which even introduces an element of dialogue into the text. The new reference to the earlier inhabitants and earlier place name of Rome, the addition of the rape of the Sabine virgins, and the introduction of an excursus on the founding of Constantinople, specifying the trick used by Constantine to secure the influx of Roman nobility into the city, also fit this picture.[65] Concurrently, a reference to a Frankish victory over the Romans, which was pleasing to French ears but undoubtedly less appealing to its new audience, is replaced by a more neutral mention of the Franks' advance along the Rhineland corridor. Interesting from a Ghibelline, pro-imperial perspective is, for example, the insertion of an exoneration of the position adopted by Caesar in the Catilinarian conspiracy. This insertion underlines that Caesar wanted to punish Catilina more, and not less, by expressing his preference for a prison sentence over the death penalty. More precisely, this exoneration is aimed to preserve the reputation of this emperor-to-be. Illustrative are also the elaborate treatment of Caesar's rise to power and the manner in which his death is discussed, namely, without any reference to the treason that provoked it. In addition, various rubrics of the historical section have been revised to stress the idea of a *translatio imperii*, which ends with the Holy German Empire.

The third component of the historical section, the one closest to the living memory of a contemporary audience, witnesses the greatest amount of scribal intervention.[66] The greater familiarity of the scribe with this period not only resulted in fewer mistaken dates or misspelled names, but the tone is immediately set by a considerable extension of the temporal coverage to 1285. As a result the historical section conveniently ends with the death of Charles of Anjou, instead of his victory at Tagliacozzo (1268). This feat did not only require a reworking of the last chapters of this section, but it also necessitated the addition of no less than fifteen new chapters. In addition, this revised part focuses no longer only on the central historical figures of the period under examination, but also includes a wider set of *personae*. Furthermore, it contains a degree of vivid detail unprecedented in the original version, for instance, in the lively descriptions of the military confrontations between Charles of Anjou and his enemies, Manfred and Conradin, on the battlefields of Benevento (1266) and Tagliacozzo (1268) respectively. In addition to this improved accuracy, extended temporal scope, and increased level of detail, the original text of the third component has been thoroughly revised in a twofold manner. First,

[65] On the significance of Roman history to communal Pisa: Buck, 'Die Bedeutung der "volgarizzamenti"', p. 14; Campopiano, 'Construction of the Text, Construction of the Past', p. 76.

[66] For a similar observation: Beltrami, 'Appunti su vicende del *Tresor*', p. 323.

its geographical focus is shifted from Florence to Pisa, and Pisa's support for the Ghibelline cause is repeatedly stressed. To start, the excommunication of Pisa as a result of its support of Frederick II is highlighted. The revised text also stresses the fierce opposition of Pisa to Charles of Anjou and its assistance to Conradin. Along the same lines, the role played by a local hero, Count Gherardo of Pisa, in the hostilities at Tagliacozzo (1268) is underlined, while the involuntary character of the truce forced upon Pisa by the French after this battle is emphasized. And last, but certainly not least, a Pisan banner, which is snatched away during an Easter procession in Sicily in 1282, is presented as the all-important catalyst of the revolt commonly known as the Sicilian Vespers.[67]

In addition to this gravitational shift of attention to Pisa, the political orientation of the section is changed into an outspoken pro-Ghibelline stance. Together with the pro-imperial examples already given above, this drastic shift in perspective can be illustrated by a fundamental revision of the way in which the conflict between the papacy and Frederick II is presented. The latter is now being portrayed as the party who is constantly being challenged and undermined by the former, instead of vice versa. The modifications are, however, not only pro-Pisan and pro-imperial, but also highly anti-Angevin in character. For instance, the altered text stresses the perfidious nature of the papal-Angevin alliance. Financial motives and national sympathies are said to be its major inspiration. Furthermore, although this collaboration eventually succeeds in conquering southern Italy, the revised text specifies that this success was only achieved as a result of widespread treason and bribery, and despite the valour exhibited by Manfred on the battlefield.[68] Similarly, Charles of Anjou, once hailed as the 'Champion of Christ' in the original version,[69] is now being portrayed as a colossal coward and an arrogant, exploitative, and untrustworthy ruler — and this, for instance, in stark contrast to the more appreciative portrayal of Conradin.[70] Finally, the Sicilian Vespers (1282) and the accidental death of Charles of Anjou (1285), who died in a humiliating fashion, crushed under his own fallen horse, are recounted with obvious delight.

[67] For the three traditional narratives on the origins of this revolt: Marshall, 'The Manuscript Tradition of Brunetto Latini's *Tresor*', pp. 96–98 (referring to the pioneering work of Michele Amari).

[68] For a similar observation: Borsa, 'Letteratura antiangioina tra Provenza', p. 418.

[69] On this image: Borsa, 'Letteratura antiangioina tra Provenza', pp. 377–78, n. 3.

[70] On the myth of the '*mauvaise seigneurie*' of Charles of Anjou: Aurell, 'Chanson et propagande politique', p. 197; Barbero, *Il mito angioino nella cultura italiana e provenzale*; Borsa, 'Letteratura antiangioina tra Provenza'.

Adjusting the Mirror: A Political Remake

This contribution confirms that the effective transmission of knowledge presupposes constant adjustment to changing circumstances. It started with putting the original composition of the *Tresor* in context. This medieval encyclopaedia was written by a Florentine notary, Brunetto Latini (*c.* 1220–94), with a specific patron in mind during his exile in France (1260–66/67). For a series of motives linked both to this deplorable personal situation and the general characteristics of the language in question, Latini preferred to write his *opus magnus* in Old French, and not in Latin or his native tongue, Old Italian. Instead of falling back on his personal experience under the *Primo Popolo* regime, Latini opted to incorporate a political manual for an Italian city magistrate, part of an existing tradition of 'mirrors for magistrates', into his encyclopaedia. Notwithstanding this specific production context, the success of this text quickly extended back to the Italian peninsula, where it sparked off an active textual tradition in Old Italian, generally known as the *Tesoro*. This immediate success of a text — even during the lifetime of its original author — is quite remarkable.

By zooming in on a specific, early exponent of this *Tesoro* tradition, namely, manuscript L^4 (1284–99), part of an intricate — and still not fully explored — web of relationships between *Tresor* and *Tesoro*, I identified, first of all, a number of contextual factors which contributed to this notable success story. More precisely, these factors were linked to the larger context in which the copying of this particular text took place. For instance, this case study has underlined the importance of maritime cities, such as Pisa or Genoa, as a conduit between different cultural spheres. The production of L^4 has also been facilitated by the fact that Pisa was already an established centre for the translation of French texts, and by the fact that exemplars of the *Tresor* to be copied and translated were readily available in Genoa. In addition, the relevance of historical events, such as the naval battle of Meloria (1284) and the resulting, prolonged imprisonment of a large number of Pisans, including an important contingent of notaries, has been stressed. The translation and copying of manuscripts became an important survival technique for those prisoners who had the necessary skills.

Second, a comparison of the incipit, prologue, and historical section of the *Tresor* and *Tesoro* (L^4) has shown the importance of another set of factors — factors which were more linked to the actual copying process, and more specifically to the person of the copyist and his intended audience. In order to allow the text to break through the boundaries of its original production context and to reach a new, untapped audience, the copyist responsible for the production of this particular manuscript, Bondì Testario, had to make a series of deliberate adjustments to this *specchio del podestà* reflecting the different historical circumstances under which this copy was made. First, he had to depersonalize the text, for instance, by erasing all references to the personal predicament of its original compiler, Brunetto Latini, and by presenting himself as the continuator of Latini's work. The affectionate and individualized dedication of the original composition also had to be broken up. In addition, Bondì Testario had to shift the focus of the text from Florence to Pisa, and he had to replace Old French with Pisan. Finally, the temporal coverage of the historical section had to be significantly prolonged, and its pro-Guelf and pro-Angevin tone had to be fundamentally reoriented to a pro-Ghibelline and anti-Angevin stance in the new copy, thus transforming this particular copy of the *Tesoro* into the political mirror image of the original *Tresor*. In short, the combination of a larger context conducive to translation and copying practices and the conscious adaptation of various aspects of the original version of the *Tresor* (both language and content) by an individual copyist was key to ensuring the effective transmission of the text across the political divide between Guelfs and Ghibellines.

Works Cited

Manuscript

Florence, Biblioteca Medicea Laurenziana, MS Pluteo 42.23

Primary Sources

Oculus pastoralis

Oculus pastoralis, ed. by Dora Franceschi, *Memorie dell'Accademia delle scienze di Torino. Classe di scienze morali, storiche, e filologiche*, 4 (1966), 1–74

Oculus pastoralis, ed. by Ludovico Muratorio, in *Antiquitates italicae mediiaevi*, IV (Mediolani: Typographia societatis palatinae, 1741), cols 93–128

'Oculus pastoralis', ed. by Terence Tunberg (unpublished doctoral dissertation, University of Toronto, 1986)

Speeches from the 'Oculus pastoralis', ed. by Terence Tunberg (Toronto: Pontifical Institute of Mediaeval Studies, 1990)

Liber de regimine civitatum

Giovanni da Viterbo, *Liber de regimine civitatum*, ed. by Gaetano Salvemini, in *Bibliotheca iuridica Medii Aevi*, III (Bononiae, 1901), pp. 215–80

De regimine et sapientia potestatis

Orfini Laudensis poema De regimine et sapientia potestatis, ed. by A. Ceruti, in *Miscellanea di storia italiana*, VII (Torino: Ex typis regiis, 1869), pp. 29–94

Orfino da Lodi, *De regimine et sapientia potestatis*, ed. by L. Castelnuovo, *Archivio storico lodigiano, 2nd ser.*, 16 (1968), 1–115

Orfino da Lodi, '*De regimine et sapientia potestatis* (Comportamento e saggezza del podestà)', ed. and trans. by Sara Pozzi, *Quaderni di studi lodigiani*, 7 (1998), 1–241

Tresor

Brunetto Latini, *The Book of the Treasure*, trans. by Paul Barrette and Spurgeon Baldwin (New York: Garland, 1993)

——, *Li Livres dou Tresor*, ed. by Francis Carmody (Berkeley: University of California, 1948)

——, *Tresor*, ed. and trans. by Pietro Beltrami and others (Torino: Einaudi, 2007)

Secondary Studies

Abulafia, David, ed., *Italy in the Central Middle Ages 1000–1300* (Oxford: Oxford University Press, 2004)

——, 'The State of Research: Charles of Anjou Reassessed', *Journal of Medieval History*, 26 (2000), 93–114

Addario, A. d', 'Brunetto Latini', in *Enciclopedia Dantesca*, III (Roma: Istituto dell'enciclopedia Italiana, 1971), pp. 578–79

Anton, Hans Hubert, *Fürstenspiegel des frühen und hohen Mittelalters* (Darmstadt: Wissenschaftliche Buchgesellschaft, 2006)

Artifoni, Enrico, 'I podestà professionali e la fondazione retorica della politica comunale', *Quaderni storici*, 63 (1986), 687–719

Aurell, Martin, 'Chanson et propagande politique: Les Troubadours gibelins (1255–1285)', in *Le forme della propaganda politica nel Due e nel Trecento*, ed. by Paolo Cammarosano (Rome: École française de Rome, 1994), pp. 183–202

Barbero, Alessandro, *Il mito angioino nella cultura italiana e provenzale fra Duecento e Trecento* (Torino: Deputazione subalpina di storia patria, 1983)

Beltrami, Pietro, 'Appunti su vicende del *Tresor*: Composizione, letture, riscritture', in *L'enciclopedismo medievale*, ed. by Michelangelo Picone (Ravenna: Longo, 1994), pp. 311–28

——, 'Introduzione', in *Tresor*, ed. by Pietro Beltrami and others (Torino: Einaudi, 2007), pp. vii–xxvi

——, 'Per il testo del *Tresor*: appunti sull'edizione di F. J. Carmody', *Annali della Scuola Normale Superiore di Pisa*, 18.3 (1988), 961–1009

Benedetti, Roberto, 'Qua fa' un santo e un cavaliere … Aspetti codicologici e note per il miniatore', in *La Grant Queste del Saint Graal* (Udine: Roberto Vattori, 1990), pp. 31–47

Berges, Wilhelm, *Die Fürstenspiegel des hohen und späten Mittelalters* (Leipzig: Karl W. Hiersemann, 1938)

Bertelli, Sandro, 'Tipologie librarie e scritture nei più antichi codici fiorentini di ser Brunetto', in *A scuola con ser Brunetto: Indagini sulla ricezione di Brunetto Latini dal Medioevo al Rinascimento (Atti del Convegno internazionale di studi, Università di Basilea, 8–10 giugno 2006)*, ed. by Irene Maffia Scariati (Firenze: Galluzzo, 2008), pp. 213–53

Bognetti, Gian Piero, *Storia del diritto italiano pubblico: Parte speciale. Appunti sul podestà* (Pisa: Pacini Mariotti, 1933–34)

Bolton Holloway, Julia, *Twice-Told Tales: Brunetto Latini and Dante Alighieri* (New York: Peter Lang, 1993)

Borsa, Paolo, 'Letteratura antiangioina tra Provenza, Italia e Catalogna: La figura di Carlo I', in *Gli Angiò nell'Italia nord-occidentale (1259–1382)*, ed. by Rinaldo Comba (Milano: Unicopli, 2006), pp. 377–432

Bratu, Cristian, 'Mirror for Princes (Western)', in *Handbook of Medieval Studies*, ed. by Albrecht Classen, 3 vols (New York: de Gruyter, 2010), III, 1921–49

Buck, August, 'Die Bedeutung der "volgarizzamenti" für die Geistes- und Literaturgeschichte', in *Studien zu den 'volgarizzamenti' römischer Autoren in der italienischen Literatur des 13. und 14. Jahrhunderts*, ed. by August Buck and Max Pfister (München: Wilhelm Fink, 1978), pp. 7–43

Campopiano, Michele, 'Construction of the Text, Construction of the Past: Historical Knowledge, Classical Myths and Ideology in a Medieval *Comune* (Pisa, Eleventh to Twelfth Centuries)', *Troianalexandria*, 9 (2009), 63–84

Canaccini, Federico, 'Restano i termini, mutano i significati: Guelfi e Ghibellini. L'evoluzione semantica dei nomi delle fazioni medioevali italiane', in *Lotta politica nell'Italia medievale* (Roma: Istituto storico italiano per il medio evo, 2010), pp. 85–94

Carmody, Francis, 'Introduction', in *Li Livres dou Tresor*, ed. by Francis Carmody (Berkeley: University of California, 1948), pp. ix–lxii

Carmody, Francis, and Françoise Fery-Hue, 'Brunetto Latini', in *Dictionnaire des lettres françaises: Édition entièrement revue et mise à jour*, ed. by Geneviève Hasenohr and Michel Zink (Paris: Fayard, 1992), pp. 213–15

Castellani, Arrigo, 'Pisano e lucchese', *Studi linguistici italiani*, 5 (1965), 97–135

——, *Saggi di linguistica e filologia italiana e romanza (1946–1976)*, 3 vols (Roma: Salerno, 1980)

——, 'La Toscana dialettale d'epoca antica', *Studi linguistici italiani*, 23 (1997), 3–46

Ceccarelli Lemut, Maria Luisa, 'I Pisani prigionieri a Genova dopo la battaglia della Meloria: la tradizione cronistica e le fonti documentarie', in *1284: L'anno della Meloria*, ed. by Renzo Mazzanti and others (Pisa: ETS, 1984), pp. 77–87

Cella, Roberta, 'Gli atti rogati da Brunetto Latini in Francia (tra politica e mercatura, con qualche implicazione letteraria)', *Nuova rivista di letteratura italiana*, 6 (2003), 367–408

Ceva, Bianca, *Brunetto Latini: L'uomo e l'opera* (Milano: Riccardo Ricciardi, 1965)

Cigni, Fabrizio, 'Copisti prigionieri (Genova, fine sec. XIII)', in *Studi di Filologia romanza offerti a Valeria Bertolucci Pizzorusso*, ed. by Pietro Beltrami and others, 2 vols (Pisa: Pacini, 2006), I, 425–39

——, 'Genova e una versione toscana della Legenda Aurea', in *Poeti e poesia, a Genova (e dintorni) nell'età medievale (Atti del convegno per Genova capitale della cultura europea 2004)*, ed. by Margherita Lecco (Alessandria: Edizioni dell'Orso, 2006), pp. 43–79

——, 'Manuscrits en français, italien, et latin entre la Toscane et la Ligurie à la fin du XIIIe siècle: Implications codicologiques, linguistiques, et évolution des genres narratifs', in *Medieval Multilingualism in England, France and Italy: The Francophone World and its Neighbours (Proceedings of the 2006 Conference at the University of Wisconsin-Madison)*, ed. by Christopher Kleinhenz and Keith Busby (Turnhout: Brepols, 2010), pp. 187–217

——, 'La ricezione medievale della letteratura francese nella Toscana nord-occidentale', in *Fra toscanità e italianità: Lingua e letteratura dagli inizi al Novecento*, ed. by Werner Edeltraud and Sabine Schwarze (Tübingen: A. Francke, 2000), pp. 71–108

——, 'I testi della prosa letteraria e i contatti col francese e col latino: Considerazioni sui modelli', in *Pisa crocevia di uomini, lingue e culture: L'età medievale (Atti del Convegno*

— *Pisa, 25–27 ottobre 2007)*, ed. by Lucia Battaglia Ricci and Roberta Cella (Roma: Aracne, 2009), pp. 157–81

Davidsohn, Robert, *Storia di Firenze*, trans. by Giovanni Battista Klein, rev. by Roberto Palmarocchi, 8 vols (Firenze: Sansoni, 1956–68)

Dessì, Rosa Maria, 'Guelfi e Ghibellini, prima e dopo la battaglia di Montaperti (1246–1358)', in *Montaperti: 1260–2010: Nella ricorrenza dei 750 anni della battaglia* (Siena: Accademia degli Intronati, 2011), pp. 21–32

Diacciati, Silvia, *Popolani e magnati: Società e politica nella Firenze del Duecento* (Spoleto: Centro italiano di studi sull'alto medioevo, 2011)

Divizia, Paolo, 'Aggiunte (e una sottrazione) al censimento dei codici delle versioni italiane del Tresor di Brunetto Latini', *Medioevo Romanzo*, 32 (2008), 377–94

——, 'Integrazioni al censimento dei codici italiani di Brunetto Latini', *Medioevo Romanzo*, 37 (2013), 184–85

Dunbabin, Jean, *Charles I of Anjou: Power, Kingship and State-Making in Thirteenth-Century Europe* (London: Longman, 1998)

Eberle, Patricia, 'Mirror of Princes', in *Dictionary of the Middle Ages*, ed. by Joseph R. Strayer (New York: Charles Scribner, 1982–89), VIII, 434–36

Fabbri, Francesca, 'Romanzi cortesi e prosa didattica a Genova alla fine del Duecento fra interscambi, coesistenze e nuove prospettive', *Studi di Storia dell'Arte*, 23 (2012), 9–32

Faini, Enrico, *Firenze nell'età romanica (1000–1211): L'espansione urbana, lo sviluppo istituzionale, il rapporto con il territorio* (Firenze: Olschki, 2010)

Fenzi, Enrico, 'Brunetto Latini, ovvero il fondamento politico dell'arte della parola e il potere dell'intellettuale', in *A scuola con ser Brunetto: Indagini sulla ricezione di Brunetto Latini dal Medioevo al Rinascimento (Atti del Convegno internazionale di studi, Università di Basilea, 8–10 giugno 2006)*, ed. by Irene Maffia Scariati (Firenze: Galluzzo, 2008), pp. 323–69

Franchini, Vittorio, *Saggio di recherche sull'instituto del podestà nel comune medievale* (Roma: Spithoever, 1912)

——, 'Trattati *De regimine civitatum* (sec. XIII–XIV)', in *Recueil de la Société Jean Bodin*, VI: *La Ville (Première partie: Institutions administratives et judiciaires)* (Brussels: La Librairie Encyclopédique, 1954), pp. 319–42

Gentile, Marco, 'Factions and Parties: Problems and Perspectives', in *The Italian Renaissance State*, ed. by Andrea Gamberini and Isabella Lazzarini (Cambridge: Cambridge University Press, 2012), pp. 304–22

Giannini, Gabriele, 'Comptes rendus: A scuola con ser Brunetto', *Romania*, 129 (2011), 235–46

——, 'Comptes rendus: Pisa crocevia di uomini, lingue e culture', *Romania*, 129 (2011), 512–15

Giola, Marco, 'Sul volgarizzamento italiano del "Tresor" di Brunetto Latini' (unpublished doctoral thesis, Università degli Studi di Ferrara, 2006)

——, *La tradizione dei volgarizzamenti toscani del 'Tresor' di Brunetto Latini* (Verona: QuiEdit, 2010)

Green, Louis, 'Florence', in *The New Cambridge Medieval History*, V: *c. 1198–c. 1300*, ed. by David Abulafia (Cambridge: Cambridge University Press, 1999), pp. 479–96

Gualtieri, Piero, *Il comune di Firenze tra Due e Trecento: Partecipazione politica e assetto istituzionale* (Firenze: Olschki, 2009)

Hanauer, Gustav, 'Das Berufspodestat im dreizehnten Jahrhundert', *Mitteilungen des Instituts für Oesterreichische Geschichtsforschung*, 23 (1902), 377–426

Herde, Peter, *Karl I von Anjou* (Stuttgart: Kohlhammer, 1979)

Hertter, Fritz, *Die Podestàliteratur Italiens im 12. und 13. Jahrhundert* (Leipzig: Teubner, 1910)

Inglese, Giorgio, 'Brunetto Latini', in *Dizionario Biografico degli Italiani* (Roma: Istituto della enciclopedia Italiana, 2005), pp. 4–12

Jones, Philip, *The Italian City-State: From Commune to Signoria* (Oxford: Clarendon Press, 1997, repr. 2004)

Luff, Robert, *Wissensvermittlung im europäischen Mittelalter* (Tübingen: Niemeyer, 1999)

Maffia Scariati, Irene, *Dal Tresor al Tesoretto: Saggi su Brunetto Latini e i suoi fiancheggiatori* (Roma: Aracne, 2010)

——, 'Gli atti rogati da Brunetto Latini in Francia (tra politica e mercatura, con qualche implicazione letteraria)', *La Rassegna della Letteratura italiana*, 109 (2005), 459–61

Maire Vigueur, Jean-Claude, ed., *I podestà dell'Italia comunale: Reclutamento e circolazione degli ufficiali forestieri (fine* XII *sec.-metà* XIV *sec.)*, 2 vols (Roma: École française de Rome, 2000)

Manni, Paola, *Storia della lingua italiana: Il Trecento toscano* (Bologna: Il Mulino, 2003)

Marshall, Jennifer, 'The Manuscript Tradition of Brunetto Latini's *Tresor* and its Italian Versions' (unpublished doctoral thesis, Royal Holloway University of London, 2001)

Mazzoni, Francesco, 'Brunetto Latini', in *Enciclopedia Dantesca*, III (Roma: Istituto dell'Enciclopedia Italiana, 1971), pp. 579–88

Meier, Christel, 'Cosmos Politicus: Der Funktionswandel der Enzyklopädie bei Brunetto Latini', *Frühmittelalterliche Studien*, 22 (1988), 315–56

——, 'Enzyklopädischer Ordo und sozialer Gebrauchsraum: Modelle der Funktionalität einer universalen Literaturform', in *Die Enzyklopädie im Wandel vom Hochmittelalter bis zur frühen Neuzeit*, ed. by Christel Meier (München: Fink, 2002), pp. 511–32

——, 'Vom Homo Coelestis zum Homo Faber', in *Pragmatische Schriftlichkeit im Mittelalter*, ed. by Hagen Keller and others (München: Fink, 1992), pp. 157–75

Miller, Anne-Hélène, 'Le Poète dans la Cité: Figures de l'intellectuel vernaculaire au tournant de la fin du Moyen Âge' (unpublished doctoral dissertation, University of Washington, 2007)

Mussafia, Adolfo, 'Sul testo del Tesoro di Brunetto Latini', in *Della vita e delle opera di Brunetto Latini*, trans. by Rodolfo Renier (Firenze: Le Monnier, 1884), pp. 279–390

Najemy, John, *A History of Florence 1200–1575* (Malden: Blackwell, 2006)

Napolitano, David, 'Brunetto Latini's *Tesoro* in Print', *Ex Historia*, 5 (2013), 19–47

——, 'Pleidooi voor een hernieuwde kennismaking met de podestà literatuur', *Incontri: Rivista europea di studi italiani*, 28.2 (2013), 22–30

——, 'The Profile and Code of Conduct of the Professional City Magistrate in Thirteenth-Century Italy' (unpublished doctoral thesis, University of Cambridge, 2014)

——, 'Verborgen schatten in een Parijse bibliotheek: Nieuw licht op de Oculus Pastoralis', *Madoc: Tijdschrift voor de Middeleeuwen*, 27 (2014), 242–49

Novati, Francesco, *La giovinezza di Coluccio Salutati (1331–1353)* (Torino: Ermanno Loescher, 1888)

Ottokar, Nicola, *Studi comunali e fiorentini* (Firenze: La Nuova Italia, 1948)

Plisson, Julien, 'Science et littérature au XIIIème siècle: Un vulgarisateur: Brunetto Latini' (unpublished doctoral thesis, Université de Provence, 1993)

Raveggi, Sergio, and others, *Ghibellini, guelfi e popolo grasso: I detentori del potere politico a Firenze nella seconda metà del Dugento* (Firenze: La Nuova Italia, 1978)

Ribémont, Bernard, 'Le Temps de l'histoire chez un encyclopédiste médiévale: L'Exemple de Brunetto Latini', in *Le Temps, sa mesure et sa perception au Moyen Âge (Actes du Colloque, Orléans, 12–13 avril 1991)*, ed. by Bernard Ribémont (Caen: Paradigme, 1992), pp. 229–42

Rosa, Daniela de, *Alle origini della repubblica fiorentina: Dai consoli al 'Primo Popolo' (1172–1260)* (Firenze: Arnaud, 1995)

Roux, Brigitte, *Mondes en miniatures: L'Iconographie du Livre du Trésor de Brunetto Latini* (Génève: Droz, 2009)

Salvemini, Gaetano, 'Il "Liber de regimine civitatum" di Giovanni da Viterbo', *Giornale storico della letteratura italiana*, 41 (1903), 284–303

Signorini, Maddalena, 'Il copista di testi volgari (secoli X–XIII): Un primo sondaggio dei fonti', *Scrittura e civiltà*, 19 (1995), 123–97

Sorbelli, Albano, 'I teorici del reggimento comunale', *Bullettino dell'Istituto storico italiano per il medio evo e Archivio Muratoriano*, 59 (1944), 31–136

Squillacioti, Paolo, 'Ricerche sulla tradizione del *Tesoro* toscano', in *A scuola con ser Brunetto: Indagini sulla ricezione di Brunetto Latini dal Medioevo al Rinascimento (Atti del Convegno internazionale di studi, Università di Basilea, 8–10 giugno 2006)*, ed. by Irene Maffia Scariati (Firenze: Galluzzo, 2008), pp. 547–63

——, 'La tradizione manoscritta delle opera di Brunetto Latini', in Brunetto Latini, *Tresor*, ed. and trans. by Pietro Beltrami and others (Torino: Einaudi, 2007), pp. xlvii–liii

Sundby, Thor, *Della vita e delle opera di Brunetto Latini*, trans. by Rodolfo Renier (Firenze: Le Monnier, 1884)

Swiggers, Pierre, *Le 'Tresor' de Brunetto Latini et l'usage du français* (Leuven: Departement Linguïstiek KUL, 1998)

Tanzini, Lorenzo, 'Tuscan States: Florence and Siena', in *The Italian Renaissance State*, ed. by Andrea Gamberini and Isabella Lazzarini (Cambridge: Cambridge University Press, 2012), pp. 90–111

Vincentiis, Amedeo de, 'Le parole di ser Brunetto', in *Atlante della letteratura italiana*, ed. by Sergio Luzzatto and Gabriele Pedullà (Torino: Einaudi, 2010–), *Volume primo: Dalle origini al Rinascimento*, ed. by Amedeo de Vincentiis, pp. 41–47

Welie-Vink, Wendelien van, 'Was Charles d'Anjou Brunetto Latini's *biaus dous amis*? De openingsminiaturen van *Li Livres dou Tresor* nader bekeken', in *Representatie: Kunsthistorische bijdragen over vorst, staatsmacht en beeldende kunst, opgedragen aan Robert W. Scheller*, ed. by Johann-Christian Klamt and Kees Veelenturf (Nijmegen: Valkhof, 2004), pp. 315–41

Zinelli, Fabio, '"Donde noi metremo lo primo in franceschoˮ I proverbi tradotti dal francese ed il loro inserimento nelle sillogi bibliche', in *La Bibbia in italiano tra Medioevo e Rinascimento (Atti del Convegno internazionale, Firenze, Certosa del Galluzzo, 8–9 novembre 1996)*, ed. by Lino Leonardi (Firenze: Sismel, 1998), pp. 145–99

——, 'Tradizione "mediterranea" e tradizione italiana del *Livre dou Tresor*', in *A scuola con ser Brunetto: Indagini sulla ricezione di Brunetto Latini dal Medioevo al Rinascimento (Atti del Convegno internazionale di studi, Università di Basilea, 8–10 giugno 2006)*, ed. by Irene Maffia Scariati (Firenze: Galluzzo, 2008), pp. 35–89

Zorzi, Andrea, *La trasformazione di un quadro político: Ricerche su politica e giustizia a Firenze dal comune allo stato territoriale* (Firenze: Firenze University Press, 2008)

Mirror Metaphors in Persian Sufi Literature

Leila Rahimi Bahmany

The mirror invites the attentive onlooker into a space where antithetical features can readily become involved in a dialectical debate. The person's seemingly dyadic nature, such as subject and object, beholder and beheld, visible and invisible, being and seeming, interior and exterior, real and illusory, present and absent, all converge on the surface of the mirror. For the onlooker, the mirror becomes a threshold through which the *this-worldly* can gain access to the *other-worldly*. This is what Jenijoy La Belle describes as 'the oxymoronic nature of the mirror'.[1] Furthermore, the mirror presents the dimensions of time and space simultaneously. A look into one's mirror not only reiterates one's present image and state of being at the time of looking, but also summons up from its depths bygone images. Moreover, it gives an outline of images still to come, triggering within the beholder a feeling of hope or sense of foreboding. Hence, the mirror can cause not only spatial dislocations but also temporal ones.

The looking glass is an ambivalent space for reality and virtuality, multiplicity and contradictions. The spatiotemporal extension of the mirror and the ongoing negotiations between the real and virtual occurring on its reflective surface make the mirror a heterotopian and even utopian space. In his comparison, Michel Foucault refers to utopia, and likewise to the mirror, in terms of a 'placeless place', 'no real place', and 'fundamentally unreal space', as they both share 'a general rela-

[1] La Belle, *Herself Beheld*, p. 161.

Leila Rahimi Bahmany (leila.rahimibahmany@gmail.com) completed her doctorate at Freie Universität Berlin, working on comparative literature and literary feminism. She is the author of *Mirrors of Entrapment and Emancipation: Forugh Farrokhzad and Sylvia Plath*.

The Mirror in Medieval and Early Modern Culture: Specular Reflections, ed. by Nancy M. Frelick, CURSOR 25 (Turnhout: Brepols, 2016), pp. 113–129 10.1484/M.CURSOR-EB.5.111044
BREPOLS PUBLISHERS

tion of direct or inverted analogy with the real space'. He argues that utopia provides society — containing the individual — with a representation 'in a perfected form' or 'turned upside down', exactly in the same manner that the mirror provides the individual with her/his image. On the other hand, Foucault sees the mirror as a heterotopia, a 'counter-site', and 'a kind of effectively enacted utopia'.[2] This utopian nature of the mirror can probably account for the numerous instances in literature where someone steps into a mirror (or desires to do so) and joins with the virtual/ideal self or the world beyond its surface. The mirror also serves as an opening, a window or threshold, inviting a person into the world of delusion and imagination. The mirror remains a virtual site of *de-realization*; within its space, the singularity of the real image is split into a plurality of images in one's mind. All these features, or shall we say powers of the mirror, enumerated hitherto, convert this instrument into a cornucopia—an inexhaustible wellspring of ambivalent meanings and symbols which have fascinated writers over the centuries, among them Persian poets. The fourteenth-century Persian Sufi poet Hāfez wittily portrays the mirror's complex ambiguities thus: 'حسن روی تو به یک جلوه که در آینه کرد | این همه نقش در آئینهٔ اوهام افتاد' (The beauty of Thy face with that one splendour that struck the mirror | All these images onto the mirror of illusion fell).[3]

Sufism is predominantly based on the doctrine of love. It defines, as well as prescribes, that man's relationship to his God is, and should be, one of love. This loving nature of the relationship is given preponderance over other types of relationships existing in Islam, such as subordination, creator-creature, and master-slave, among others. The relationship between man and God, including all His manifestations, in the form of His creations, is that of lover and the Beloved. This Eros-oriented nature of Sufism has turned Sufi literature into a highly metaphoric literature of love. On the one hand, Sufis believe that language and words are incapable of conveying the complex and profound truths of the human condition. Words are unable to reflect a portrait of the Ultimate Beloved and the nature of man's relationship to his divine Beloved. Sufis are urged to resort to a metaphoric language that would enable them to communicate the ineffable truths. How can a tongue or a pen convey the fire of love without being consumed? Hāfez says, 'قلم را آن زبان نبود که سرّ عشق گوید باز | ورای حدّ تقریرست شرح آرزومندی' (The reed pen does not have that tongue to reveal the mystery of love | Beyond the limit of narration, is the explanation of longing).[4]

[2] Foucault, 'Of Other Spaces', trans. by Miskowiec, p. 24.

[3] Hāfez, *Divān*, ed. by Qazvini and Ghani, p. 86 (translations are all mine).

[4] Hāfez, *Divān*, ed. by Qazvini and Ghani, p. 341. The Persian word *zabān* signifies both language and tongue.

On the other hand, Sufis also believe that the language of Sufi poetry, which seeks to convey the beauty of the Beloved, should be highly figurative, having many layers of meaning. These layers of meaning allow the reader to enjoy the sense of the poem according to her/his spiritual state. Yet the ambiguity of the poem engages the reader in a mental struggle in the world of meanings. The epiphanic mode of Sufi poetry, through its metaphors and symbolisms, and by its very brevity, provides the reader with moments of insight into the essential meaning of things. Hence, Sufi poetry is in itself a mirror, engaging the reader in laboriously polishing and unveiling the text for the higher meanings to be reflected.

The present article intends to address three of the most recurrent mirror metaphors in Persian classical literature: *āyena-ye del* (the mirror of the heart), *jām-e jam* (the mythical cup of Jamshid), and *harrāqa* or *āyena-ye suzān* (the burning mirror). These three key mirror metaphors have been used frequently and interchangeably, to the extent that almost any evocation of the word mirror in Persian Sufi literature is loaded with one, two, or even all three of these metaphors combined.

The Mirror of the Heart

The 'mirror of the heart' is probably the most dominant mirror metaphor in classical Persian literature. It has been a recurring metaphor in its non-mystical and mystical sense since at least the eleventh century. In the non-mystical usage of the metaphor, the receptivity and sincerity of the mirror in reflecting an image has been employed to liken the heart to a mirror; in itself, the heart is a metaphor for something beyond physical existence, for the human soul. In experiencing worldly love, the heart should be as a polished mirror, reflecting nothing but the beauty of the beloved. Persian Sufi poets, however, have adopted the mirror in their texts to explicate, as well as to prescribe, the nature of man's relationship with his Ultimate Beloved. The mirror's ability to bring the subject and object together, as well as its unyielding obedience and passive reflectivity, were frequently drawn upon by many Sufi poets to highlight their doctrine of the supremacy of the experience of divine receptivity over reason.[5] This state of passivity, reflecting things while remaining unaffected by them, represents the serene non-activity of the wise man. The thirteenth-century Persian Sufi poet Rumi describes the metaphor of 'the mirror of the heart' thus:

[5] When the Sufis attempted to assign a more active agency to the part of the Sufi seeker, they used the term *kashf* (unveiling) instead.

آينهٔ دل چون شود صافی و پاک
نقشها بينی برون از آب و خاک

هم ببينی نقش و هم نقّاش را
فرش دولت را و هم فرّاش را

(When the mirror of the heart becomes pure and clean
You will behold images outside of [the world of] water and earth

You will behold both the painting and the painter
Both the carpet of kingdom and the carpet spreader.)[6]

When a man succeeds in polishing his heart, cleansing it of all the tarnish and rust of worldly sins that veil it from reflectivity, he can see on its reflective surface all of creation and its Creator.

In this metaphor, man's heart is depicted as a mirror reflecting God's *imago* and His different manifestations. It is believed that the pure, untarnished heart can reflect the divine light of God. Through self-indulgence, however, this mirror gathers rust, hampering its ability to reflect the divine light. In order to polish the mirror and restore its reflectivity, one has to refrain from involvement in the self and its desires. The perfect, spotless polish of this mirror is maintained only when *sālek* (path seeker) succeeds in attaining the state of absolute *fanā* (self-annihilation). It is at this stage that *tajalli* (theophany) occurs to the Sufi's heart, and he revels ecstatically in the *Unio Mystica*: the divine receptivity engraved upon his heart. Thereby, for the Sufi, the heart as mirror becomes a space where microcosm and macrocosm mutually reflect each other and man becomes one with the universe and its Creator. To put it in literary terms, Sufis use the vehicles of mirror and heart to convey the tenor of the nature of their love relationship. As understood in *Unio Mystica*, when the soul is turned into a mirror, it is not merely reflecting but also contemplating and participating in the image and the beauty to which it is exposed. By assimilating the image, the soul-mirror undergoes metamorphosis and ultimately becomes part of that beauty.

The mirror figure represents the schism between two ostensibly identical worlds or two modes of being: while it can serve as a bridge from one level of reality to the higher level of reality, from the known to the unknown, it can also become a locus of illusion and deception. Hence in Sufism, it is believed that before the ultimate stage of perfect annihilation of the self in God, there is a stage in which the two presences of the man-as-a-mirror and the Truth or

[6] Mawlavi [=Rumi], *Masnavi Ma'navi*, ed. by Sorush, II, 186 (translations mine).

God coexist. Some Sufis assert that this stage is marked by imperfection and deficiency, because the subject is still in dubious suspense between these two presences, such that his view fluctuates between them.[7] It is also suggested that man can never comprehend the Truth in its totality. The whole universe is comprised of mirrors which reflect various radiations of the one being. Man can only grasp incomplete representations of the Divine Truth, as though through the ongoing reflections from pieces of a shattered mirror.

The metaphorical image of the 'mirror of the heart' can be found in ancient Neoplatonic thought, as can the twofold division of macrocosm and microcosm, which includes the division of the world into visible and invisible, human soul and the Divine Being, where one mutually reflects the other. The metaphor of the human soul as a mirror for God's reflection has been a culturally significant mirror metaphor in Western thought, as it is omnipresent in the spiritual literature of the Middle Ages.[8] Muslim Sufis seem to have inherited this metaphor from the Hellenistic tradition, by way of Neoplatonic thinkers, particularly Plotinus. In his study of mirror symbolism in Islamic mysticism, Titus Burckhardt observes that the mirror has long been the most appropriate symbol for demonstrating the essential nature of Islamic mysticism, which is 'purely spiritual', as well as for conveying 'the spiritual vision, the *contemplatio*, and more generally the knowledge, because through it, the approximation of subject and object is revealed'.[9] This appears to be particularly valid when one is referring to intuitive knowledge.

[7] The parable of the mirror has been recurrently used in the explanation and criticism of *shathiyyāt* (the ecstatic utterances) of some mystics. For example, Hosayn ebn-e Mansur al-Hallāj (executed in 922) declares 'I am the Truth', and Bāyazid Bastāmi (d. *c.* 874) utters in ecstasy 'Glory be to me! How great is my majesty!' and 'I am He'. See: Purjavādi, *Eshrāq o 'erfān*, pp. 26–28.

[8] Theodore Ziolkowski argues that this metaphor 'descends from Plato and the Bible by way of Christian Platonism. This syncretic analogy, which was first publicized by Augustine and his commentators, is based principally on two passages in the New Testament' (1 Corinthians 13.12 and James 1. 23–24); Ziolkowski, *Disenchanted Images*, p. 152.

Melchior-Bonnet discusses the origin and development of this metaphoric thought: 'The mirror became part of the religious vocabulary of the Middle Ages, which developed its symbolic meanings from scriptural writings, Neoplatonic texts, and the patristic tradition (the writings of the church fathers)'; Melchior-Bonnet, *The Mirror: A History*, p. 108. See also: Leisegang, 'Die Erkenntnis Gottes im Spiegel der Seele und der Natur'.

[9] Burckhardt, 'Die Symbolik des Spiegels in der islamischen Mystik', p. 12 (translation mine). Compare with the original: 'weil es [das Sinnbild des Spiegels] wie kaum ein anderes dazu geeignet ist, das Wesen dieser Mystik, ihren vornehmlich erkenntnishaften Charakter zu zeigen;

In Persian Sufism, the metaphor of the heart can be traced in the writings of Muslim philosophers such as Avicenna (d. 1037), Abu Hāmed Ghazāli (d. 1111), his younger brother Ahmad Ghazāli (d. 1123 or 1126), and Sohrawardi (d. 1191).[10] This metaphor was then appropriated by Persian Sufi poets like 'Attār (d. *c.* 1221), Rumi (d. 1273), Fakhr od-Din 'Erāqi (d. 1289), Sa'di (d. *c.* 1292), Hāfez (d. 1389), and Bidel (d. 1720), among others.[11] Writing about kings, Rumi in his *Masnavi* says in a didactic tone:

صوفیان را پیش رو موضع دهند
کاینهٔ جانند و ز آیینه بهند

سینه صیقلها زده در ذکر و فکر
تا پذیرد آینهٔ دل نقش بکر

هر که او از صُلب فطرت خوب زاد
آینه در پیش او باید نهاد

(The kings place the Sufis before their countenance
For the Sufis are mirrors for the soul, even better than a mirror

der Spiegel ist in der Tat das unmittelbarste Sinnbild der geistigen Schau, der *contemplatio*, und ganz allgemein der Erkenntnis, denn durch ihn wird die Angleichung von Subjekt und Objekt veranschaulich.' It should be recalled that the philosophical polarity between subject and object did not exist at the time. The idea was to be developed by Descartes and later thinkers. The relationship between the mirror and the reality it reflected was understood as a form of 'speculation'.

[10] For a succinct study of this mirror metaphor in the works of Avicenna and Sohrawardi see: Purjavādi, *Eshrāq o 'erfān*, pp. 25–28. For mirror metaphor in the works of Abu Hamed Ghazāli see: Lazarus-Yafeh, *Studies in Al-Ghazzali*, pp. 314–20. Lazarus-Yafeh, in this thematic study of Ghazāli's Arabic works, devotes a section to 'The Parable of the Mirror', compiling the passages in which the parable of the mirror in the different Arabic works of Ghazāli appears.

[11] For a further discussion of the mirror of the heart in classical Persian literature, see: Zipoli, 'Semiotics and the Tradition of the Image'. In this study, Zipoli traces the development of mirror image in the works of three stylistically representative authors in the history of classical Persian poetry: Farrokhi, Hāfez, and Sā'eb. See also: Musavi, 'Āyena dar shāhkār-hā-ye adabi tā qarn-e hashtom', pp. 75–86. Bidel Dehlavi's poetry, famous for its stylistic opacity and obfuscatory ambiguities, is a plethora of mirror imagery. Mohammad Rezā Shafi'i Kadkani aptly titles his anthology of Bidel's poems as 'The Poet of the Mirrors'. At the end of this book Shafi'i Kadkani provides us with a glossary of the poet's mirror associations and interpretation, understandably far from comprehensive. See: Shafi'i Kadkani, *Shā'er-e ā'ine-hā*. For the study of Bidel's mirror imagery, see also: Meneghini, 'A New Approach to Analyzing the Use of the Word *Āyine* (Mirror)'; Zipoli, 'Āyine (Mirror) in Bidel's Ghazals'; Zipoli, 'A Computer-Assisted Analysis of Bidel's *Tur-e Ma'refat*'.

They have so thoroughly polished their breasts
in commemoration (*zekr*) and meditation (*fekr*)
That the mirror of the heart may receive the virgin image

Whoever is born good from the loins of nature
A mirror must be placed in front of him.)[12]

In Islamic Sufism, the human being can become a *Seelenspiegel*, a living soul-mirror for the soul of the fellow man, the friend, or the beloved. The mirror can serve as a symbol for the perfect human being, for the prophet of Islam, for the Sufi master, and the Sufi path seeker, all versions of the ideal self. A famous *hadis* (prophetic tradition) bespeaks this reciprocal awareness of the soul mirror, 'المؤمن مرآة المؤمن' (the faithful is the mirror of the faithful). Alluding to the same *hadis*, Rumi says: 'چونک مؤمن آینهٔ مؤمن بود | روی او ز آلودگی ایمن بود' (Since the faithful is the mirror of the faithful | His face is immune to any defilement).[13]

Furthermore, as God and His divine light may be perceived by the Sufi, His divine voice may be heard in his interior, or, as they express it, by his *gush-e jān* (ear of the soul). It is through the master Sufi that the path seekers can see and hear their beloved God. The mirror metaphor in the latter sense abounds in Persian literature. For example, the twelfth-century exegete of the *Qur'ān*, Abo'l-Fazl Rashid od-Din Maybodi, says: 'The *'āref* [Sufi] is a mirror; he who looks upon it sees his Lord.'[14] The heterotopic-utopian feature of the mirror makes many Sufi-poets utter their desire to turn into a mirror themselves; a perfect man, a utopian 'placeless place' and a 'counter-site' to this rough, earthen prison world and the prison body. Indeed, many of these Sufi poets have described the world and their bodies as a *hejab* (veil) covering the mirror of their self. Rumi believes that it is with constant polishing of the breast, that is, the soul, through continual remembrance of, and meditation on, the Ultimate Beloved, that the heart becomes receptive to His divine *imago*, invisible to the physical eye. Sa'di, too, adopts the mirror metaphor to convey the same Sufi principle: 'تأمّل در آيينهٔ دل كنى | صفايى به تدريج حاصل كنى' (Reflect in the mirror of the heart | And gradually you will acquire purity).[15]

[12] Mawlavi [=Rumi], *Masnavi Ma'navi*, ed. by Sorush, I, 142.

[13] Mawlavi [=Rumi], *Masnavi Ma'navi*, ed. by Sorush, II, 184. For the sources of this *hadis* see: Foruzānfar, *Ahādis o qesas-e Masnavi*, pp. 155–56.

[14] Maybudi, *Kashf al-asrār wa-'oddat al-abrār*, ed. by Hekmat and others, III, 788; See: Keeler, *Sufi Hermeneutics*, p. 308, n. 102.

[15] Sa'di, *Kolliyāt-e Sa'di*, ed. Foroughi, p. 187 (translation mine).

Persian Sufi thinkers and poets have merged this key metaphor, the 'mirror of the heart', with other popular specular images in Persian literature, such as the mythological 'cup of Jamshid' and 'the burning mirror', in order to aesthetically convey the complexities of their doctrine in highly figurative language; a language which in itself needs constant polishing so as to mirror its meanings. These concepts and meanings require a succession of interpretations (in the form of unveilings) to be fully grasped. Here the two topoi of 'cup of Jamshid' and 'burning mirror' are succinctly addressed.

The Cup of Jamshid

The Sufi metaphor of divinity, 'the mirror of the heart', was further augmented with another figurative *topos* adopted from ancient pre-Islamic Persian mythology, the *jām-e jam* or the cup of Jamshid. *Jām* literally means a chalice, cup, glass, or mirror, but it also denotes 'clean' and 'pure', and *jam* etymologically means a great king or ruler, again denoting purity. In addition, *jam* refers to the legendary Persian priest-king Jamshid, who figures prominently in Ferdawsi's *The Book of Kings*, the Persians' most significant national epic, written between *c.* 977 and 1010 AD. Jamshid is in turn based on the mythological figure of the world-ruling king Yima at the time of the golden age in the *Avesta* and other sacred texts of Zoroastrianism. According to these texts, Jamshid's kingdom, lasting several hundred years, brought prosperity, peaceful felicity, and immortality to the lands under his rule.

King Jamshid is credited with many inventions, among them wine. The cup of Jamshid is the container from which Jamshid used to drink his wine and from which he was granted wisdom. It is, however, also depicted as a surreal, magical, 'all-knowing' mirror; a spatially and temporally clairvoyant reflecting surface which provided its owner with the knowledge he desired. Catoptromancy, the divination performed with the help of a magic mirror, is one of the most ancient forms of divination and is said to have originated in Persia.[16] Catoptromantic mirrors are indeed cognizant magic mirrors. As a reliable and articulate source of information, *Jām-e Jam* is even today used as the title for Iran's national TV channel.

Functioning as a microcosmic space onto which a macrocosm is reflected, this mirror granted its owner, King Jamshid, a view of the whole universe, transcending the limits of time and space. This cup of divination or scrying is said to have been formed from seven bands or rings and was filled with the elixir of immortality. It allowed its owner to perceive at once the seven heavens of

[16] De Givry cites Varro mentioning that catoptromancy originated in Persia; de Givry, *Witchcraft, Magic, and Alchemy*, p. 304.

the universe, thereby granting the owner dominion over the world. Therefore it was credited by many authors with the empowerment of the Persian Empire. In the world of Persian poetry, a world-revealing cup has been attributed to other kings too, such as Solomon, Kay Khosraw, Alexander, and others. Due to the broad similarities and overlapping of the legends present in classical sources, some hold that King Jamshid, King Solomon, and King Kay Khosraw in fact all referred to the same person.

Over time the catoptromantic cup of Jamshid acquired even richer metaphorical connotations, all revolving around the notion of reflection, whether mental or visual. Many different interpretations were in fact made of this cup, among the most popular identifications being: water, an astrolabe (also called *speculum cosmographicum* or cosmographical mirror), a mirror, a globe, a map, a compass, a magic lantern, wine, the heart, the perfect human, and the world. It has even been envisioned as a crystal ball in Persian literature.[17]

Persian Sufi poets have recurrently incorporated this mythological cup into their conception of the 'mirror of the heart', making it a rich and highly multifaceted metaphor. The Sufi, through his total self-abnegation and constant polishing of the heart to remove the dirt and stain of the self, is granted a reflective surface in which he can see his Beloved and become one with Him. In the Sufi's reinterpretation, the heart becomes the cup of Jamshid, through which he is granted a view of the whole of creation and its Creator. It becomes a space where the Sufi can traverse the limits of time and space. The mirror of the heart is a heterotopic, even a utopic, microcosmic space over which the Sufi can exercise some levels of influence, even to the point of giving him, ideally, *tasarrof* (total ascendancy) over the higher realities of the macrocosm. Ahmad Ghazāli says:

تا جام جهان نمای بر دست من است
از روی خرد چرخ برین پست من است

تا کعبهٔ نیست قبلهٔ هست من است
هشیارترین خلق جهان مست من است

(So long as the world-revealing cup is in my hand
Out of my intellect, the higher wheel [of heaven] is under my rule

[17] The other equivalent terms used in literature to refer to the same cup are: *jām-e jahān namā*, *jām-e giti namā*, *jām-e jahān bin*, *jām-e 'ālam bin*, *jām-e jahān ārā* (all meaning the world-revealing cup), *jām-e Jamshid*, *jām-e Kay Khosraw*, *āyena-ye Soleymān* (Solomon's mirror), *āyena-ye Eskandar* (*Sekandar* or *Sekandari*).

For a study of this topos in Persian literature see: Mo'in, '*Jām-e jahān namā*'; Musavi, 'Āyena dar shāhkār-hā-ye adabi', pp. 48–53; see also: Omidsalar, 'Jamšid ii. In Persian Literature'. For more about Jamshid see: Skjærvø, 'Jamšid i. Myth of Jamšid'.

> So long as the *Ka'be* of non-existence is the *qibla* of my existence
> The soberest person in the world seems drunk to me.)[18]

As long as the poet possesses Jamshid's cup, that is, the reflecting heart, he has ascendancy over the higher wheel. Furthermore, drawing on the image of the *Ka'be* in Mecca as the direction towards which Muslims pray, he emphasizes his nonexistence to an extreme degree while also stressing his conviction in pursuing it. The sentence appears paradoxical in the sense that the Sufi-poet believes he will gain *baqā* (immortality) only if he reaches *fanā* (annihilation). This reflecting surface, by means of connecting the lover to his Beloved and to the entire universe, further connects him to unlimited knowledge and becomes the source of his wisdom, power and peacefulness.

In Persian Sufism, mirror, wine, as well as poetry, have been used in close association with each other. They were often used interchangeably, because they are all supposed to induce a state of psychological exaltation, de-realization, illusion, bewilderment, and ecstasy known as *sukr*, *masti*, or *khumāri* (inebriation). Mirror, wine, and poetry provide a person with a heterotopic-utopic space for imagination and dreams, granting him intellectual freedom. Moreover, Persian Sufis referred to the seven rings of Jamshid's cup to explain the different spiritual states or stations of Sufism (*ahwāl* or *maqāmāt*) that the soul can seek to attain. It also refers to the different levels of capacity for wine drinking and intoxication, again metaphorically referring to the differences in capacity for being inebriated by divine love. 'Attār (d. *c.* 1221), the Persian poet, hagiographer, and theoretician of Sufism, elucidates the mystical meaning of Jamshid's cup in these lines:

نشسته بوذ کیخسرو چو جمشیذ
نهاذه جامِ جم در پیشِ خورشیذ

نگه می‌کرد سرّ هفت کشور
وز آنجا شذ به سَیر هفت اختر

نماند از نیک و بذ چیزی نهانش
که نه درجام جم می‌شذ عیانش

طلب بوذش که جامِ جم به بینذ
همه عالم دمی درهم به بینذ

اگرچه جملهٔ عالم همی دیذ
ولی درجام جام جم نمی دیذ

بسی زیر و زبر آمذ در آن راز
حجابی می نشذ از پیشِ او باز

[18] Ghazāli, *Savāneh*, ed. by Purjavādi, p. 19 (translation mine).

(Kay Khosraw was sitting like Jamshid
Positioning the cup of Jamshid in front of the sun

Observing the mysteries of the seven countries
From there to the mysteries of the seven stars

Nothing, vice or virtue, was hidden from him
That was not revealed to him in the cup of Jamshid

He desired to have a better view of the cup of Jamshid
To see the whole world all together in an instant

Although he was watching the whole world
He could not see the cup of Jamshid in the cup

He searched a lot up and down in that mystery
No veil was lifted from in front of him.)[19]

In this allegorical story, ʿAttār weaves together both of the popular metaphors—the divine mirror of the heart and the cup of Jamshid—in order to explicate the abstract meanings central to Sufism. In the opening lines, Kay Khosraw, the legendary Persian king of the Kayāniān dynasty and a messiah figure in Zoroastrianism, who is praised by many Sufis for his spiritual glory, is seated by the poet in the manner of King Jamshid, holding his cup to the sun. With this cup, or mirror of divination, not only could he see the mysteries of the seven lands of this earthen world, but he could also traverse the heavenly bodies, referring to the seven *maqāmāt* (spiritual stations) of Sufism. He could see everything in this omniscient mirror, except for the cup of Jamshid—representing in this story the master Sufi and the perfect human. After much spiritual exertion an image reveals itself in an epiphanic moment:

بآخر گشت نقشی آشکارا
که در ما کی توانی دیذ ما را

چو ما فانی شذیم از خویشتن پاک
که بینذ نقشِ ما در عالم خاک

چو فانی گشت از ما جسم و جان هم
ز ما نه نام ماند و نه نشان هم

تو باشی هرچه بینی ما نباشیم
که ما هرگز دگر پیذا نباشیم

[19] ʿAttār, *Elāhi-nāma*, ed. by Ritter, pp. 184–85 (translation mine).

چو نقش ما به بی نقشی بَدَل شذ
چه جوئی نقش ما چون با ازل شذ

همه چیزی بما زان می‌توان دیذ
که ممکن نیست ما را در میان دیذ

(At last an image revealed itself
[Saying] how can you see me within me?

Since I have been thoroughly annihilated from my self
Who can see my image in the earthen world?

Since my body and my spirit are both perished
No name and no trace remains of me

Whatever you see is you, not me
For never again will I be manifest

As my image is turned imageless
Why are you searching for my image, as it has united with eternity?

You can see everything in me
Only because it is impossible to see me in between.)[20]

An image starts talking to Kay Khosraw within this mirror-cup, addressing by extension the listener or reader. It tells him that Jamshid's mirroring cup cannot be seen, since it has been turned into a mirror; thoroughly annihilated and joined with eternity. The image confides in him that whatever he sees in this cup is nothing but his own image. In fact, the image reveals to Kay Khosraw, that he is seeing himself in the mirror-cup, not the cup. Ironically, one can see everything in the mirroring cup only because it is impossible to see it, the mirror itself. Here 'Attār employs many rhetorical questions to assert the certainty of the information and its acknowledgement by the reader. Through this extended metaphor, held in these early lines of the poem, 'Attār manages to convey the mystical concept of *fanā* (self-annihilation), when *sālek* (the path-seeker) gains reflectivity and is united with the macrocosmic reality. He becomes a mirror, reflecting the terrestrial and celestial worlds, thus turning into the perfected man, the ideal self and the Beloved. Hāfez employs the same metaphor of the cup of Jamshid to convey the same meaning: 'سالها دل طلب جام جم از ما می‌کرد | وآنچه خود داشت ز بیگانه تمنا می‌کرد' (For years my heart was seeking the cup of Jamshid

[20] 'Attār, *Elāhi-nāma*, ed. by Ritter, p. 185.

from me | Begging from strangers that which it itself possessed).[21] For many years, the poet yearned to have the cup of Jamshid, begging for it from others, while all the time he was carrying it within himself. He yearned for the mirroring heart in which he could catch a glimpse of the higher realities, through which he could step into eternal paradise and be united with his Beloved. The poet has been yearning to take hold of that omniscient mirror — one which could give him unrestrained terrestrial, celestial and divine knowledge ; thereby, becoming a fountainhead of (his) blissful contentment. A sip from the wine of this mirroring cup would give him immortality and love-drunkenness accompanied by the bliss of union.

Burning Mirrors

A burning mirror is actually a concave reflecting surface that concentrates the rays of the sun with enough intensity that it can set on fire anything that is placed under its focal point. In Persian literature, there are various narrations of the famous Greek legend of how a huge concave mirror was constructed and installed atop a mountain to observe the enemy from afar or to set their vessels ablaze in the sea. These stories seem to have captured the imagination of the Persians. In Persian texts, the design and construction of this burning mirror was sometimes attributed to Archimedes and even to Plato, but it is mostly called 'Alexander's mirror'. 'Alexander's mirror' always had a magical and surreal aura about it. There are some early texts in Arabic and Persian recounting the story of Alexander's mirror and how it was ruined. These works, reaching back to the tenth century, appear to be rather fictitious. Various Persian texts narrate different stories and different uses for the burning mirror, which ranged from a defensive instrument, used against enemy ships, against a sea monster with a 'lethal look' (reminiscent of the Medusa myths), serving as a lighthouse or beacon, or even in some cases functioning as an astrological instrument for observing the heavenly bodies.[22]

Persian Sufi poets and thinkers merged the popular images of the flat mirror of the heart and Jamshid's cup with the concave burning mirror, to explicate the

[21] Hāfez, *Divān*, ed. by Qazvini and Ghani, p. 109.

[22] For a study of burning mirrors in Persian literature and travelogues, see: Ma'sumi Hamadāni, 'Āyena-ye suzān-e aflātun'. Mohammad Mo'in, in his article 'Alexander's Mirror' has collected some of the old Arabic and Persian texts recounting the story of Alexander's mirror. Mo'in, 'Āyena-ye Sekandar'. See also: Musavi, 'Āyena dar shāhkār-hā-ye adabi', pp. 35–37 and 42–53, and Schimmel, *A Two-Colored Brocade*, p. 115.

Sufi's state of total self-annihilation through self-immolation. The Sufi's heart is not merely a flat surface on which the divine Beloved is reflected: it is also Jamshid's cup, giving the owner wine-drunkenness and its esoteric knowledge, often veiled from the sober. Furthermore, the Sufi's heart eventually turns into a burning mirror, consuming the self totally, so that what is left is only the Beloved.

In a letter to his Sufi friend and disciple, 'Eyn ol-Qozāt-e Hamadāni (d. 1131), Ahmad Ghazāli (d. 1123 or 1126), the Persian Sufi master and metaphysician of love, explicates the Sufi state of total self-annihilation in a highly metaphorical language by resorting to the mirror image. In this letter, Ghazāli imagines a debate between philosophers of India and *Rum* (Anatolia) and a talking mirror. These scholars assemble at the threshold of the mirror. First, they show their veneration towards the mirror by claiming loyalty to his sincerity and purity. Then they reveal their bewilderment by asking:

گفتند: 'ما در کار تو سرگردانیم'، به حق وفا بر صفای او دادند. او گفت:
'مرا زبانی هست، اما با کسی گوید که روی من بی روی خود بیند. اما مرا
ترجمانی هست. رویی چون روی بی رویی نیست و و روی من بی روی خود
نمی توانید دید. به درگاه ترجمان روید'. گفتند: 'آن کیست؟'
گفت: 'حرّاقه'. به درگاه او رفتند. هرکه در سایه دید، خود ندید؛ و هر که
در آفتاب دید، خود نماند. گفت: 'آری، طمس طمس الغیب است. تا روی را
غارت کند، بی رحمتی صفت اوست. آن که بوَد نیابد و آن که نبوَد چون یابد؟'

(The scholars said: 'We are wondering at your work.' Truly they stressed their devotion to the mirror's purity.

The mirror said: 'I have a tongue, but it would only talk to someone who can see my face without seeing his own face. However, I have an interpreter. Since there is no face like a faceless face and my face might not be seen without your own face. Go to the threshold of the interpreter.'

They inquired: 'Who is he?'

The mirror said: 'The burning mirror.'

They went to the threshold of the burning mirror. Anyone who looked at it in the shadow did not see himself; and anyone who looked at it in the sunlight, no self was left.

The mirror said: 'Yes, indeed the effacement is a divine effacement. Until the face is plundered, mercilessness is his attribute. The one who exists, would not find, while the one who does not exist — how can he find?')[23]

[23] Ghazāli, *Mokātabāt-e khʷāja Ahmad Ghazāli bā 'Eyn ol-Qozāt-e Hamadāni*, ed. by Purjavādi, pp. 27–28 (translation mine).

By anthropomorphizing the mirror, Ghazāli gives it an authoritative subjectivity. It is a knowledgeable mirror which has answers for the human soul. It also possesses the human means, the ability to speak, of conveying that knowledge. He gets involved in a scholarly discussion to convey his complex message; a message so complex that it requires an interpreter. However, the answer that the burning mirror gives is an ironic one, provided through a rhetorical question. Hāfez vows that he will attain that mirroring heart, regardless of whether it incinerates him or not: 'من آن آیینه را روزی بدست آرم سکندروار | اگر می گیرد این آتش زمانی، ور نمی گیرد' (One day, I will take hold of that mirror Alexander-like | Whether this fire catches sometime, or whether it does not).[24]

By recurrently bringing these mirror metaphors together — the mirror of the heart, Jamshid's cup, and the burning mirror — and by using them interchangeably, the Persian Sufi thinkers and poets transformed mirror imagery into a cornucopia of meanings and symbols. After constant polishing to remove this-worldly stains, man's heart turns into a utopic space, a mirror which provides him a glimpse of the divine paradise and of union with the Beloved. Likewise, Persian Sufi poetry is a mirror, a threshold to that celestial space where a soul can withdraw for peace, contentment and dialogue with the Beloved. Yet according to the Sufis, the ascension to this space is not that easy. It requires experience in love and engagement in the mental/spiritual struggle to move beyond the bars of one's *this-worldly* prison and the language of the temporal realm. The metaphor of the mirror of divinity remains a key metaphor in Persian Sufi literature, its function transcending beyond mere description and becoming constitutive. Through appropriation of this metaphor, the Sufis configure and describe their entire *Weltanschauung* and mould their cognitive ethos.

[24] Hāfez, *Divān*, ed. by Qazvini and Ghani, p. 116.

Works Cited

Primary Sources

'Attār, Farid od-Din, *Elāhi-nāma*, ed. by Hellmut Ritter (Istanbul: Matba'a-ye ma'āref, 1940)

Ghazāli, Ahmad, *Mokātabāt-e khʷāja Ahmad Ghazāli bā 'Eyn ol-Qozāt-e Hamadāni*, ed. by Nasrollāh Purjavādi (Tehran: Khānqāh-e Ne'mat ollāhi, 1356/1977–78)

——, *Savāneh*, ed. by Nasrollāh Purjavādi (Tehran: Enteshārāt-e bonyād-e farhang, 1359/1980–81)

Hāfez, *Divān-e Hāfez-e Shirāzi*, ed. by Mohammad Qazvini and Qāsem Ghani (Tehran: Anjoman-e khoshnevisān-e Irān, 1368/1989–90)

Mawlavi [=Rumi], Jalāl od-Din Mohammad ebn-e Mohammad, *Masnavi Ma'navi*, ed. by 'Abdol-Karim Sorush, 2 vols (Tehran: Sherkat-e enteshārāt-e 'elmi o farhangi, 1386/2007–08)

Maybudi, Abo'l-Fazl Rashid od-Din, *Kashf al-asrār wa-'oddat al-abrār*, ed. by 'Ali Asghar Hekmat and others, 10 vols (Tehran: Amir Kabir, 1331–39/1952–60)

Sa'di, Mosleh ebn-e Abd ollāh, *Kolliyāt-e Sa'di*, ed. by Mohammad 'Ali Foroughi (Tehran: Enteshārāt-e zavvār, 1381/2002–03)

Secondary Studies

Burckhardt, Titus, 'Die Symbolik des Spiegels in der islamischen Mystik', *Symbolon*, 1 (1960), 12–16

Foucault, Michel, 'Of Other Spaces', trans. by Jay Miskowiec, *Diacritics*, 16 (1986), 22–27

Foruzānfar, Badi'oz-Zamān, *Ahādis o qesas-e Masnavi: Talfiqi az do ketāb-e Ahādis-e Masnavi o Ma' ākhez-e qesas o tamsilāt-e Masnavi* (Tehran: Enteshārāt-e Amir Kabir, 1381/2002–03)

Givry, Grillot de, *Witchcraft, Magic, and Alchemy*, trans. by J. Courtenay Locke (New York: Dover, 1971)

Keeler, Annabel, *Sufi Hermeneutics: The Qur'ān Commentary of Rashīd al-Dīn Maybudī* (New York: Oxford University Press, 2006)

La Belle, Jenijoy, *Herself Beheld: The Literature of the Looking Glass* (Ithaca: Cornell University Press, 1988)

Lazarus-Yafeh, Hava, *Studies in Al-Ghazzali* (Jerusalem: The Magnes Press, 1975)

Leisegang, Hans, 'Die Erkenntnis Gottes im Spiegel der Seele und der Natur', *Zeitschrift für philosophische Forschung*, 4 (1950), 163–83

Ma'sumi Hamadāni, Hoseyn, 'Āyena-ye suzān-e Aflātun', *Nashr-e dānesh*, 7.1 (1379/2000), 3–15

Melchior-Bonnet, Sabine, *The Mirror: A History*, trans. by Katharine H. Jewett (New York: Routledge, 2001)

Meneghini, Daniela, 'A New Approach to Analyzing the Use of the Word *Âyine* (Mirror) in Bidel's Ghazals', *Annali di Ca' Foscari*, 43, serie orientale, 35 (2004), 157–71

Mo'in, Mohammad, 'Āyena-ye Sekandar', in *Majmu'eh maqālāt-e doktor Mohammad Mo'in*, ed. by Māhdokht Mo'in, 2 vols (Tehran: Enteshārāt-e Mo'in, 1367/1988–89), II, 465–94

——, '*Jām-e jahān namā*', in *Majmu'eh maqālāt-e doktor Mohammad Mo'in*, ed. by Māhdokht Mo'in, 2 vols (Tehran: Enteshārāt-e Mo 'in, 1368/1989–90), I, 345–66

Musavi, Seyede Faribā, 'Āyena dar shāhkār-hā-ye adabi tā qarn-e hashtom' (unpublished master's thesis, Tarbiat Moallem University, Tehran, 1374/1995–96)

Omidsalar, Mahmoud, 'Jamšid ii. In Persian Literature', in *Encyclopædia Iranica*, ed. by Ehsan Yarshater, XIV (New York: Encyclopædia Iranica Foundation, 2008), pp. 522–28 <http://www.iranicaonline.org/articles/jamsid-ii> [accessed 16 December 2012]

Purjavādi, Nasrollāh, *Eshrāq o 'erfān: maqāleh-hā o naqd-ha* (Tehran: Markaz-e nashr-e dāneshgāhi, 1380/2001–02)

Shafi'i Kadkani, Mohammad Rezā, *Shā'er-e ā'ine-hā: Barresi-ye sabk-e Hendi o she'r-e Bidel*, (Tehran: Mo'asese-ye enteshārāt-e āgāh, 1371/1992–93)

Schimmel, Annemarie, *A Two-Colored Brocade: The Imagery of Persian Poetry* (Chapel Hill: University of North Carolina Press, 1992)

Skjærvø, Prods Oktor, 'Jamšid i. Myth of Jamšid', in *Encyclopædia Iranica*, ed. by Ehsan Yarshater, XIV (New York: Encyclopædia Iranica Foundation, 2008), pp. 501–22 <http://www.iranicaonline.org/articles/jamsid-i> [accessed 16 December 2012]

Ziolkowski, Theodore, *Disenchanted Images: A Literary Iconology* (Princeton, New Jersey: Princeton University Press, 1977)

Zipoli, Riccardo, 'A Computer-Assisted Analysis of Bidel's *Tur-e Ma'refat*', *Annali di Ca' Foscari*, 44.3 (2005), 123–38

——, 'Āyine (Mirror) in Bidel's Ghazals. Lexical Solidarities: Āyine (Mirror) and Dāġ (Brand)', *Annali di Ca' Foscari*, 47.3 (2008), 115–29

——,'Semiotics and the Tradition of the Image', *Persica*, 20 (2005), 155–72

Language as Mirror: Semiramis and Alexander in a Late Byzantine Romance

Alison Beringer

Looks can kill. The classic literary proof of this frightening axiom is provided by the story of the Gorgon Medusa, whose very glance is said to have turned men and beasts to stone.[1] It took the bravery of a Perseus and the battlefield wisdom of an Athena to defeat the monster's gaze: the hero, as Apollodorus tells us in his second-century BCE *Library*, held a glistening shield in his hand, and used the reflection from that impromptu mirror to decapitate his snake-tressed foe.[2]

A millennium and a half after Apollodorus, an anonymous Byzantine poet produced another story about a deadly female overcome by a clever hero. The full title of this fourteenth- or early fifteenth-century Greek romance is *The Narrative of Alexander and Semiramis, the Queen of Syria and Concerning*

[1] An easily accessible overview of the Medusa legend is provided in *The Medusa Reader*, ed. by Garber and Vickers. This collection includes English translations of primary sources on Medusa as well as seminal secondary works that draw on the figure. Creative engagement with the legend is not limited to literary production, but rather includes a variety of aesthetic responses.

[2] Apollodorus, *The Library*, in *The Medusa Reader*, ed. by Garber and Vickers, pp. 23–25.

Alison Beringer (beringera@montclair.edu) is Associate Professor of Classics and General Humanities at Montclair State University. She works on medieval German literature, translation studies and reception, as well as women and gender. She is the author of *The Sight of Semiramis: Medieval and Early Modern Narratives of the Queen of Babylon*.

The Mirror in Medieval and Early Modern Culture: Specular Reflections, ed. by Nancy M. Frelick, CURSOR 25 (Turnhout: Brepols, 2016), pp. 131–147 10.1484/M.CURSOR-EB.5.111045
BREPOLS PUBLISHERS

the Eleven Riddles (I adopt the abbreviated title current in scholarship: *The Narrative of Alexander and Semiramis*).[3] The narrative of the encounter between the two titular antagonists recalls — reflects — the structure of the tale of the Medusa, but transforms it — refracts it — with great and revealing subtlety. Here, too, looks can kill, yet it is not the terrible gaze of a monster but the overwhelming sight of physical beauty that overcomes the masculine viewer; and this story ends not with the death of the female but with a different conquest; namely, her subjection in marriage. And the mirror, a physical artifact of bright shining steel in the story of Perseus and the Gorgon sisters, is transmuted here into a verbal phenomenon. Words, like mirrors, offer a reflection and only a reflection of the object to which they refer; they are remote from what they reflect. Alexander's eventual victory over Semiramis in the Byzantine romance is assured by his more skillful use of the mirror of language, his superior ability to manipulate the verbal reflections of objects while averting his gaze from the dreadful objects themselves.

The text at the core of this essay is extant in two seventeenth-century manuscripts, Meteora, Monastery of St. Barlaam, Codex 197 ('B') and Mount Sinai, Monastery of St. Catherine, Cod. gr. 2122 ('S').[4] In 2004 Ulrich Moennig published a critical edition with a German translation based on manuscript B.[5] Moennig's work, which includes an extensive discussion of the text's literary-historical context as well as careful analysis of narrative strategies and linguistic elements, considers the relationship between these two textual versions.[6] Prior to Moennig's groundbreaking study, this Greek verse narrative of about 1400 lines had received limited scholarly attention.[7]

[3] The Greek title is Διήγησις Ἀλλεξάνδρου μετὰ Σεμίραμης βασίλισσας Συρίας περὶ τῶν ἕνδεκα ἐρωτημάτων. Ulrich Moennig used the abbreviated title referred to above in his critical edition and extensive study of the text: *Die Erzählung von Alexander und Semiramis*, ed. by Moennig.

[4] Manuscript descriptions are found in *Die Erzählung von Alexander und Semiramis*, ed. by Moennig, pp. 7–14.

[5] See n. 3 above. For the most part, I have based my reading on Moennig's German translation (pp. 35–69), which is based on his edition of manuscript B. In the interests of space in my paper, therefore, I quote from the German text, which I translate into English.

[6] Moennig postulates one of the centres of Byzantine culture already under Ottoman influence before the fall of Constantinople as a likely place of origin for the text; *Die Erzählung von Alexander und Semiramis*, ed. by Moennig, pp. 35–36. On dating the text, see pp. 35–69.

[7] Manuscript B contains 1411 lines, and S contains 1329, plus an additional two lines stating that this is the end of the text. Earlier plans for editions were not realized, which no

The romantic — and historically impossible — pairing of Alexander and Semiramis is to my knowledge not found in narratives of Semiramis produced in the Latin West;[8] when the two rulers are mentioned in the same context, it is a military one, in which Semiramis figures as the only person before Alexander to have conquered India.[9] While a study of the narratives of Alexander goes beyond the bounds of this paper, it should be noted that the sixth-century Byzantine chronicler John Malalas records that Alexander met Candace of Meroe, a later ruler of Semiramis's empire, and a woman whom he would marry, because he wanted to see 'the world-renowned palace of Semiramis', which Candace now inhabited.[10] Thus, though Semiramis and Alexander did not themselves become a pair, her fame led Alexander to one of his wives.[11]

doubt contributed to the general absence of scholarship on this text: 'Unterdessen fehlt in den aktuellen Geschichten der byzantinischen und neugriechischen Literatur sowie in der Bibliographie zum spätbyzantinischen (Liebes-) Roman — die *Erzählung von Alexander und Semiramis* gehört in diese Gattung — so gut wie jeder Hinweis auf diesen Text'; *Die Erzählung von Alexander und Semiramis*, ed. by Moennig, p. 1. On page 2, Moennig discusses the work of Christos Dimitrulopulos; Dimitrulopulos included a synoptic edition of the two manuscripts in his 1999 publication Διήγησις 'Αλεξάνδρου μετά Σεμίραμης βασίλισσας Συρίας. An English translation of the romance remains a *desideratum*.

8 In the Western European literary tradition, Semiramis's first husband is Onnes (=Menon), and her second is Ninus. These husbands are recorded by Diodorus Siculus, *Library of History*, trans. by Oldfather, pp. 357–61. In the anonymous *Annolied*, an Early Middle High German text assumed to be written between 1077 and 1101, Semiramis is Ninus's wife and is depicted in her common role as builder of Babylon. The passage is included in *Sovereignty and Salvation in the Vernacular*, trans. by Schultz, pp. 64–65.

9 For example, in Rudolf von Ems's thirteenth-century world chronicle: Rudolf von Ems, *Weltchronik aus der Wernigeroder Handschrift*, ed. by Ehrismann, ll. 3584–91. Pfaffe Lamprecht's *Alexanderlied*, the earliest German Alexander poem, based on the French *Roman d'Alexandre* and known in three versions, does not mention Semiramis. The Vorauer version (*c.* 1150) does not record Alexander's marriage; the Strassburger one (*c.* 1170) recounts his marriage to Darius's daughter, Roxanne (ll. 3863–64 and 3982–84) as does the *c.* 1270s Basler one (Roxanne is called 'Reksaman', l. 2807 and ll. 2869–87).

10 See Richard Stoneman, who cites Malalas: Stoneman, *Alexander the Great*, pp. 134–35. Stoneman's book is the best place to start for studying the Alexander narrative material.

11 In stories of the Latin West about Semiramis, she is rarely the object of romantic pursuit — the exception is a now fragmentary anonymous Greek novel written in the first century CE which records a romance between Ninos [Ninus] and his unnamed female cousin, believed by scholars to be Semiramis, as her mother's name, Derkeia, appears in the fragments. The fragments are published and translated, with an introduction, in *Ancient Greek Novels*, ed. by Stephens and Winkler, pp. 23–71.

Semiramis, as the anonymous Byzantine romance begins, is the mighty queen of Syria, renowned for her beauty and her large armies. The events of the romance are set in motion by a familiar trope: the kingdom requires a consort and an heir, but Semiramis is unwilling to submit to any man in marriage. And so she issues the proclamation that any young man who comes, firmly withstands her gaze, and answers a riddle of her posing will become ruler of Syria and of her person. Those who fail in their quest will be put to death in the presence of her courtiers and scholars (ll. 19–24). Predictably, many princes try, and many princes fail.[12]

There are several notable aspects to this beginning but the most interesting for our purposes is the odd complexity of the challenge issued by the Syrian queen. The trial Semiramis sets for her suitors combines the visual dangers posed by Medusa — they must behold her face — with the verbal test demanded by the sphinx — they must solve a riddle. These twin challenges are redundant, and the conditions that Semiramis constructs for possession of her kingdom and her body are strangely overdetermined, as if she were unwilling to risk defeat in only one contest.

Meanwhile, we learn that Alexander, the Macedonian king, has suffered defeat at the hands of the Persian Darius and has fled his kingdom, naked and alone. After fifty days on the road, Alexander arrives in Cairo, where by chance he finds the escaped falcon of the king. Upon returning the falcon to its owner, Alexander is rewarded with clothing, food, drink, a horse, shoes, and money; in other words, he has the outer trappings of a would-be suitor. After another thirty days, Alexander arrives in Syria, where he takes up lodging in the house of an old woman without, however, revealing his true identity. On his first day, he happens to witness the execution of one of Semiramis's unhappy suitors and in the evening questions his landlady about this event. She informs him of the challenge Semiramis has set and of the many princes whose heads have been displayed on the gates. The landlady emphasizes Semiramis's beauty, a beauty which causes men to lose their mind, and thus their head. In repeating the challenge that Semiramis issued, the landlady initially omits the stipulation that the suitor must look at Semiramis and relates instead only that the suitor

[12] The core narrative material is probably most familiar to modern readers from Giacomo Puccini's 1926 opera, *Turandot*; the story is also found in Aarne and Thompson, *The Types of Folktale*, where it is entitled both 'The Princess's Hand Is Won' and 'The Princess Who Cannot Solve the Riddle'. I have taken this information from Moennig, who explores the source of the Byzantine narrative extensively; *Die Erzählung von Alexander und Semiramis*, ed. by Moennig, pp. 19–34; here p. 19.

who speaks to Semiramis and answers the riddle she poses will become ruler of Syria (ll. 80–86). Alexander reflects on the landlady's words and the next morning determines to try his luck; notably, he has implicitly understood the requirement to look at the Queen and hear her riddles: 'Mother dear, I, too, will go and see what can be done, to look upon the Queen and have her ask me a riddle' (ll. 91–92).[13] Not surprisingly, the old woman tries to dissuade her young lodger, emphasizing that whoever looks at the Queen loses his mind. She concludes with an uncomplimentary character description of Semiramis: the Queen is bloodthirsty, lacks sympathy, decapitates young men, and has no pity for anyone (ll. 97–98).

There would be no romance if Alexander heeded this warning; for his part, he tells his landlady that he has lost all his power and all that he had, including armies and cities (ll. 102–03). As a result, he has lost his purpose and so he determines to risk his head to gain a kingdom (ll. 104–05). Despite this revealing information about his past, Alexander does not name himself, nor does the landlady ask any questions; thus his identity remains unknown in Syria. Alexander heads to the palace, where first the guards and later Semiramis's courtiers urge him not to expose himself to danger, but he declares himself ready for the challenge (ll. 122–243). The Queen, informed of this new suitor, determines to set him three riddles, though first she attempts to dissuade him from his plan.

This initial attempt to dissuade Alexander takes the form of a twenty-eight-line speech. In the first fifteen lines the Queen concentrates on the suitor and appeals to his intelligence — he should not trust to his luck, but should withdraw and save his life (ll. 268–82). The second part of the speech reads as an attempt to define the Queen and begins with the words, 'For I am the angel who leads away the souls of young men' (l. 283).[14] The next ten lines (ll. 285–94) all begin 'I am'.[15] This anaphoric structure produces a type of aural reflection for the listener: each verse appears to mirror the one before it, only to end differently and so to disrupt the linguistic reflection. As Moennig has observed, the unusual content of this description — Semiramis casts herself as harbinger of death, at one point even identifying herself *as* death (l. 286), and draws on more expected comparisons of herself as the sun (l. 285) — reflects

[13] '"Mütterchen, auch ich werde gehen und schauen, was sich machen läßt, | die Königin anzuschauen und mich ihren Rätselfragen zu stellen."'

[14] '"Denn ich bin der Engel, der hinwegführt die Seelen der jungen Männer"'.

[15] '"Ich bin."'

the conflation of two roles found in the genre of the late Byzantine romance. In this romance, Semiramis is simultaneously the beloved, the object of the protagonist's pursuit, and the adversary of that protagonist, the one who seeks to prevent the union of lover and beloved.[16]

Regardless of content and of rhetorical structure, the speech fails in its intention: Alexander remains. The audience now receives the first sign that perhaps Alexander will succeed in his self-imposed mission. Alexander responds to Semiramis with a flowery speech cleanly divided in subject matter between the Queen and himself. Of twenty-four lines (ll. 299–322), he devotes the first twelve to Semiramis, then in the thirteenth makes a clear shift to himself, 'I am not [...]' (l. 311).[17] This careful attention to symmetry offers another example of reflection, this time on a structural level. Using language intended to flatter his royal addressee, Alexander explains to her and her gathered courtiers that he knows how she has vanquished all those princes who have made the attempt before him: it is not the riddle that proved their undoing, but the other part of her challenge, the requirement that they look her full in the face. Because Semiramis is beautiful, clever, and powerful, the men who have gazed on her lost their minds and their capacity to reason (ll. 307–09). As Moennig shows, losing one's reason at the sight of the beloved is not uncommon behaviour for protagonists of romances; however, the treatment of the motif here is exceptionally drastic, given that the nature of the suitors' test rests on mental acuity. Loss of reason at the sight of the beloved leads *immediately* and *directly* to death.[18] Alexander has understood the danger of the power of seeing Semiramis, and in his speech to her, he manages to de-emphasize the requirement that he look at her. He compliments the Queen on her beauty, her fame, and the identity of her suitors — sons of rulers — and contrasts his own poor self with these suitors: 'I am not son of a ruler, nor scion of a king, | I am a lonely foreigner, and have come from far away' (ll. 311–12).[19]

Having focused the attention on his identity — or rather lack thereof — Alexander then asks that he be allowed to take up the challenge, to hear —

[16] *Die Erzählung von Alexander und Semiramis*, ed. by Moennig, pp. 97–98.

[17] '"Ich bin nicht [...]"'.

[18] *Die Erzählung von Alexander und Semiramis*, ed. by Moennig, pp. 90–92.

[19] '"Ich bin nicht Sohn eines Herrschers, kein Sproß eines Königs, | ich bin ein einsamer Fremder, von weit her bin ich gekommen."' Alexander's notion of the truth is rather malleable: though it is true that he had been *king*, rather than *son of the king* before Darius defeated him, given that his father, Philip II, had also been king of Macedonia, technically he was a king's son. Alexander's ambiguous use of language is a topic I explore in *The Sight of Semiramis*, chap. 4.

and to answer if God and his fate so desire — the royal riddles (ll. 313–15). Through linguistic skill, Alexander has shifted the focus of the challenge to the riddle, a shift that does not go unnoticed and causes the Queen later to reiterate the visual demand. For the moment, however, the Queen proceeds and Alexander refuses to look at her, instead focusing his eyes on the floor in a show of humility and an act of self-preservation.

By untangling the two components of Semiramis's deadly challenge, Alexander reveals her as not just the sphinx-like poser of riddles but also and simultaneously as a new Gorgon, able to kill at a glance. If this were Perseus, a mirrored shield would be at hand; but Alexander has already introduced himself to the Queen and her court as a foreigner without contacts ('lonely'). Alexander resorts instead to a different kind of mirror, a verbal expertise that exploits the ontological gap between words and things to solve the Queen's riddles while keeping the deadly danger she poses at a distance.

Semiramis now states her three riddles, perhaps better described as learned questions. Even Alexander does not respond immediately, as he must gather his thoughts, but, having done so, he correctly answers all three.[20] Angered at her suitor's success, the Queen responds with more riddles — six in total — all of which Alexander successfully solves (ll. 407–515). After nine riddles — eight more than the Queen's original proclamation — the interview is ended: Semiramis withdraws to confer with her handmaids and eunuchs while Alexander returns to his lodging.[21]

[20] I include here a succinct version of the first three riddles to provide a taste of the type of question; for the fuller version, as well as the subsequent ones, the reader is referred to Moennig's text. The first three riddles and their answers are found in lines 409–33. In answer to the first riddle, 'Where do you stand?' Alexander replies 'between life and death', thereby showing that he can think beyond the surface meaning of words. The second, 'What is the land that saw the sun one single time and never again and will not see it again until the Last Trump?', allows Alexander to reveal his scriptural knowledge: the answer is the Red Sea, exposed to the sun when Moses led the Israelites out of Egypt. (Fritz Loewenthal's early twentieth-century study of the Germanic riddle includes a similar riddle with the answer of the crossing of the Red Sea. Lowenthal traces this answer back to the eleventh-century Byzantine poet, Psellos. Loewenthal, *Studien zum germanischen Rätsel*, p. 63.) Semiramis's third riddle asks what wonderful tree has twelve branches, each with thirty black and thirty white leaves, and whose shadow reaches throughout the world. The answer is time: the branches are months, the leaves are nights and days. Here Alexander displays his ability to think allegorically.

[21] For 'handmaids', I follow Moennig's translation, 'Zofen' not the manuscript's 'Hofräte'; see *Die Erzählung von Alexander und Semiramis*, ed. by Moennig, p. 334, l. 536.

The riddle is the perfect rhetorical form for the exchange between Alexander and the woman he dare not behold, and he uses them just as Perseus used his mirrored shield. Like a mirror, a riddle offers the reflection of an object — its answer — without explicitly containing that object; just as what is lacking from a reflection is the thing reflected, so too what is lacking from a riddle is the thing riddled about. The answer is there, but it is not *here* in the riddle; the reflected object is there, but it is not *here* in the mirror. The riddle, like the mirror, evokes what it reflects without attempting to contain it physically or linguistically. Perseus is safe as long as he regards only the reflection of the Gorgon, and Alexander is safe as long as he restricts his interaction with Semiramis to language in the form of the riddle.

When Semiramis leaves the site of their first verbal sparring, she admits to her entourage that she has lost her heart and her mind to this new suitor, but she continues to resist him. She views her state radically: 'My heart is enflamed, I am lost to this world' (l. 548).[22] She demands that her entourage advise her on how she can still execute Alexander and create a sensation (ll. 550–51); the importance of spectacle resonates with the public nature of the execution of unsuccessful suitors formulated in the initial challenge. Despite the Queen's request for help in defeating this suitor, her retinue extols Alexander's success, warns their Queen that he has the support of the Syrian people, and asserts that it is the divine order for a woman to bow to a man (ll. 574–76).[23] What really seems to bother Semiramis here is not the expected subjugation to a male — after all, she herself had introduced that idea in the initial formulation of the wager — nor even that she feels love for the as yet unnamed young suitor, but rather that this young man, by his own admission, is a foreigner and an orphan,[24] a nobody in other words, who has succeeded where the highest princes had failed (ll. 588–96).[25] Continuing to resist the prospect of subjuga-

[22] '"Mein Herz ist entflammt, ich gehe dieser Welt verloren."' Such a metaphorical death is a motif not generally associated with Semiramis in Western narratives, but the idea that (sexual) love renders the female powerless and thus 'dead' in a sense resonates with other female figures, for example Brunhilde in the *Nibelungendlied* who loses her superior strength along with her virginity. See *Das Nibelungenlied*, ed. by de Boor, stanzas 681–82, p. 117.

[23] This misogynistic assertion is part of a discourse on the relationship between male and female which Moennig explores in greater depth: *Die Erzählung von Alexander und Semiramis*, ed. by Moennig, pp. 72–76.

[24] Alexander calls himself an orphan a little later, in line 708: '"and I am a foreigner and an orphan among orphans"' (und ich bin ein Fremder und der Waise unter den Waisen).

[25] See *Die Erzählung von Alexander und Semiramis*, ed. by Moennig, pp. 75–76.

tion to any man, especially to this nobody, Semiramis resolves to pose two more riddles. If Alexander can solve these, then she will place herself under his sway.

Naturally, Alexander solves these riddles without difficulty, and Semiramis finally yields (ll. 676–87). In his happiness, Alexander pronounces an involved encomium on the Queen (ll. 690–708), and then, almost without motivation, offers her a chance to escape the bondage to which she has pledged herself: now he will pose a riddle, and if Semiramis can solve it, he will submit to execution. If not, he will claim his just reward from her. That Alexander should give the Queen a second deadly chance initially strains credibility, but his offer is not entirely unmotivated at the level of the romance's narrative. By turning the tables on the original poser of riddles, the text creates a new narrative impulse and introduces another opportunity for Semiramis to display her own linguistic abilities — or rather, significantly, her relative lack of them.

For the reader, the riddle posed by Alexander is transparently autobiographical:

> Who is, noble lady, that king,
> who possessed innumerable armies, who feared no one?
> Fortune drove him away and became weary of him:
> he waged war against another king,
> that one conquered him, and he fled, so as not to be caught.
> Fifty days he fled, naked and with disheveled hair
> he found nothing to eat during these days,
> only wild herbs and the tips of trees.
> The first death found him, he fought against it,
> and God helped him, he escaped and fled.
> The bird of fortune came and he plays with him;
> he is skillful, he catches him, holds him in his hand;
> and so he ate and drank and continued alone.
> And the second death found him.
> And again God helped him, a second time;
> he conquered death, in addition he won the crown,
> he became powerful once more, more powerful than before.
> (ll. 726–42)[26]

[26] '"Wer ist, edle Dame, jener König | der unzählige Heere besaß, niemanden fürchtete? | Das Glück vertrieb ihn und wurde seiner überdrüssig: | Er führte einen Krieg mit einem andern König, | jener besiegte ihn, und er floh, damit man ihn nicht fange. | Fünfzig Tage floh er, nackt und mit zerzaustem Haar, | zu essen fand er nichts in diesen Tagen, | nur wilde Kräuter und Baumspitzen. | Es fand ihn der erste Tod, er kämpft gegen ihn, | und Gott hilft ihm, er entkommt und flieht. | Es kam der Vogel des Fatums und er spielt mit ihm; | er ist geschickt, fängt ihn, hält ihn in seiner Hand; | er aß also und trank und allein zieht er dahin. | Und es fand

Unable to answer the riddle immediately, Semiramis begs Alexander for time to reflect. She withdraws, and the courtiers, sure of Alexander's triumph, greet him as ruler and king of both Semiramis and the Syrians; again, this welcome speech is heavily anaphoric with the initial eight lines (ll. 758–65) all beginning with 'Be welcomed'.[27] Upon returning to his lodging, Alexander finds that his rooms are filled with gifts from the courtiers, even from the Queen. Given Semiramis's subsequent behaviour, it is a little odd that she sends gifts — behaviour that possibly disturbed the scribe of manuscript S, too, as there the gifts all come from the courtiers.[28]

Regardless of the gifts, Semiramis continues to resist, a resistance motivated less by an explicit dislike of Alexander than by wounded pride (ll. 830–37): Alexander has answered eleven of her riddles and she cannot answer one of his. As she cannot solve the riddle honestly, the Queen turns to deceit. Taking her handmaid, food, and wine, she seeks Alexander out in his lodgings and introduces herself as a wealthy relative of the Queen's. She and the Queen have quarrelled and she has herself fallen in love with the stranger, inspired by what she has heard about his great virtue and other advantages. She claims she herself is a wealthy orphan, wealthier in fact than Semiramis. The anonymous king and the incognito queen share two meals: first, one offered by the landlady, and second, one that the visitor has brought. With the second meal, the visitor urges Alexander to drink the wine she has brought,[29] but he refuses, telling her that he will not drink wine again until he has that which his heart desires — namely, Semiramis. Disappointed and angry, the young woman prepares to leave, but the owner of Alexander's lodging intervenes, telling him that his behaviour is improper and that it is right for him to entertain himself with the guest, 'tonight it is proper that you entertain yourself with her' (l. 948)[30] and

jenen der zweite Tod. | Und wiederum hilft Gott ihm, zum zweiten Mal; | er besiegte den Tod, erlangte zudem die Krone, | er ward wieder mächtig, mächtiger als zuvor."' The ambiguity of line 736 — does the bird play with Alexander or Alexander with the bird? — is present in the Greek, too.

[27] '"Sei gegrüßt."' Moennig discusses this speech in *Die Erzählung von Alexander und Semiramis*, ed. by Moennig, p. 103.

[28] See *Die Erzählung von Alexander und Semiramis*, ed. by Moennig, p. 341 n. 5.

[29] Manuscript B has a lacuna of one line at this crucial point, but Moennig has supplied a replacement based on manuscript S. See *Die Erzählung von Alexander und Semiramis*, ed. by Moennig, p. 257, l. 920a as well as the German translation on p. 344, l. 920a.

[30] '[H]eute abend ist es gerecht, dich mit ihr zu unterhalten.' The Greek phrase for 'zu unterhalten' is *νὰ ποίσῃς ὁμιλίαν*; see *Die Erzählung von Alexander und Semiramis*, ed. by Moennig, p. 481.

that tomorrow morning he can be sensible again (ll. 945–49). Important here is the sexual connotation present in the Greek phrasing of this idea: the noun *ὁμιλία* can mean both company and sexual intercourse.[31]

Alexander remembers his manners (ll. 952–54), and the young woman returns to the table, where they continue to eat, drink, and entertain each other. As Alexander drinks more wine, he attempts to touch his visitor; she fends him off, but tells him that he may approach her once they are in bed. Confident that she has him where she wants him, the young woman asks the question that has motivated the entire charade: what is the answer to the riddle he posed Semiramis?

Alexander replies at first that they will approach the answer together, in bed, but on the lady's insistence, he reveals that the answer is himself, king of Macedonia, Alexander Paris. Hiding her great joy, the lady attempts to convince Alexander that Semiramis does not want him, to which the loyal suitor responds that he would rather die than be without the Queen (l. 1007). Then, in what seems at first to be a very strange blind motif, the woman removes her clothing and leaves, promising to return to sleep with him (l. 1010). She does not.

The contrast between the two protagonists and their treatment of reflective language is stark: Alexander has now answered a total of eleven riddles without hesitation, while Semiramis, unable to see what is alluded to but absent, has had to stoop to deception to solve the King's single riddle. What is more important is that Alexander's cunning use of language as a mirror, a device for seeing what is there but not immediately here, has also overcome the first part of the Queen's challenge. Unbeknownst to him — he thought that he was receiving Semiramis's less inhibited cousin — Alexander has spent the evening dining with the Queen herself, looking her full in the face. Despite a brief initial incapacity, where Alexander is stunned by the beauty of his guest and does, in fact, lose his sense and his ability to speak (ll. 861–63), Alexander has survived, as surely as if he had had Perseus's shield to help him.

The next day Alexander arrives at the palace, late, where Semiramis publically announces that the suitor is the answer to his own riddle, though she does not identify him by name (ll. 1036–42). Alexander, in turn, recognizes in the Queen his visitor of the evening before, though the text remains ambiguous about the source of his recognition. Does Alexander visually recognize

[31] Liddell and Scott, *A Greek-English Lexikon*, s.v. *ὁμιλ -ία*, 1 and 2. Byzantine Greek is of course much later than classical Greek, but in view of the development in the narrative, it is plausible that the sexual connotation of this word is still present in the later form of Greek.

Semiramis? In this case, the visual effect exercised by Semiramis would ultimately and ironically lead to *her* loss of the wager, rather than to the suitor's. Does he recognize her voice? Or does he deduce her identity based on her knowledge of his identity (ll. 1043–44)? If the last possibility is correct and his recognition follows from the Queen's linguistic assertions about his identity, this suggests that Alexander is a character who focuses on signifiers, rather than signifieds: on language rather than on objects. It is to his linguistic skill — riddles — that he now turns in order to save himself.

The Macedonian king narrates a story about a beautiful bird that flew to his lodging and that he caught and prepared, but just as he was about to dine upon it, that bird escaped, leaving its feathers behind. This bird has caused Alexander's tardiness. If the Queen wishes, Alexander will fetch the feathers and display them to the court at large. The identity of the bird, as the Queen well knows, is herself. Semiramis politely declines Alexander's offer, stating that, 'Humans should not reveal all that they see, but nor should they deceptively engage in intrigues' (ll. 1060–61),[32] and, demanding that he never repeat what he has said, she promises that he will, eventually, dine on the aforementioned bird.

The answers to both riddles that Alexander has posed — one directly and one indirectly through the story of the bird — mirror each other: Alexander Paris, king of Macedonia, and Semiramis, queen of Syria. Neither answer (name) is publically announced; it is enough that the names are privately shared. Alexander has won, and his superior linguistic abilities have disarmed both the Gorgon and the Sphinx in Semiramis, who becomes his wife.

This Byzantine romance is full of allusions to mirroring and reflecting. Most concretely, such allusions are found in the rhetorical strategy of anaphora, where the words literally mirror each other. I referred above to the anaphoric 'I am' (ll. 285–94) of Semiramis in her introduction of herself to Alexander and to the anaphoric 'Be welcomed' (ll. 758–65) in the court advisors' welcome and recognition of Alexander as their future king. In addition to their respective use of anaphora, these two speeches can also be seen as mirroring each other in terms of content: each is devoted to description of one of the two protagonists. Mirroring of the two protagonists and their actions is also present in the practice of posing riddles with the threat of death for the wrong — or

[32] 'Nicht darf der Mensch alles verraten, was er sieht, | aber auch darf man nicht hinterlistig intrigieren.'

for no — answer, an activity that Semiramis introduces, and Alexander later reproduces.[33]

These various uses of the mirror trope become more prominent when the narrative as a whole is read as a reflection of the Medusa legend. Like Perseus, Alexander must face a woman with a gaze that kills. Unlike the mythic hero, the historical one saves himself not with a tangible mirror, but rather with intangible language: specifically, language in the form of riddles. For Alexander in this romance, riddles mimic both the structure and the function of a mirror. Figurative language reflects its object, but expressly lacks the defining presence of that object. Just as Perseus's mirror produced a reflection that lacked the deadly essence of Medusa, so did Alexander's focus on the Syrian queen's riddles allow him to avoid the perils posed by her physical beauty. Knowledge of the riddles' answers saves Alexander's life eleven times; his ability to decode the riddles and provide the missing words suggests a superior linguistic understanding and a flexible cognitive ability.

Finally, I would like to turn to the one episode in which Alexander's dedication to the linguistic realm blinds him to reality — a blindness that, in an ironic twist, saves him. The context is the visit by Semiramis's unnamed cousin. The visitor's true identity can be seen as an (unvoiced) twelfth riddle, 'Who am I?' The answer to that riddle, the name Semiramis, has the potential to save Alexander. When the nightly visitor attempts to ply her host with wine, Alexander refuses, asserting that 'for Semiramis my soul burns, with her are my heart and my senses' (ll. 929–30).[34] But for his landlady, Alexander would not have pursued the company of this visitor any further; his devotion to his beloved would have withstood the temptations and the wiles of the beautiful woman before him. Alexander's determined faithfulness to the woman he has not seen — expressed through her name, 'Semiramis' — would have prevented him from divulging the answer to the riddle he had publicly posed the Queen. Consequently the Queen would have remained unable to answer the riddle on the following day, and Alexander would not have been in danger of his life. As we know, however, thanks to the landlady's interference the young woman didn't leave, and Alexander, indulging in wine, divulged the answer to his riddle.

[33] There are other instances of mirroring not discussed in this paper; for example, two sets of correspondence between the protagonists occur, each initiated by a different spouse (ll. 1142 and 1292).

[34] '"Für Semiramis brennt meine Seele, | bei ihr ist mein Herz, ist mein Sinnen."'

Alexander's puzzlement at his guest's subsequent stark naked departure from the house is recorded by the narrative, but he does not yet put name and body together. Alexander excels in the linguistic realm, in (arbitrary) signifiers, in linguistic terms that were created to reflect the object. As he has never knowingly seen the object (the Queen), he is unable to recognize it (her) and clings instead to its linguistic counterpart, the name 'Semiramis'.

When the young suitor does finally match signifier to signified — the name 'Semiramis' to the woman Semiramis — it is again the signifier, 'Semiramis', that saves him, though this time the salvific effect of the name lies in his *not* giving voice to it. Rather, he spins the story of the beautifully feathered bird and its discarded plumage, thereby showing complete mastery over both name and object. Alexander has managed to bear witness to his love for the Queen through his careful use of her name: first, by announcing it to his visitor in the private context of his lodgings; and then by keeping it silent in the public context of the court.

What repeatedly saves Alexander, then, is his expertise in the use of language: his knowledge of *how and when* to speak, his ability to decode words and discover the object (or concept) they reflect, and conversely to encode words with a private meaning, allowing him to convey publically what should remain private; in sum, his ability to use the riddle as a linguistic mirror.

I began this essay with the idea that the Byzantine romance reflects the Gorgon legend, and that Alexander, in facing his Gorgon, transforms the shining mirror of Perseus into a linguistic phenomenon. Given the prevailing image of Medusa as a horrifying monster with serpentine hair, the idea of aligning her with the outstandingly beautiful Semiramis might initially strain the imagination. Yet Medusa (like Semiramis) has a myriad of tales disseminated about her, one of which proves illuminating for my purposes.

In Book IV of his *Metamorphoses*, in the context of Perseus's wedding with Andromeda, Ovid provides the story behind Medusa's horrific head. In his account Medusa is the only one of the three Gorgons with snakes as hair (in Apollodorus the siblings share this coiffure), and he informs us that this was not always the case. Indeed, Medusa was originally strikingly beautiful — her hair was her best asset — and she was sought after by many suitors. Neptune violated her[35] — no explanation for his action is given — in Minerva's temple. The virginal daughter of Jupiter reacted to this event by changing Medusa's hair to snakes; the young woman's status as victim of both rape and punishment

[35] The verb used is 'vitiasse'. See Ovid, *Metamorphoses*, IV, 797.

seems to have had no effect on Minerva. Like Semiramis, Medusa was once beautiful and desired by many suitors.

The literary motif of Medusa as beautiful is present in other texts, too, though it is perhaps less familiar to the modern reader. A noteworthy example, which invites an interesting reflection by the Byzantine Romance's heroine, is found in *The Hall*, written by the second-century CE Greek writer Lucian. Here, the author refers to the beauty of all the Gorgon sisters and asserts that visual power surpasses auditory power. He describes the Sirens' charm as merely delaying voyagers (a voyager could potentially ignore the Sirens' music); in contrast, the beauty of the sisters 'being extremely powerful and affecting the very vitals of the soul, stunned its beholders and made them speechless'.[36] This aphasic reaction leads to the common assertion that people 'turned to stone in wonder'.[37] Medusa stunned people *with her beauty*, exactly the way that Semiramis's beauty robs her pre-Alexander suitors of their cognitive abilities, and this is precisely the reason that Alexander refuses to look at her before he has answered the riddles. Like her legendary precursor, Semiramis, too, is a figure who poses a visual threat through her beauty, a threat that Alexander overcomes by using the metaphorical mirror of riddles.

[36] Lucian, *The Hall*, in *The Medusa Reader*, ed. by Garber and Vickers, p. 43.

[37] Lucian, *The Hall*, in *The Medusa Reader*, ed. by Garber and Vickers, p. 43.

Works Cited

Primary Sources

Ancient Greek Novels: The Fragments. Introduction, Text, Translation, and Commentary, ed. by Susan A. Stephens and John J. Winkler (Princeton: Princeton University Press, 1995)

Diodorus Siculus, *Library of History*, trans. by C. H. Oldfather, Loeb Classical Library, 279 (Cambridge, MA: Harvard University Press, 1933)

Die Erzählung von Alexander und Semiramis: Kritische Ausgabe mit einer Einleitung, Übersetzung und einem Wörterverzeichnis, ed. by Ulrich Moennig, Supplementa Byzantina Texte und Untersuchungen, 7 (Berlin/New York: de Gruyter, 2004)

The Medusa Reader, ed. by Marjorie Garber and Nancy J. Vickers, Culture Work Series (New York: Routledge, 2003)

Das Nibelungenlied, ed. by Helmut de Boor, based on the edn by Karl Bartsch, 22nd rev. edn by Roswitha Wisniewski, Deutsche Klassiker des Mittelalters (Mannheim: Brockhaus, 1988)

Ovid, *Metamorphoses*; published as *Die Metamorphosen des P. Ovidius Naso*, ed. by Rudolf Ehwald and others, 2 vols (Berlin: Weidmann, 1903–16); available online at <http://ovid.lib.virginia.edu/metamorphoseonUVA.html> [accessed 30 July 2013]

Pfaffe Lamprecht, *Alexanderlied*, ed. by Karl Kinzel, Germanistische Handbibliothek, 6 (Halle: Buchhandlung des Waisenhauses, 1884) <http://scrineum.unipv.it/wight/alexsv.htm> [accessed 30 July 2013] [Vorauer and Strassburger texts]

——, *Die Basler Bearbeitung von Lamprechts Alexander*, ed. by Richard Maria Werner (Tübingen, 1881) <http://titus.uni-frankfurt.de/texte/etcs/germ/mhd/a_basler/a_bas.htm> [accessed 30 July 2013] [Basler text]

Rudolf von Ems, *Weltchronik aus der Wernigeroder Handschrift*, ed. by Gustav Ehrismann, Deutsche Texte des Mittelalters, 20 (Berlin: Weidmann, 1915) <http://archive.org/details/RudolfsVonEmsWeltchronikAusDerWernigeroderHandschrift> [accessed 30 July 2013]

Sovereignty and Salvation in the Vernacular, 1050–1150: Das Ezzolied, Das Annolied, Die Kaiserchronik, vv. 247–667, Das Lob Salomons, Historia Judith, trans. by James A. Schultz, Medieval German Texts in Bilingual Editions, 1 (Kalamazoo: Medieval Institute Publications, Western Michigan University, 2000)

Secondary Studies

Aarne, Antti, and Stith Thompson, *The Types of Folktale: A Classification and Bibliography*, Folklore Fellows, 184 (Helsinki: Suomalainen Tiedeakatemia, 1964)

Beringer, Alison L., *The Sight of Semiramis: Medieval and Early Modern Narratives of the Babylonian Queen*, Medieval and Renaissance Texts, 487 (Tempe: Arizona State University Press, 2016)

Dimitrulopulos, Christos, Διήγησις Αλεξάνδρου μετά Σεμίραμης βασίλισσας Συρίας. Ένα ανέκδοτο πρωτονεοελληνικό ποίημα για τον Μέγα Αλέξανδρο. Έκδοση κριτική με εισαγωγή, σχόλια και λεξιλόγιο (Athens: [n. pub.], 1999)

Liddell, Henry Georg, and Robert Scott, *A Greek-English Lexikon* 2 vols, reprint of the 9th edn (Oxford: Clarendon Press, 1925–40)

Loewenthal, Fritz, *Studien zum germanischen Rätsel*, Germanistische Arbeiten, 1 (Heidelberg: Winters, 1914)

Stoneman, Richard, *Alexander the Great: A Life in Legend*, (New Haven: Yale University Press, 2010)

Sight Lines: The Mirror of the Mind in Medieval Poetics

Suzanne Conklin Akbari

> For herself, she lingered in the soundless saloon long after the fire had gone out. There was no danger of her feeling the cold; she was in a fever. She heard the small hours strike, and then the great ones, but her vigil took no heed of time. Her mind, assailed by visions, was in a state of extraordinary activity, and her visions might as well come to her there, where she sat up to meet them, as on her pillow, to make a mockery of rest [...]. When the clock struck four she got up; she was going to bed at last, for the lamp had long since gone out and the candles burned down to their sockets. But even then she stopped again in the middle of the room and stood there gazing at a remembered vision — that of her husband and Madame Merle unconsciously and familiarly associated.
>
> Henry James, *Portrait of a Lady*[1]

It may seem odd, in an essay on medieval poetry, to begin with a passage from the 1908 New York edition of Henry James's novel. Yet the experience recounted in this prose passage highlights the intimate proximity of sight and knowledge, in which the sudden experience of recognition — of abruptly coming to understand what had previously been obscure — is expressed through

[1] James, *The Portrait of a Lady*, ed. by Bamberg, p. 364. On seeing and knowing in this scene, see Marshall, *The Turn of the Mind*, pp. 138–40; see also Cameron, *Thinking in Henry James*.

Suzanne Conklin Akbari (s.akbari@utoronto.ca) is Professor of English and Medieval Studies at the University of Toronto. She has written on optics and allegory, European views of Islam and the Orient, and travel literature. Among others, she is the author of *Seeing through the Veil: Optical Theory and Medieval Allegory*.

The Mirror in Medieval and Early Modern Culture: Specular Reflections, ed. by Nancy M. Frelick, CURSOR 25 (Turnhout: Brepols, 2016), pp. 149–169 10.1484/M.CURSOR-EB.5.111046
BREPOLS PUBLISHERS

the language of vision. To see is to know. Such alignment of sight and knowledge has a long history, some of which I have explored in an earlier book on optical theory and allegory in the Middle Ages.[2] In the medieval optical tradition, the sense of sight is thought to be a two-part relationship between seeing subject and the object seen, and is theorized in two fundamentally different ways. In extramission theories of optics, the visual ray is thought to emanate from the one who sees and then to encounter the thing that is seen, so that the power of vision can be said to be located in the seeing subject; conversely, in intromission theories of optics, the thing that is seen is thought to emit forms that can be apprehended by the one who sees, so that the power of vision — or, at least, the active component of the act — can be said to be located in the object itself.

In the present essay, I will turn to a different way in which medieval texts depict the relationship of seeing and knowing, where instead of the two-part visual exchange of seeing subject and object that is seen, we find a model of vision that articulates the nature of the subject through the very act of seeing, producing a moment in which the subject itself is constituted through the act of self-reflective vision. In some of these moments, as I will demonstrate, the seeing subject is actually defined through the ability to see, rather than through the completed act of sight, with its concomitant linkage of seeing subject and object seen, even where that object is a reflection of the self. The examples I will focus on are drawn from two prolific writers, both working around the year 1400, one in French and one in English: in Christine de Pizan's *Livre de la Mutacion de Fortune*, the seeing subject finds herself in the mirror of history; while in Chaucer's *Troilus and Criseyde*, the seeing subject finds herself in the self-reflective gaze of inner vision. For both of these late medieval writers, the subject is constructed in terms of solitude rather than union, and it comes fully into being not through erotic intimacy but through a withdrawal or an abstraction from the earlier, apparently transcendent experience of the mutual gaze. After an introductory survey of reflective vision in medieval culture, including examples from other works by Chaucer and Christine de Pizan, I'll turn to a more detailed reading of the ways in which the seeing solitary subject is constituted in Christine's *Mutacion de Fortune* and Chaucer's *Troilus and Criseyde*.

Before turning to these works, however, it will be useful to lay out in some detail the ways in which medieval writers described how individual subjectivity might be constituted through the act of vision. In devotional and theological writing, in scientific and philosophical writing, in lyric and in epic, vision appears consistently as the fundamental analogical language through which

[2] Akbari, *Seeing through the Veil*.

the ability of the subject to know the world around him is articulated: to put it another way, to 'see' is to 'know'. At times the reliability and accuracy of vision is emphasized, along with its ability to mediate between subject and object; at other times, the deceptive, distorting nature of vision is at the fore. The ability of vision to mediate between subject and object has an especially profound force in devotional literature, in which the viewer's contemplative gaze is the means of achieving nearness to God. This contemplative gaze can have many different proximate objects — a physical statue or icon; a manuscript illumination; an image conjured up in the mind by a pious text that is read (or simply remembered).[3] The non-proximate, distant object of the devotional gaze, however, is always God who is (within the terms of Christian doctrine) the only proper object of desire. The devotional gaze is often highly mediated, as when the viewer imaginatively gazes upon the suffering Christ, achieving an affective link with the divine not through direct contemplation of the Passion, but through indirect contemplation of Mary's compassion with the suffering Christ. Here, vision acts as the medium binding the seeing subject to the object of his gaze, not directly but through a convoluted web of visual referents, including not only the statue or painted image and the image held in the mind, but also the mediating vision of Mary as primary witness of the Passion.[4] Paradoxically, it is the highly mediated act of vision that most closely approximates the unmediated, perfect experience of heaven, in which the subject finally comes to 'see face to face'.

There is a tendency to artificially separate such devotional and theological modes of vision from the scientific and philosophical writings that emphasize the precision and essential knowability of vision.[5] In medieval texts on optics, as in contemporary works of literature that rely on optical theory, vision — whether direct or distorted — is always represented as being quantifiable, geometric, and precise. The extent both of verity and of falsehood can be measured. The former, devotional mode of seeing is essentially affective; the latter, quantifiable mode of seeing is essentially intellectual. Now, the frailty and instability of this binary opposition will, I am sure, be immediately evident, and nowhere is this binary opposition so fragile as in literary descriptions of vision in the context of erotic love. As seen, for example, in the breathtaking moment of recognition that Narcissus experiences as he gazes into the water

[3] On visionary experience, see Newman, 'What Did It Mean to Say "I Saw"?', esp. pp. 10–14.

[4] On Mary as visual mediator, see Snow, '*Maria Mediatrix*'.

[5] I myself have been guilty of this tendency: see Akbari, 'The Object of Devotion', esp. pp. 304–07.

and exclaims 'iste ego sum' ('It is I'; *Metamorphoses* III.463),[6] the moment of erotic desire is situated at the point of sudden convergence of two very different experiences of vision: the mutual gaze of the lover and beloved, as Narcissus gazes at the 'imaginis umbra' ('shadow', III.434; 'imago', III.463) that gazes back at him from the water, and the singular, self-reflective gaze that marks the shattering moment of anamnesis.

Much more could be said about the role of mythography in premodern constructions of the seeing subject;[7] here, however, I will merely point out the extent to which depictions of vision in love, both in late medieval and early modern examples, juxtapose (or, more precisely, place into a complex relation) the moment of the singular, self-reflective gaze and the shared mutual gaze. This juxtaposition of the dual and singular gaze lies at the heart of Sarah Kay's brilliant Lacanian reading of the *Roman de la Rose*, in which she argues that the fountain of Narcissus, as presented in Guillaume de Lorris's poem, reveals that 'love for another is the projection outwards of what starts as an investment of energy in the self [...]. [L]ove [...] invests desire in something patterned on the model of the self'.[8] In devotional literature, desire for the divine is based precisely on the vision of the self in the other, in recognizing the humanity of Christ, and, through that recognition, participating in the creation of an affective bridge that links God and man through contemplative experience. In non-devotional literature, whether the *Roman de la Rose* or the works of Chaucer, the expression of desire also begins with the vision of the self in the other; it differs, however, from devotional seeing in its embrace of idleness. Unlike devotional practice, which requires the exercise of disciplined visual contemplation, desire arises from vision in very undisciplined, lazy ways: it is no coincidence that the garden in which Guillaume de Lorris's narcissistic narrator falls in love has a gateway guarded by Oiseuse, the personification of Idleness.

A similar emphasis on lazy vision appears in Chaucer's *Merchant's Tale*. The elderly January is determined to find a wife regardless of what his advisors

[6] Quotations are from Ovid, *Metamorphoses*, trans. by Miller, and are cited in the text by book and line number.

[7] On the role of Narcissus in medieval constructions of the subject, see Goldin, *The Mirror of Narcissus in the Courtly Love Lyric*; Poirion, 'Narcisse et Pygmalion dans le *Roman de la Rose*'; Hult, *Self-Fulfilling Prophecies*; Akbari, *Seeing through the Veil*, chap. 3, pp. 45–77.

[8] Kay, 'The *Roman de la Rose* and the Inverted Bouquet', p. 308. On the relationship of vision, ethics, and desire in the *Rose*, see Rosenfeld, 'Narcissus after Aristotle', esp. pp. 10–14 and pp. 34–39.

counsel him to do; he wants a young one, not just because she will be of an age to give him heirs, but because (he thinks) a young wife is easier to control:

> Heigh fantasye and curious bisynesse
> Fro day to day gan in the soule impresse
> Of Januarie aboute his mariage.
> Many fair shap and many a fair visage
> Ther passeth thurgh his herte nyght by nyght,
> As whoso tooke a mirour, polisshed bryght,
> And sette it in a commune market-place,
> Thanne sholde he se ful many a figure pace
> By his mirour [...].
> [W]han that he hymself concluded hadde,
> Hym thoughte ech oother mannes wit so badde
> That inpossible it were to repplye
> Agayn his choys; this was his fantasye.[9]

In this passage, the imaginative faculty of the mind (the 'fantasye', or Latin 'phantasia') acts as a kind of 'brightly polished mirror' upon which the remembered images of local women can be reflected, made repeatedly available to be seen again.[10] Rather than a disciplined mode of thought, January's mode of thought here is casual, characterized as 'curious bisynesse'. The capriciousness of his imaginative mind is likened to the random quality of images that would appear in a mirror set up on the market place, catching glimpses of passersby. The same capriciousness affects his quality of judgement, so that his power of 'choys' is, precisely, 'his fantasye'.[11]

The idleness that motivates the subject — or, perhaps, the lack of motivation that itself constitutes idle thought — substantially affects what the subject sees within the mirror of his mind. Idleness colours not only the imagination but also the other mental faculties of judgement and will. But what about when the subject rises above idleness into a state of focused contemplation? The aim of raising the reader into such a state is central to Christine de Pizan's earliest allegory, the *Epistre Othea* or 'Letter of Othea to Hector', a mythographic mirror for princes. Here, the potential idleness of youth is redirected toward a

[9] Chaucer, *Merchant's Tale*, ll. 1577–85, 1607–10. Quotations from Chaucer's works are from *The Riverside Chaucer*, ed. by Benson, and are cited by line number within the text.

[10] On medieval theories of the imagination, see Karnes, *Imagination, Meditation, and Cognition in the Middle Ages*.

[11] For a more detailed analysis of this scene in the context of Chaucer's allegory, see Akbari, *Seeing through the Veil*, pp. 222–33.

more focused object: from the opening chapters, the youthful noble reader is urged to gaze upon Hercules, or gaze upon Perseus, in order to come to resemble them, just in the way that a devout reader might contemplate an image of a saint, or of the Passion, in order to reform the self. For example, Othea declares to the young Hector, 'Se tu de grant vertu fais force, | Vers Herculés te faut virer | Et ses vaillances remirer' ('If you set store by great virtue, you must turn towards Hercules and gaze upon his worthiness'; chap. 3, 204).[12] Two chapters later, Othea urges Hector to turn his gaze upon another hero, Perseus: 'Après te mire en Perseüs' (chap. 5, 209). This phrase seemingly instructs Hector to gaze upon Perseus just the way he had gazed upon Hercules; but Othea tells Hector 'te mire' — that is, gaze upon yourself in Perseus. In other words, Perseus becomes a mirror in which the young Hector is able to see himself.

The significance of this transformation lies in the story that Christine does not tell explicitly in this chapter, but which was central to mythographic accounts of Perseus, such as that found in the *Ovide moralisé*: that is, Perseus's defeat of the Gorgon. Perseus avoids Medusa's petrifying effect by looking at her not directly but only as she is reflected in his shield. In the *Epistre Othea*, immediately after telling Hector to gaze at himself in Perseus, the goddess Othea then tells him to shift that gaze to the shield of Perseus: she says, 'Si te mires en son escu | Luisant, qui plusieurs a vaincu; | De son fauchon soyes armé, | Si seras fort et affermé' ('So look at yourself in his shining shield, which has vanquished many; be armed with his sword, and so you will be strong and steadfast'; chap. 5, 165.21–24). By looking at his shield, Hector avoids the harmful petrifaction caused by Medusa; instead, he experiences a desirable petrifaction, becoming 'affirmé', powerful, and overwhelmingly potent.

While several of the chapters in the *Epistre Othea* encourage the reader to gaze upon mythical figures, others warn against looking upon them. For example, Othea cautions Hector, 'Narcisus ne vueilles sembler' ('You do not want to resemble Narcissus'; chap. 16, 226), a warning that is reiterated in Christine's gloss on the passage: 'Pour ce deffent au bon chevalier que il ne se mire point en ses bienfais par quoy il en soit oultrecuidez' ('For this reason, a good knight must beware that he never admire [*mirer*, gaze upon] himself in his good deeds, because that will make him overly proud'; chap. 16, 226). The narcissistic gaze, for Christine, is to be avoided; but gazing at the self is nonetheless repeatedly enjoined upon the reader, whether in seeing oneself in another (in Hercules, or in Perseus) or even more directly, as in Christine's return to the figure of Perseus

[12] Quotations from Christine's *Epistre Othea* are from the edition by Gabriella Parussa, and are cited in the text first by chapter and then by page number; translations are my own.

later in the *Epistre Othea*. Christine alters her source in the *Ovide moralisé* by stating that Perseus protected himself from Medusa's petrifying effect not by looking at her reflected image in his shield but by looking at himself reflected in the shield. Christine's gloss interprets this act by saying that 'Perseus se mira en son escu, c'est a dire en la force et chevalerie' ('Perseus gazed at himself in his shield, that is to say, in strength and chivalry'; chap. 55, 277). In the higher level of gloss, or, as she calls it, the 'allegorie', Christine adds that one should 'soy mirer en l'escu de l'estat de perfeccion' ('gaze at oneself in the shield of the state of perfection'; chap. 55, 278). This counsel seems to run counter to the earlier warning in the chapter on Narcissus, which states that a good knight 'ne se mire point en ses bienfais' ('does not admire [gaze upon] himself at all in his good deeds'). The chapter on Narcissus does not warn the knight to avoid contemplation of his good deeds, but to avoid seeing himself in them — that is, he must look beyond the surface of the mirror, avoiding the error of Narcissus. According to this reading, it would be possible for the second account of Perseus to be consistent in telling the noble reader to gaze at himself in the shield of Perseus, as long as what he sees in it is something more than the sterile insubstantial image. But what does he see, as he attempts to look below the surface? Or is it the act itself — the act of seeing, rather than the object seen — that Christine seeks to elevate?

This problem of how to express the coming into being of the subject through self-reflective vision, while still avoiding the narcissistic trap, persists beyond Christine's *Epistre Othea* into her hybrid combination of allegory and world history, the *Livre de la Mutacion de Fortune*. This work has a peculiar place in the canon of Christine's works, because while many readers are familiar with its opening book — in which Christine recounts her allegorical transformation from a woman into a man — very few tend to read the later books, which contain a versified universal history adapted largely from the expansive *Histoire ancienne jusqu'à César*, a universal history that grafts French national and imperial history onto the rootstock of Orosius's fifth-century *Seven Books of History against the Pagans (Historiarum Adversum Paganos Libri* VII*)*. In general terms, the *Mutacion de Fortune* confronts the intersection of poetics and history by transposing a universal history mainly in prose into verse; in addition, it deals with the intersection of poetics and history in a highly focused and specific way, as the elaborate 'sale merveilleuse' (or 'marvellous great room') housed within the Castle of Fortune serves as the point of junction between the allegorical opening books of the *Mutacion de Fortune* and the universal history that dominates the latter books.[13]

[13] On the highly visual presentation of history in the *Mutacion de Fortune*, placing it in the context of ancient and medieval scenes of ekphrasis including Virgil's *Aeneid*, the *Roman*

In structural terms, the *Mutacion de Fortune* is a fusion of Orosian historiography with a Boethian view of the role of Fortune in the life of the individual, and of Providence in the unfolding of time itself. Historiographical and philosophical models of change are integrated throughout the work, expressed through the figures of Fortune and Providence who act as guiding principles within the effort to understand the nature of the changes that take place both on the level of the individual life and on the level of kingdoms and empires.[14] Book I is, as I mentioned, the allegorical autobiography, which integrates Ovidian and Boethian models of change;[15] Books II and III are ekphrastic accounts of the Castle of Fortune, its walls, gates, and pathways, and the inhabitants located in and around the castle. Book IV is a transitional book, containing an elaborate, ekphrastic account of the 'sale merveilleuse' or marvellous great room whose walls are richly decorated with images of the entire sequence of world history. Books V, VI, and VII are a sustained exposition of universal history, elaborating in words the images depicted on the walls of the 'sale merveilleuse', moving from the empires of Assyria and Babylon to conclude with imperial Rome and rulers of European nations in Christine's own day.[16]

The autobiographical allegory of Book I enacts the process of self-examination from the perspective of the narrator, positioning her as an authoritative figure whose own 'mutacion' enables her to recount the 'grandes mutacions' (l. 1460) of history.[17] The counterpart of this authorial self-examination appears in the seventh, final book of the *Mutacion de Fortune*, which disrupts the conventional order of imperial succession by moving the story of Alexander the Great's world conquest from its usual place, in Orosian chronicles, just after the rule of Babylon, and then using the figure of Alexander as a model for rule — both the rule of others, and the proper rule of one's own self. The reader is encouraged to see the figure of Alexander — and, by extension, all history — as

de la Rose, the *Prose Lancelot*, and Dante's *Commedia*, see Brownlee, 'The Image of History in Christine de Pizan's *Livre de la Mutacion de Fortune*'.

[14] On the personification of Fortune in the *Mutacion de Fortune*, see Attwood, *Fortune la contrefaite*.

[15] On the integration of Boethian and Ovidian metamorphosis in book one of the *Mutacion de Fortune*, see Akbari, 'Metaphor and Metamorphosis in the *Ovide moralisé* and Christine de Pizan's *Mutacion de Fortune*'.

[16] For a detailed account of the historical ekphrases of the *Mutacion de Fortune*, see Akbari, 'Ekphrasis and Stasis in the Allegories of Christine de Pizan'.

[17] Quotations are from *Le Livre de la Mutacion de Fortune*, ed. by Suzanne Solente, and are cited in the text by line number; translations are my own.

a mirror reflecting the self, balancing the autobiographical account of the first book with the exemplary biography in the last book. As Christine puts it in the closing passages of her Alexander narrative, 'Mire toy, mire en ceste istoire' ('Look at yourself, look within this history', l. 23274). Here we find the same kind of reflective vision that the goddess Othea urged on Hector — and which the narrator of the *Epistre Othea* urges on her noble young reader — where the youth was urged to 'te mire[r]' (look at yourself) in the image of heroes such as Hercules and Perseus. Here, however, not just the individual exemplary figure of the hero but the entire span of history itself is the mirror into which the subject is urged to gaze, as Alexander appears as the culmination of the sequence of exemplary histories. The reader is thus urged to find himself not only in the historical mirror that reflects Alexander but in the mirror of history itself.

The *Mutacion de Fortune* has a symmetrical structure, with an opening focus (in Book I) on the narrator's self-examination counterbalanced by a closing focus (in Book VII) on the reader's self-examination, and the generally allegorical framework of the opening three books balanced by the generally historiographical framework of the closing three books. The junction or hinge that links the two halves of this symmetrical structure lies in Book IV of the *Mutacion*. While we can refer to the work as being divided into 'two halves', since this bipartite division accurately represents the number of books devoted to allegory and history-writing, this division does not accurately represent the overall balance of the work: of its approximately 24,000 lines, more than two-thirds of the work is comprised of the historical chronicle. Allegory serves, therefore, as the preliminary stage or foundation for the exposition of history, with ekphrasis acting as the mediating principle that enables the movement between these two modes. We see this process unfold in Book IV of the *Mutacion*, as the ekphrasis of the 'sale merveilleuse' located within the Castle of Fortune moves the reader from the external perspective of the architectural allegory to the intimate, internal perspective of the reader of narrative ekphrasis. This narrative ekphrasis begins with the Boethian ladder of Philosophy, ranging from theoretical to practical knowledge, through an engagement with all the various branches of knowledge including the Seven Liberal Arts, to an exposition of world history from creation through the first age of mankind.

Christine's opening description of the 'sale merveilleuse' emphasizes its enormous scale and its geometrical form: it is 'reonde' (or 'round', l. 7090), its perfect circularity emphasizing its essentially microcosmic nature. It is 'belle, clere, grande et haulte' (l. 7094), a 'strong piece of work' ('fort ouvrage', l. 7095) in spite of the fact that it, like the whole of the Castle of Fortune, is constantly in motion ('toudis tremble', l. 7095). The great chamber is 'painte moult richement | D'or et d'azur' ('painted richly with gold and azure', ll. 7104–05), and illustrated with pictorial narratives of the history of the world: 'Si sont escriptes les gestes | Des grans princes et les conquests | De tous les regnes, qu'ilz acquistrent' ('And the "gestes" are also written there, | Of the great princes and of the conquests | Of all the kingdoms that they acquired', ll. 7107–09). Here, the pictorial quality of the images that are said to be 'painted' (l. 7104; cf. 'pourtraict', l. 7117) is intertwined with the narrative quality of the 'gestes' that are said to be 'written' or 'escriptes' upon the walls: in other words, text and image are mutually constitutive, united in the ekphrastic writing.

The 'sale merveilleuse' anchors the universal history that follows, as the narration of the sequence of great empires flows out of the intense ekphrastic imagery that adorns the inner walls of Fortune's castle. The account of universal history continues through Books V, VI, and VII, almost until the end of the entire work; it is halted only in the final book, when an abrupt return to the ekphrastic framework punctuates the annals of world history, reminding us of its original visual context in the 'sale merveilleuse', and introduces a concluding section on Alexander the Great. This turn to Alexander is at once historiographical and prescriptive, a mirror for princes that seeks to provide guidance to the reader, whatever rank of society he (or she) comes from. In this story of Alexander, as in all narratives recounted in the *Mutacion de Fortune*, the reader can see himself in the mirror of history.[18] Like everyone living in the sublunary realm, Alexander lives at the whim of Fortune, who is sometimes his beloved 'amie', sometimes his hateful 'ennemye'. It is this very mutability that makes Alexander an appropriate focus for the reader, a point made explicitly in the lines that conclude the Alexander narrative of the *Mutacion de Fortune*:

> O tout homme, ou maint vaine gloire,
> Mire toy, mire en ceste istoire,
> Vois se Fortune la perverse,
> En peu d'eure, de moult hault verse!
>
> (Oh, every man, in whom there is so much vainglory,
> Look at yourself, look within this history [or story],
> See how Fortune, the perverse one,
> In short time, from high above, throws down!) (ll. 23273–76)

These lines evoke two crucial moments in Guillaume de Lorris's *Roman de la Rose*: the narrator's identification of the fountain of Narcissus, and his lament

[18] On other exemplary heroic narratives in the *Mutacion de Fortune*, see Dulac, 'Le Chevalier Hercule de l'*Ovide moralisé*'.

concerning Fortune. In the first of these two moments in Guillaume's *Rose*, the narrator recognizes the dangerous nature of the mirroring fountain in the garden: 'C'est li miroërs perilleus | ou Narcisus, li orgueilleus, | mira sa face et ses ieuz vers, | dont il jut puis morz toz envers' ('It is the perilous mirror where Narcissus, the proud one, looked at his face and his gray eyes, for which reason he then fell down dead', ll. 1569–72).[19] Don't gaze at the fountain of Narcissus, Christine warns; instead, 'Mire toy, mire en ceste istoire'. Look at this, she says, and see yourself as you might become. The Narcissus passage has its counterpart, in Guillaume's *Rose*, in the narrator's closing lament regarding Fortune:

> Ele a une roe qui torne
> et, quant ele veut, ele met
> le plus bas amont ou somet,
> et celui qui est sor la roe
> reverse a un tor en la boue.
> Et je sui cil qui est versez!
>
> (She has a wheel that turns
> and, when she wishes, she places
> the one who is the lowest high up at the top,
> and the one who is on top of the wheel
> she throws in one turn into the mud.
> And I am he who is turned!) (*Roman de la Rose*, ll. 3960–65)

In this passage, as David Hult has persuasively argued, 'versez' means both to be turned on the wheel of Fortune and to be immured within poetic verse, to become exemplary for those who will come afterward. A similar pun appears in the earlier passage from the *Rose*, where Narcissus is said to fall down dead ('envers') or, alternatively, 'in verse' ('en vers').[20] In the *Mutacion de Fortune*, Fortune similarly 'de moult hault verse', throws people down from on high. They fall; but they too become immured in verse, transformed into examples for the one who can learn from them. The historical mirror of the *Mutacion de Fortune* is, we might say, the good mirror of Narcissus: by gazing at the 'vrayes histoires' recounted in Christine's verse, it is possible for the reader to make out how he might similarly be tossed on the tides of change.

[19] Quotations from *Le Roman de la Rose* are from the edition of Félix Lecoy, and are cited in the text by line number; translations are my own.

[20] Hult, *Self-Fulfilling Prophecies*, p. 297.

The concluding turn to Alexander, as a mirror for the reader, is introduced by a return to the stunningly visual ekphrastic mode, with another description of the images depicted on the walls of the 'sale merveilleuse', in a final return to the general form of history just as it was presented in the opening historiographical passages of Book VI. In the earlier book, the ages of mankind served as a kind of epitome or temporal overview of the shape of time; in this final book of the *Mutacion de Fortune*, the concluding ekphrastic description of the parts of the world serves a similar ordering purpose, here providing a geographical overview that corresponds to the temporal overview that opened the initial move into the historiographical mode in Book IV. The moments of ekphrastic intensity, in Christine's universal history, suspend the forward movement of the exposition in temporary stasis. In Book IV, at the opening of her ekphrastic description of Philosophy and the various branches of knowledge that flow from her, Christine describes this state of being in terms of 'abstraction':

Par les escriptures, qu'y vy,
Mon esperit y fu ravy
Et astract, si que supposay
D'elle, ainsi qu'icy le posay.
Si vous en diray mon rapport,
Ainsi qu'ay de l'escript recort.

(By the engravings that I saw there
My spirit was ravished from that place
And abstracted, so that I imagined these things concerning
Her, just as I present them here.
So I will tell you my report
Just as I recorded it from the engravings.) (ll. 7203–08)

This state of being lifted outward from the present moment, in a movement of abstraction (the narrator's spirit is 'astract') that marks a temporal pause, where the narrator is lifted out of her body — the same body whose physical transformation from male to female was recounted in Book I — and then returned to continue the sequence of ekphrastic exposition. This return is marked by the direct address to the reader in which Christine renews her commitment to recount her experiences accurately, 'just as I recorded it' from the images on the wall.

In a thoughtful survey of ekphrasis from antiquity to the present, Valentine Cunningham describes ekphrases as 'pausings for thought', in which 'the linear flow of narrative slows or even stops'.[21] Such pauses appear repeatedly in the

[21] Cunningham, 'Why Ekphrasis?', p. 61.

Mutacion de Fortune, initially in the form of a sequence of repeated pauses for ekphrastic description and then movements of return to historical chronology, found throughout Books II and III and much of Book IV. But once the move to historiography takes place in the latter parts of Book IV and the ekphrastic image retreats into the background, the movement of abstraction is transposed from the state of ecstatic wonder induced by ekphrasis into synoptic moments that summarize, with brevity and intensity, the sequence of imperial succession. In these synoptic moments, great patterns of history are made visible as if they were contained within a single glimpse. These include the synoptic view of the ages of mankind, in a temporal moment of stasis, and the synoptic view of the empires of the world, in a geographical moment of stasis. These highly concentrated moments make linear time visible in a single glance, making the river of history into a gleaming mirror.

* * *

As a counterpart to the moments of visual, reflective encounter we have examined in the poetry of Christine de Pizan, I will turn to Geoffrey Chaucer's *Troilus and Criseyde*. In this long narrative poem, we find a comparably detailed engagement with how vision mediates knowledge, and how the subject is constituted not through the erotic mutual gaze but by the aftermath of that experience. The depiction of visual experience in Chaucer's *Troilus and Criseyde* has been a kind of magnet for readings based in psychoanalytic theory and film theory, as well as in medieval optics.[22] Here, however, I will focus particularly on the process through which vision mediates knowledge, and on the interrelation of intellect, affect, and desire. I use the word 'process' deliberately, for vision in the experience of love, as described in *Troilus*, has a double nature: it is sudden, happening all at once in a single moment of rupture; and it is gradual, unfolding over time. We can see this double nature in Chaucer's account of Troilus's initial look at his soon-to-be-beloved Criseyde:

> Withinne the temple he wente hym forth pleyinge,
> This Troilus, of every wight aboute, [person]
> On this lady, and now on that, lokynge,
> Wher so she were of town or of withoute; [whether]
> And upon cas bifel that thorugh a route [by chance it happened] [crowd]

[22] Stanbury, 'The Lover's Gaze in *Troilus and Criseyde*'; Stanbury, 'The Voyeur and the Private Life in *Troilus and Criseyde*'; Stanbury, 'Women's Letters and Private Space in Chaucer'. See also Holley, 'Medieval Optics and the Framed Narrative in Chaucer's *Troilus and Criseyde*'; Hill, '*She, this in blak*'.

His eye percede, and so depe it wente,
Til on Criseyde it smot, and ther it stente. [struck] [stuck]

And sodeynly he wax therwith astoned, [astonished]
And gan hir bet biholde in thrifty wise. [began] [better] [careful way]
[...]
And of hire look in him ther gan to quyken [began to stir]
So gret desir and such affeccioun,
That in his herte botme gan to stiken [bottom of his heart] [began]
Of hir his fixe and depe impressioun. (*Troilus and Criseyde*, 1.267–75, 295–98)

Here, the sudden temporal rupture is plainly visible as Troilus's gaze strikes and rests upon Criseyde, in a moment of abrupt termination of the visual beam. For the one who sees, this metamorphic moment is not only a reenactment of the experience of Narcissus, but also of the victims of Medusa: Troilus is 'sodeynly [...] astoned', both astonished and, in a Middle English pun, turned to stone, transfixed by this transformative object of the gaze.

Yet this singular moment soon proves to be more liminal than climactic: it is the entryway into a process of seeing that does not reach its final termination until the poem's closing stanzas at the end of Book v. Troilus begins to 'behold her more closely', looking at her 'in thrifty wise'. This phrase, difficult to translate, is echoed later on in Book ii when the manipulative Pandarus calculates how best to phrase his appeal to Criseyde while looking at her in a 'bysi wyse' (ii.274). It suggests an intensity of gaze, an attempt to see further, looking beyond the visible surface of things. This 'thrifty' gaze elicits a returning gaze from the eyes of Criseyde, precipitating the narcissistic gaze of love, in which desire is born. Criseyde's 'look' causes the 'quickening' — literally, the coming to life, the conception — of desire.

Yet the look of love is still not complete; when Troilus withdraws to his private room, the process of looking continues:

And whan that he in chambre was allone,
He doun upon his beddes feet hym sette,
And first he gan to sike, and eft to grone, [sigh]
And thought ay on hire so, withouten lette, [ever] [pause]
That, as he sat and wook, his spirit mette [stayed awake] [dreamed]
That he hire saugh a-temple, and al the wise
Right of hire look, and gan it newe avise. [examine]

Thus gan he make a mirour of his mynde [began]
In which he saugh al holly hire figure, [wholly]
And that he wel koude in his herte fynde. (*Troilus and Criseyde*, i.358–67)

Note how closely the wording of this passage parallels the phrasing we saw in the first passage, taken from the parodic and darkly ironic *Merchant's Tale*. In both cases, the subject makes a 'mirror of his mind', reviewing the images of the waking day in the quiet night-time solitude of his private chamber. This image of Criseyde, like the women's images reviewed by January, seems to be the product of idleness: Troilus experiences a waking dream (he 'mette') in which he 'newly regards' the 'look' of Criseyde. Earlier, we contrasted the focused vision of devotional practice, in which the subject regards the divine object of desire through mediating forms, with the idle vision of January; but can we simply identify Troilus's vision as belonging to the latter category of idle vision? Although the daydream setting suggests idleness, other aspects of this scene align it with the devotional language of visual contemplation: for example, it is surely significant that this reflective, after-the-fact vision causes Troilus to see Criseyde 'all wholly', in an entirely full image of the beloved object of desire, in contrast to the fragmented, random images of women that fill January's idle thoughts.

The extended period of seeing the object of desire continues through the subsequent books of *Troilus and Criseyde*. In this excerpt from Book III, which appears just after the lovers have parted after spending their first night together, not just the visual but all the sensory experiences of the beloved are rehearsed, repeated, and re-experienced in the solitude of the private room:

He softe into his bed gan for to slynke, [began to creep]
To slepe longe, as he was wont to doone. [in the habit of doing]
But al for nought, he may wel ligge and wynke, [lie awake]
But slep ne may ther in his herte synke,
Thynkyng how she for whom desir hym brende [burned]
A thousand fold was worth more than he wende. [believed]

And in his thought gan up and down to wynde [began]
Hire wordes alle, and every countenaunce,
And fermely impressen in his mynde
The leeste point that to him was plesaunce; [pleasing]
And verraylich of thilke remembraunce [verily, truly] [that memory]
Desir al newe hym brende, and lust to brede [burned]
Gan more than erst, and yet took he non hede. [began] [at first]

(*Troilus and Criseyde*, III.1535–47)

This passage repeats some of the visual terminology that we have already seen, in which the impact of the object of vision is described as an 'impression' upon the seeing subject. Here, the receptive soul is figured as a kind of wax that takes on the form of the impressing object. More specifically, however, this passage recalls the lines examined above, in which Troilus's first glance at Criseyde resulted in the conception of desire (I.267–75, 295–98). In that passage, Criseyde's look back at Troilus was fertile, 'quickening' his own impulse to love; in this passage, the 'remembrance' of her words and her 'countenance' causes him to 'burn' with desire, and causes 'lust to breed'. In some ways, as we have seen, the mutual gaze of Troilus and Criseyde is a narcissistic moment, in which the passionate engagement of the seeing subject is engendered by the returning look: to put it another way, the arrows of Love that wound Troilus proceed not from their apparent, proximate source — the eyes of Criseyde — but from a more distant source, that is, the dynamic eye-beams of the seeing subject himself. To put it another way, the eyes of the beloved function as mirrors that reflect the narcissistic image of the lover back to him. Yet, paradoxically, the narcissistic origin of desire is characterized not in terms of sterility, as is the norm in medieval literary tradition, but in terms of its opposite: exuberant fertility. From Alain de Lille's *De planctu naturae* to the *Roman de la Rose*, and throughout the tradition of Ovidian commentary, the sterile love of Narcissus is repeatedly opposed to models of fertile love, particularly as expressed in the counterexample of Pygmalion. In *Troilus and Criseyde*, however, narcissistic love has its own generative fecundity.

Thus far, we have focused on the visual experience of Troilus in the construction of the subject in Chaucer's poem; the visual experience of Criseyde, however, is at least as complex as Troilus's, although it proceeds along rather different lines. Like Troilus, Criseyde withdraws into the private space of her 'chamber' in order to review earlier events; unlike Troilus, however, who gazes back upon the image that has been deeply impressed upon his soul, Criseyde engages in rational thought, even logical argumentation:

> [S]he gan in hire thought argue [began]
> In this matere of which I have yow told,
> And what to doone best were, and what eschue, [eschew, avoid]
> That plited she ful ofte in many fold. [pleated]
> Now was hire herte warm, now was it cold.
> [...]
> But right as when the sonne shyneth brighte
> In March, that chaungeth ofte tyme his face,
> And that a cloude is put with wynd to flighte,
> Which oversprat the sonne as for a space, [spreads over]
> A cloudy thought gan thorugh hire soule pace, [began]
> That overspradde hire brighte thoughtes alle,
> So that for feere almost she gan to falle. [began]
> [...]

And after that, hire thought gan for to clere, [began]
And seide, 'He which that nothing undertaketh,
Nothyng n'acheveth, be hym looth or deere'. [reluctant or eager]
And with an other thought hire herte quaketh;
Than slepeth hope, and after drede awaketh; [fear]
Now hoot, now cold; but thus, bitwixen tweye, [between the two]
She rist hire up, and wente hire for to pleye. [rose]

(*Troilus and Criseyde*, bk II, 694–98, 764–70, 806–12)

We might be tempted to express the difference between the interiority of Troilus and that of Criseyde by means of the binary opposition of affective and intellectual modes of seeing, the former based in devotional literature and theology, the latter based in scientific and philosophical discourse. To do so, however, would elide the profoundly affective nature of Criseyde's experience.

While she begins by offering herself arguments and counter-arguments, she ends by 'quaking' with 'dread' or fear. Clearly, Criseyde's intellectual engagement does not preserve her from emotional lability; moreover, the very language used to describe her rational processes — itself drawn from contemporary scientific theories of motion along latitudes — is also used to describe her emotional turmoil. On the intellectual scale, Criseyde moves between degrees of doubt and certainty just as temperature moves between the extremes of hot and cold; but she also moves along the affective scale, torn between hope and dread, ultimately caught 'bitwixen tweye' (between two). Finally, we should note one striking aspect of the visual language in this passage: instead of positing any one single specific object of vision, Criseyde's very process of thinking is itself obscured as a 'cloudy thought' enters her mind. This cloudy thought might be understood in terms of the passions of the body, in which the affective states of fear and hope rise up to impair the processes of the mind. Unlike Troilus's contemplative vision of the form of Criseyde impressed upon his soul, Criseyde's inner vision is self-reflective and intellectual, yet remains vulnerable to the affective states of the body. We might call this a solipsistic form of interiority, entirely self-contained yet never completely satisfied in itself.

The private space of Criseyde's mind, on one hand, is filled with possible alternatives, potential movements in one direction or another; the private space of Troilus's mind, on the other hand, which is experienced most fully in the private space of Troilus's room, is filled with the image of Criseyde, repeatedly reviewed over the time following his initial sight of her. Interestingly, although we are told that Troilus sees her 'figure' in his mind, that figure is not literally described in the text until near the end of the fifth, final book of the poem, in a sequence of three portraits describing the three members of the love triangle poised about Criseyde: herself, Troilus, and his Greek successor, Diomede. The portraits are sometimes described as ekphrastic, not because they are actual descriptions of works of art, but because the elaborate visuality of the language aligns them with the ekphrastic tradition. We can see the portrait of Criseyde, which appears between the portraits of the other two:

Criseyde mene was of hire stature;	[mean, middling]
Therto of shap, of face, and ek of cheere,	[the same] [also] [countenance]
Ther myghte ben no fairer creature.	
And ofte tymes this was hire manere:	
To gon ytressed with hire heres clere	[to go with her bright hair dressed]
Doun by hire coler at hire bak byhynde,	[collar]
Which with a thred of gold she wolde bynde.	
And save hire browes joyneden yfeere,	[except]
Ther nas no lak, in aught I kan espien.	[was no lack] [anything I can see]
But for to speken of hire eyen cleere,	
Lo, trewely, they writen that hire syen	[they that saw her wrote]
That Paradis stood formed in hire yën.	[eyes]
And with hire riche beaute evere more	
Strof love in hire ay, which of hem was more.	[strove] [also] [them]

(*Troilus and Criseyde* v.806–19)

The peculiar position of these portraits, situated at the end of the work rather than (as would be usual) at the outset, is striking: it positions, first, the narrator and, second, the reader in the position of seeing Criseyde in something like the way that Troilus sees her — more precisely, the ways in which he repeatedly sees her within the mirror of his mind, as we saw earlier. In this portrait passage, we learn how Criseyde looked: not simply how she looked at one particular moment in time, but how she looked 'often times' — that is, how she usually looked, how she was seen over time by those who knew her. These lucky few saw not only her outer features but also looked more deeply — perhaps even, like Troilus, looking at her in 'thrifty ways' — to see that 'paradise stood formed in her eyes'. The position of the narrator is peculiar here, because while he does not associate himself with those who 'hire syen' (who saw her), he does imply visual knowledge of her in saying that she lacked nothing 'in aught I kan espien' (as far as I can see).

In the narrator's mediated visual knowledge of Criseyde, focused through the mediating gaze of Troilus, we might perhaps see an analogue to the role of Mary in the visual contemplation of the Passion in devotional texts and imagery of the period. The comparison to the devotional context is useful precisely because it illuminates the position of the viewer who is not privy to the originary moment of desire. The narrator's vision of Criseyde is not so much like Troilus's initial view of Criseyde at the temple as like his later, retrospective, interior views of her within his mind. Like Troilus, the narrator views Criseyde as her image appears in its habitual form, as she usually looked, flashed up on the mirror of his mind. Unlike Troilus, however, the narrator's desire for Criseyde is grafted upon the stock of Troilus's love; it is secondary, mediated through Troilus's desire in something like the way the devotional reader's desire for Christ is mediated through the desire of Mary.

In the end, the common ground that links Christine de Pizan's *Mutacion de Fortune* and Chaucer's *Troilus and Criseyde* is not simply their common focus on the ability of vision to mediate knowledge, and their repeated recourse to specular reflection in describing how the subject comes into being. They also share a common ground in investigating the temporality of the visual act: that is, the way in which the act of seeing unfolds in linear time, yet also provides a means of knowing that is synoptic, that takes place all at once in a singular moment of vision. In Christine's *Mutacion de Fortune*, the ekphrastic imagery of the painted walls of the 'sale merveilleuse', like the ekphrastic portraits of Chaucer's *Troilus*, provide moments of suspension where the reader is compelled to pause to experience the contemplative sense of wonder engendered by ekphrasis. Through the medium of the ekphrastic pause, the reader shares in the narrator's remembered moment of temporal stasis, the period during which she is (in Christine's words) 'abstracted' from the body. In other words, the moment of temporal stasis experienced by the reader is itself a reflection of the moment of interiority experienced by the seeing subject within the text.

In Chaucer's *Troilus*, the reader is invited into the interior space of Criseyde's mind, to witness her own act of self-reflective contemplation that takes place within her chamber — an interior space that is literally her private room but which also consists of the figurative chamber of her mind. This passage expresses Criseyde's experience of interior vision as she turns over the images she has seen and processes them intellectively; in turn, the reader witnesses the sight of the thinking self. In some ways, this interior vision — experienced by Criseyde, and witnessed by the reader — is comparable to the introspective moments of vision that we saw in Christine de Pizan's *Epistre Othea* and, especially, in the *Mutacion de Fortune*. There, the reader was repeatedly urged to 'soi mirer', to gaze upon oneself, in the mirror of another. In the *Epistre Othea*, the exemplary figures of Hercules and Perseus serve as mirrors; through gazing into these mirrors, the subject can, so to speak, find himself as he is reformed to correspond to the model of what is visible in the mirror. Even more richly in the *Mutacion the Fortune*, the reader's visual experience is focused upon the introspective gaze into the mirror of history. No longer limited to individual heroic exemplars, as in the *Epistre Othea*, the reader now gazes into the mirror of history, finding in that sight not simply a sequence of historical facts, but something far more profound: here, the reader is able to gaze upon the phenomenon of 'mutacion', of change itself.

Works Cited

Primary Sources

Christine de Pizan, *Epistre Othea*, ed. by Gabriella Parussa, Textes littéraires français, 517 (Geneva: Droz, 1999)

——, *Le Livre de la Mutacion de Fortune*, ed. by Suzanne Solente, Société des anciens textes français, 4 vols (Paris: Picard, 1959–66)

Guillaume de Loris and Jean de Meun, *Le Roman de la Rose* ed. by Félix Lecoy, 3 vols (Paris: Champion, 1965–70)

James, Henry, *The Portrait of a Lady*, ed. by Robert D. Bamberg, 2nd edn (New York: Norton, 1995)

Ovid, *Metamorphoses*, trans. by Frank Justus Miller, Loeb Classical Library, 42–43, 2nd edn, 2 vols (Cambridge, MA: Harvard University Press, 1984)

The Riverside Chaucer, ed. by Larry D. Benson, 3rd edn (Boston: Houghton Mifflin, 1987)

Secondary Studies

Akbari, Suzanne C., 'Ekphrasis and Stasis in the Allegories of Christine de Pizan', in *The Art of Vision: Ekphrasis in Medieval Literature and Culture*, ed. by Andrew James Johnson, Ethan Knapp, and Margitta Rouse (Columbus: Ohio State University Press, 2015), pp. 184–205

——, 'Metaphor and Metamorphosis in the *Ovide moralisé* and Christine de Pizan's *Mutacion de Fortune*', in *Metamorphosis: The Changing Face of Ovid in Medieval and Early Modern Europe*, ed. by Alison Keith and Stephen James Rupp (Toronto: Victoria University Centre for Reformation and Renaissance Studies, 2007), pp. 77–90

——, 'The Object of Devotion: Fundamentalist Perspectives on the Medieval Past', *Religion and Literature*, 42.1 (2010), 299–315

——, *Seeing through the Veil: Optical Theory and Medieval Allegory* (Toronto: University of Toronto Press, 2004)

Attwood, Catherine, *Fortune la contrefaite: L'Envers de l'écriture médiévale* (Paris: Champion, 2007)

Brownlee, Kevin, 'The Image of History in Christine de Pizan's *Livre de la Mutacion de Fortune*', in *Contexts: Style and Values in Medieval Art and Literature*, ed. by Daniel Poirion and Nancy Freeman Regalado (= *Yale French Studies*, special issue (1991)), pp. 44–56

Cameron, Sharon, *Thinking in Henry James* (Chicago: University of Chicago Press, 1989)

Cunningham, Valentine, 'Why Ekphrasis?' *Classical Philology*, 102 (2007), 57–71

Dulac, Liliane, 'Le Chevalier Hercule de l'*Ovide moralisé* au *Livre de la Mutacion de fortune* de Christine de Pizan', *Cahiers de recherches médiévales*, 9 (2002), 115–30

Goldin, Frederick, *The Mirror of Narcissus in the Courtly Love Lyric* (Ithaca: Cornell University Press, 1967)

Hill, Thomas E., *'She, this in blak': Vision, Truth, and Will in Geoffrey Chaucer's 'Troilus and Criseyde'* (Oxford: Taylor and Francis, 2008)

Holley, Linda T., 'Medieval Optics and the Framed Narrative in Chaucer's *Troilus and Criseyde*', *Chaucer Review*, 21 (1986), 26–44

Hult, David F., *Self-Fulfilling Prophecies: Readership and Authority in the First 'Roman de la Rose'* (Cambridge: Cambridge University Press, 1986)

Karnes, Michelle, *Imagination, Meditation, and Cognition in the Middle Ages* (Chicago: University of Chicago Press, 2011)

Kay, Sarah, 'The *Roman de la Rose* and the Inverted Bouquet: Reflections of Love', in *Mythes à la cour, mythes pour la cour (Courtly Mythologies)*, ed. by Alain Corbellari and others, Publications romanes et françaises, 248 (Geneva: Droz, 2010), pp. 295–310

Marshall, Adré, *The Turn of the Mind: Constituting Consciousness in Henry James* (Madison, NJ: Fairleigh Dickinson University Press, 1998)

Newman, Barbara, 'What Did It Mean to Say "I Saw"? The Clash between Theory and Practice in Medieval Visionary Culture', *Speculum*, 80 (2005), 1–43

Poirion, Daniel, 'Narcisse et Pygmalion dans le *Roman de la Rose*', in *Essays in Honor of Louis Francis Solano*, ed. by Raymond J. Cormier and Urban T. Holmes (Chapel Hill: University of North Carolina Press, 1970), pp. 153–65

Rosenfeld, Jessica, 'Narcissus after Aristotle: Love and Ethics in *Le Roman de la Rose*', *New Medieval Literatures*, 9 (2007), 1–39

Snow, Clare S., '*Maria Mediatrix*: Mediating the Divine in the Devotional Literature of Late Medieval England' (unpublished doctoral dissertation, University of Toronto, 2012)

Stanbury, Sarah, 'The Lover's Gaze in *Troilus and Criseyde*', in *Chaucer's 'Troilus and Criseyde': 'Subgit to alle Poesye': Essays in Criticism*, ed. by R. A. Shoaf, Medieval and Renaissance Texts and Studies, 104 (Binghamton: Pegasus, 1992), pp. 224–38

——, 'The Voyeur and the Private Life in *Troilus and Criseyde*,' *Studies in the Age of Chaucer*, 13 (1991), 141–58

——, 'Women's Letters and Private Space in Chaucer,' *Exemplaria*, 6 (1994), 271–85

Comme dans un miroir, comme dans un adage : chatoiements et reflets de la reprise et de l'amitié

Hélène Cazes

Mirage ou concept, tout à la fois illusion et abstraction, le miroir offre l'image sans donner la chose et toujours dans la dualité du reflet : déclinaison du dédoublement dans l'insensible différence, le miroir semble pour l'auteur des *Adages*, maître de la reconnaissance de soi dans le chatoiement des emprunts et lieux communs, l'emblème d'une écriture seconde et d'une connaissance médiate. En effet, dans la forêt des *Adages* qu'Érasme commence d'assembler vers 1500 et continue jusque la fin de sa vie, dans cette sylve sans cesse bruissante et mouvante, la pluralité des sentences, des voix et de leurs échos se donne à lire et penser par le discours sur le miroir, justement. Dès lors, emblème du dédoublement et de la reconnaissance, le miroir devient le symbole de la médiation et le reflet de l'amitié.

Dialogue entre citation et commentaire, l'œuvre des *Adages* met en scène les rapports du Moi et de l'Autre, du sujet et du reflet, dans sa constitution même : le texte commenté est une manière de se contempler dans l'emprunt du lieu commun et l'explication de l'apophthegme ou du proverbe ouvre, dans une structure de dialogue, la reconnaissance de l'écriture personnelle dans le geste du compilateur. Relevée dans l'organisation interne des annotations formant la continuation de chaque adage, cette affinité de l'écrivain et de ses citations,

Hélène Cazes (hcazes@uvic.ca) est professeur au Département de français de l'Université de Victoria. Elle travaille sur les miroirs de l'amitié et sur la culture humaniste de la tradition. Entre autres, elle est directrice adjointe de *Renaissance et Réforme/Renaissance and Reformation* et co-directrice du volume collectif *Miroirs de la mélancolie*.

The Mirror in Medieval and Early Modern Culture: Specular Reflections, ed. by Nancy M. Frelick, CURSOR 25 (Turnhout: Brepols, 2016), pp. 171–189 10.1484/M.CURSOR-EB.5.111047
BREPOLS PUBLISHERS

se donne également à lire dans un réseau thématique du double et des reflets : nombre d'adages traitent explicitement de la similarité (les gouttes d'eau, les jumeaux), de la similitude (les tableaux en trompe-l'œil), de l'écho (les résonances harmoniques) ou de l'affinité (la ressemblance familiale, l'amitié). Or ces thèmes, présentés dans les adages comme les catégories générales sous lesquelles se rangent et s'écrivent les recherches et rêveries sur la démultiplication de l'identité, se reconnaissent également dans l'agencement et la poétique de l'œuvre : allitération, métaphore, métonymie, les figures du 'même' inspirent le passage d'un adage à l'autre, d'une citation à l'autre, d'un sujet à l'autre. De fait, le miroir est, au sens propre, une réflexion sur l'écriture ; modèle d'une reconnaissance de soi dans l'objet, il sert une poétique citationnelle, construite comme une conversation. Dès lors, c'est l'ami, cet 'autre moi' qui se découvre dans le miroir en miroir, comme dédicataire, source d'inspiration, ou lecteur. Les multiples visages de l'amitié se dévoilent au miroir des adages, thèmes, structures et poétiques de la similarité.

Chiliades de miroirs

Collection de commentaires et essais organisée autour d'adages anciens et modernes — le terme 'adage' désignant, ainsi que le définit la préface, proverbes, expressions, sentences, lieux communs et même symboles pourvu que la formulation en soit quelque peu énigmatique ou ambigüe —, l'œuvre d'Érasme connue sous le titre *Adagiorum Chiliades*, tisse les traditions et y insère maints commentaires spéculaires sur ces traditions (antique, classique, chrétienne, humaniste) et sur la manière de s'y regarder. Construite comme une galerie de miroirs, la collection d'adages donne en effet à lire l'emprunt et l'appropriation de formules antiques pour une éthique moderne ; elle propose ainsi, dans la multiplicité des fragments citationnels et des énonciations, tant un reflet de soi (pour l'auteur) qu'un reflet du lecteur autour du lieu commun, dans le dialogue du connu et de l'inconnu. Dans ce regard porté sur le reflet et le partage de soi, le miroir apparaît comme l'emblème tant de la représentation que de l'élucidation : dans une éthique du dédoublement fondateur, Érasme propose une rêverie philologique sur le *speculum*, outil de l'examen de soi-même (réflexivité), de la connaissance du monde (spéculation) et de la reconnaissance de la similarité. Car le miroir de la coquette ou du fanfaron, absorbés tous deux dans la solitude de leur vanité, peut également être le miroir de la conscience et de la science.

L'éloge érasmien des miroirs, inattendu chez le moraliste mais bien à sa place dans une poétique de la reprise philologique, est poétiquement étayé sur un éloge de l'amitié, reconnaissance du moi et de l'autre tout à la fois qui permet de voir plus clairement, dans le reflet, la vérité — plus profonde dans ce regard second que dans la perception première. Pure du désordre des perceptions et de l'aveuglement des émotions, la représentation idéalement abstraite du reflet dans le miroir permet alors la pensée et Érasme ne dédaigne pas entrer dans la spécularité par la surface du miroir occasionnellement apparu dans un adage ancien. Or la réflexion sur le thème même de la spécularité ressemble fort à une définition, proprement poétique, du projet des adages, citationnel et dialogique tout à la fois. Symbole du dédoublement et du regard second, le miroir est aussi le lieu où se cherche et se trouve, dans une ontologie de l'amitié, la compagnie des textes et de soi.

Gouttes d'eau et gouttes de lait

Le premier miroir des *Adages* est une affaire de frères se ressemblant comme deux gouttes d'eau, au premier livre dans sa première mouture. 'Rien ne ressemble plus à une goutte d'eau qu'une goutte d'eau', ainsi que l'épelle un vers de Plaute, puis deux, puis trois :

> **Non tam aqua similis aquae**
> Eamdem vim obtinet, quod usurpat Plautus in *Menaechmis* : Nec aqua aquae, nec lacte lacti, crede mihi, usquam similius Quam hic tui est, tuque hujus.
> Ad hanc formam et illud ex eadem scena : Tam consimilis est quam potest speculum tuum. Rursum in *Milite* :
> Nam e summo puteo similior numquam potest Aqua aquae sumi.[1]

Le miroir n'est pas ici au titre, mais au milieu du chapitre et fait le lien entre les citations jumelles des portraits crachés, se répétant en sosies, puis revenant aux gouttes d'eau.

> **Comme deux gouttes d'eau**
> Pareille force d'expression se lit chez Plaute aux *Ménechmes* : 'Il y a plus loin de la goutte d'eau à la goutte d'eau, de la goutte de lait à la goutte de lait, crois-moi, que de toi à lui ou de lui à toi' [*Ménechmes*, 1088–89]. Ce qui ressemble tout à fait à la fin de cette scène à : 'Il te ressemble tant qu'il pourrait te servir de miroir'.

[1] *Adages* I, V, 12, p. 403, dans Érasme, *Les Adages*, trad. et éd. par J.-C. Saladin, Collection 'Le Miroir des humanistes' (Paris : Belles Lettres, 2011). Cet adage figure déjà dans les *Collectanea* de 1500, ancêtre des *Adages* assemblé à l'usage du jeune Lord Mountjoy. Sur ce premier recueil, voir *Adagiorum Collectanea*, éd. par Heinimann et van Poll-van de Lisdonk. La seconde citation est ajoutée dans l'édition de 1523. Voir pour les identifications de références : Erasmus, *Collected Works of Erasmus. 31 : Adages*, trad. par Phillips, annot. par Mynors, p. 394. On verra également *Adagia*, éd. par Le Clerc. L'édition savante de référence, également consultée, est la suivante : *Adagiorum Chiliades*, éd. par van Poll-van de Lisdonk et d'autres.

On trouve encore au *Miles* : 'Jamais, du plus profond des puits, eau ne pourra être puisée qui ressemble plus à de l'eau' [*Miles gloriosus*, 551–52].[2]

De fait, ce court excursus sur la ressemblance met en son centre le miroir : entre le début et la fin de la scène de Plaute ici citée, se trouve un vers consistant en une exclamation nominale : 'Ton miroir' (*Speculum tuum !*) que la première citation introduit et la seconde commente. Invisible dans le titre, escamoté dans la citation, le miroir est ainsi au centre de l'adage consacré à la ressemblance.[3] Se répondent, symétriquement de part et d'autre de l'évocation du miroir, les images, ou reflets, de la ressemblance : les gouttes d'eau jumelles. Ces gouttes se dédoublent elles-mêmes d'abord en gouttes de lait. Puis l'eau appelle le miroir de la surface des puits. La galerie des ressemblances et similitudes se construit dans la structure de l'adage comme dans sa thématique. Métaphore de la ressemblance parfaite, semblable au dédoublement, le miroir est le 'portrait' de l'ami, du frère.

Or ce premier miroir explicite se lit d'abord dans la répétition : le titre de l'adage latin *Non tam aqua similis aquæ* répète comme en un bégaiement le mot *aqua* mais il évoque également la location latine bien connue *tam quam*, ici présente sous forme anagrammatique et comme cachée à la première lecture. L'effet est redoublé par la répétition de l'adjectif *similis* du titre par son composé *consimilis* dans la première citation de Plaute. Le texte se mire dans ses propres reprises, jusqu'au vertige.

Voire, cet adage sur les gouttes d'eau reflète celui qui le précède, sur les gouttes de lait :

Non tam lac lacti simile.
Idem pollet, et a Plauto usurpatur in Amphitryone :
Neque lacte lacti magis est simile, quam ille ego similis est mei. Dicitur a Sosia de Mercurio, qui hujus assumpserat figuram. Idem in Milite : Tam simile, quam lacte lacti est.[4]

[2] Ma traduction. Ailleurs dans l'article, sauf indication expresse, la traduction française est empruntée à Érasme, *Les Adages*, trad. (Lat. et Gr.) et éd. dirigées par Saladin. Je ne m'écarte de cette version que pour restituer la relative lourdeur des répétitions du latin.

[3] On me pardonnera, tout au long de cet article, d'utiliser 'adage' dans le sens métonymique 'chapitre des adages', usage certes impropre, toutefois passé dans la langue et sûrement plus élégant que la formulation répétée 'chapitre consacré à l'adage'.

[4] *Adages* I, v, 11 (éd. cit. p. 402).

Bien : l'eau et le lait s'appellent l'un l'autre et l'adage sur les gouttes de lait se continue, après une transition citationnelle liant eau et lait, en un adage sur les gouttes d'eau. Miroirs, miroirs.

> **Comme deux gouttes de lait.**
> Pareille efficacité se lit dans l'expression qu'emprunte Plaute dans son *Amphitryon* [601] : 'Deux gouttes de lait ne se ressemblent plus fortement que je ne me ressemble à lui'. C'est Sosie qui parle de Mercure, dont il a pris l'apparence. Et encore, dans le *Miles Gloriosus* [240] : 'Ils se ressemblent comme deux gouttes de lait.'[5]

Ici encore, la seconde citation est ajoutée en 1523, comme si la relecture pour réédition ouvrait encore un pan de miroirs dans la chambre des citations. Or le jeu des reflets et continuations se poursuit, de part et d'autre, encore, du miroir central du douzième adage. Avant les gouttes de lait et les gouttes d'eau, ce sont les œufs qui se ressemblent ; et après les gouttes, ce seront les abeilles. Et dans l'un comme l'autre cas, le titre-adage se plaît à la déclinaison d'un même mot répété, tout comme *aqua* était repris par *aquæ* et *lac* par *lacti*. Car la petite série des ressemblances commençait par l'adage *Non tam ovum ovo simile*, première polyptote de l'unité thématique sur la ressemblance. Dans ce premier adage — lui aussi déjà présent dès les *Collectanea* de 1500 —, la thématique et, de fait, la série étaient annoncées, en première phrase, par la promesse d'une série d'adages 'sur la similitude' : *Aliquot similitudinis adagia*, où le mot *similitudinis* était repris dès la ligne suivante ('À propos de similitudes interdisant la distinction').

> **Non tam ovum ovo simile**
> Extant apud auctores aliquot similitudinis adagia, quorum de numero est : Non tam ovum ovo simile, de rebus indiscretae similitudinis. Cicero Academicarum quaestionum libro secundo : Videsne, ut in proverbio sit ovorum inter se similitudo ? [...][6]

> **Comme deux œufs**
> On trouve chez les classiques plusieurs adages sur la similitude, dont : comme deux œufs, comme signifier des choses si semblables qu'on ne peut les distinguer l'une de l'autre. Cicéron, dans les *Questions académiques*, au second livre [*Academica priora*, 2, 18, 57], demandait : vois-tu comment ce proverbe exprime la similitude entre un œuf et un autre ? [...][7]

[5] Ma traduction.

[6] *Adages* I, v, 10 (éd. cit. p. 402).

[7] Ma traduction.

Placée sous le signe des *Questions académiques* de Cicéron, la thématique de la ressemblance se décline avec Quintilien, puis Sénèque au cours de cette ouverture.

Elle se termine, en 1528, par l'ajout de l'adage *Quam apes apum similes*, qui ferme la boucle en reprenant et terminant la citation cicéronienne du premier adage et amenant la conjonction *quam* appelé par les *tam* de tous les titres de la série, attendue jusque cette résolution. Surtout, développant la similitude par des *simile*, Érasme propose comme clausule le rappel puis une reprise homophonique du premier adage de la série : les œufs font place aux abeilles, tandis que le dernier proverbe fait entrer les moutons, dont la désignation latine (*ovis* et *ovi*) semble répéter *ovum* et *ovo*.

> **Quam apes apum similes**
> His M. Tullius illud etiam adjungit in Academicis quaestionibus : Ut similia sunt, et ova ovorum, et apes apum. Ubi jocatur in Democriti mundos usquequaque similes inter sese. Simili forma dicunt : Quam ovis est ovi similis.[8]

> **Ainsi des abeilles ressemblent aux autres abeilles**
> Cicéron continue ces adages dans les *Questions académiques* par un autre adage : 'Comme un œuf ressemble à un œuf, ou des abeilles aux autres abeilles'. C'est dans le passage où il plaisante à propos des mondes de Démocrite, indiscernables les uns des autres. Semblablement, on dit : 'Comme un mouton ressemble à un autre mouton'.[9]

Symétries et échos de la ressemblance

Le jeu des miroirs polyptiques et des comparaisons de la ressemblance se continue, toujours de part et d'autre du miroir caché au creux de l'adage sur les gouttes d'eau, central dans la série des similitudes. En fait, tandis qu'homophonies, reprises et structures se conjuguent pour construire, par le texte, la ressemblance (phonétique, thématique, rhétorique), la disposition des adages met en place une architecture du dédoublement : tout comme œufs et moutons, très similaires du point de vue phonétique en latin, encadraient le premier dispositif des répétitions, l'élargissement aux deux adages voisins de la série établit une symétrie nouvelle. En effet, l'adage qui précède nos indiscernables œufs est consacré à la musique, plus précisément à la fausse note :

[8] *Adages* I, V, 13 (éd. cit. p. 403).

[9] Ma traduction.

> **Eadem oberrare chorda**
> Horatius in *Arte poetica* sententiam haud dissimilem dissimili metaphora extulit,
> Et citharoedus, inquiens, Ridetur, chorda qui semper oberrat eadem.
> [...] Primus enim lapsus vel casui datur vel imprudentiae, iteratus stultitiae tribuitur aut inscitiae.[10]

> **Refaire la même fausse note.**
> L'idée qu'exprime Horace, dans l'*Art poétique* [255–56], n'est pas trop différente sous une métaphore différente :
> Car on se moque, dit-il,
> Du citharède qui refait toujours la même fausse note. [...] La première fois, ce peut-être un coup du hasard ou de la précipitation, mais la fois suivante s'explique ou par la bêtise ou par le manque d'expérience.[11]

Cet adage musical reprenait lui-même un adage sur la maladresse,[12] lui aussi présent dès les *Collectanea*, qui explorait 'différence et répétition' dans le bégaiement des erreurs réitérées. La musique sert alors de comparant, en tête de la série sur les ressemblances, à une dynamique dans la reprise : première occurrence et seconde occurrence diffèrent radicalement. Le discours revient donc, par la métaphore, aux similitudes des similitudes, par le détour des répétitions qui transforment la valeur de l'événement, interprété différemment selon qu'il se produit pour la première ou pour une seconde fois. La série d'éléments indiscernables les uns des autres se développe ainsi comme une autre perspective, toujours dynamique mais non chronologique cette fois : certaines ressemblances empêchent distinction ou même spécificité. La structure en reflet porte néanmoins un mouvement poétique et philosophique : le miroir caché au centre des ressemblances permet de faire succéder le *quam* au *tam*, de compléter la conjonction de comparaison.

Cet achèvement de la similitude se redit dans la continuation, toujours dès les *Collectanea*, de l'architecture thématique : à l'adage sur la fausse note, répond, en écho, l'adage sur les cigales d'Acanthe. Continuant l'adage sur les abeilles, l'analyse sur les cigales muettes de la ville d'Acanthe introduit une exégèse d'histoire naturelle, empruntée à Stéphane de Byzance, Pline, Pausanias et Strabon, pour conclure sur une anecdote où une cigale supplée la note juste dans un concours musical.

[10] *Adages* I, v, 9 (éd. cit. p. 402).

[11] Ma traduction.

[12] *Adages* I, v, 8 Iterum eundem ad lapidem offendere : Trébucher deux fois sur la même pierre (éd. cit. p. 401).

Acanthia cicada
Ἀκάνθιος τέττιξ, id est, Acanthia cicada, in indoctos atque infantes aut musices ignaros torqueri proverbio solitum. Auctor est Stephanus Byzantius juxta oppidum Aetoliae Acanthum cicadas alibi vocales mutas esse, atque hinc ortum adagium, cujus auctorem citat Simonidem. Plinius Naturalis historiae libro undecimo capite vigesimoseptimo scribit in agro Rhegino cicadas omnes silere, ultra flumen in Locris canere. Idem testatur Pausanias libro rerum Eliacarum secundo. Strabo libro Geographiae sexto refert Rheginum agrum a Locrensi dirimi fluvio, cui nomen Alex, cicadas autem, quae in Locrensi versentur ripa, sonantius stridere, cum in Rhegina sint mutae.[13]

Dans l'érudition et la collection de références, l'adage dit le silence, mystérieux, infamant, des cigales d'Acanthe. L'accumulation des citations ainsi que la longueur des indications bibliographique crée un effet cumulatif, non pas d'authenticité — puisque la ville d'Acanthe devient un champ de Rhegium —, mais d'insistance.

Les Cigales d'Acanthe
Ἀκάνθιος τέττιξ, c'est-à-dire, cigale d'Acanthe, expression proverbiale utilisée à l'encontre de ceux qui sont ignorants, inexpérimentés ou encore sans talent pour la musique. On tient de Stéphane de Byzance que, dans la ville d'Acanthe, en Étolie, les cigales sont muettes, alors que partout ailleurs elles chantent. De ce fait est né l'adage, qu'il attribue à Simonide [Fragment 105 de l'édition Page]. Pline l'Ancien, du livre onze chapitre vingt-sept de l'*Histoire Naturelle* [11.95], écrit que toutes les cigales du district de Rhegium restent sans voix, tandis que, de l'autre côté du fleuve, chez les Locriens, elles chantent. Pausanias atteste ce même fait au second livre de ses études sur Elis [6, 6, 4]. Strabon, au sixième livre de la *Géographie* [6, 1, 9], fait référence à un champ de Rhegium, qui jouxte le pays locrien, dont seule le sépare une rivière nommée Alex : du côté locrien, les cigales chantent bruyamment, mais du côté de Rhegium, elles sont muettes.[14]

Or de la similitude, c'est la simple répétition, muette, stérile, qui prévaut. Elle rappelle la faute récurrente du mauvais musicien. Or, de l'autre côté du miroir, voici qu'une cigale, l'animal qui symbolise l'âme des poètes, vient au secours d'un citharède. À la fin de l'adage, une anecdote remarquable rapportée par Strabon fait d'une cigale le personnage crucial d'un concours musical : faisant résonner la note juste lorsque la corde du citharède s'est brisée, elle conclut, achève, accomplit le morceau de musique. Tout comme le pourvoyeur d'adages,

[13] *Adages* I, V, 14 (éd. cit. p. 403–04).

[14] Ma traduction.

qui semble répéter les similitudes sur la ressemblance, fournit la note finale, en écho, de la série.

> [...] cum in Eunomi cithara una inter canendum chorda frangeretur, cicada supervolans astitit, ac vocem alioqui defuturam suo cantu supplevit. Atque ita victor declaratus statuam citharoedi posuit cum cicada citharae insidente. Hujusmodi ferme tradit Strabo.[15]
> Une corde de la cithare d'Eunomus se brisa et une cigale, volant au dessus de lui, lui fournit par son chant la note qui autrement lui aurait manqué. Ainsi, il fut proclamé vainqueur et il fit ériger la statue d'un citharède portant sur sa cithare une cigale. Voilà ce qu'en dit, en substance, Strabon.[16]

Cigale individuelle, unique, immortalisée par la statue que fait ériger Eunomus, la figure dernière du dispositif en miroir sur les similitudes sort du silence et sort le musicien du silence, tout comme l'auteur des *Adages* sort de la discrète et indiscernable répétition pour offrir la note manquante, magique, qui termine le morceau de musique. Figure de l'auteur, la cigale est alors le comparant pour l'auteur qui, d'abord compilateur comme les abeilles de l'adage précédent, devient en passant de l'autre côté du miroir, artiste.

Miroirs aux alouettes et miroirs d'aveugles

La transfiguration de la répétition en mouvement, la dialectique de la reprise du compilateur devenue création du musicien, repose sur la conscience de la ressemblance. Entre goutte d'eau et goutte d'eau, il y a similitude, non identité. Entre le personnage et son sosie, pareillement. Et entre le sujet et son reflet dans le miroir, la similitude ne doit pas effacer la distinction.

Les *Adages* se plaisent à prévenir la confusion du moi et de ses reflets, dénoncée comme une illusion où la réflexion dans le miroir est prise pour la réalité. Sous le nom d'Akkô, est dessiné, par un couple d'acteurs allégoriques, un portrait des mauvais usages du portrait de soi : la coquette aveuglée par son reflet et le soldat fanfaron ne savent pas distinguer l'identité de la ressemblance.

Le premier personnage entre sur scène devant son miroir : Akkô, Accus sous son nom latinisé, ne sait pas ce qu'elle veut, car elle ne sait pas qui elle est. Sa stupidité tient à son incapacité de percevoir la ressemblance en tant que telle : elle prend le reflet pour l'original, elle ignore la dualité du miroir et l'unité du sujet.[17]

[15] *Adages* I, V, 14 (éd. cit. p. 404).

[16] Ma traduction.

[17] *Adages* II, II, 99 (éd. cit. p. 997–98) : **Accissare** / Ἀκκίζειν, id est accissare, Graeco

> **Faire son Akkô**
> Par ce proverbe grec, on désignait ceux qui feignaient de refuser une chose qu'ils désiraient pourtant ardemment recevoir. Cette habitude perdure encore chez bien des contemporains. On lit qu'il y avait une femme nommée Akkô, d'une insigne stupidité. Elle avait pour habitude de s'entretenir devant son miroir avec son propre reflet comme si c'était une autre femme. De là, on en est venu à dire couramment que ceux qui font quelque chose de stupide ou d'inepte 'font leur Akkô' et on les appelle du nom d'Akkô. Il apparaît aussi que cette femme avait pour autre trait de caractère de refuser ce qu'elle désirait pourtant. [...][18]

Cette Akkô sans distinction, au sens propre du mot, se reflète dans les traits du soldat fanfaron, qui aime son reflet sous l'uniforme sans savoir, lui non plus, qui il est ni ce qu'il voit.[19]

> **Sous les armes il prend la pose**
> Comme un homme épris de lui-même et qui se contemple comme il le ferait de sa propre statue en armes. Conviendra à qui est sottement épris de sa propre personne. Est rapporté avec bien d'autres dans l'*Étymologique*. J'ai parlé ailleurs d'Akkô, la femme qui minaudait sottement devant son miroir dans l'adage 'Faire son Akkô'.[20]

L'un comme l'autre épris d'eux-mêmes, ces personnages sont aveuglés par l'immédiateté. Tout comme les oiseaux topiques qui tentent de picorer les raisins peints par Zeuxis,[21] les coquets ne savent pas distinguer l'image de la chose. Ils sont des aveugles à qui le miroir ne sert de rien, pour citer l'adage sur 'le miroir de l'aveugle'[22] :

proverbio dicebantur, qui, cum maxime cuperent accipere quippiam, tamen ficte recusabant ; qui mos hodie quoque multis dutat. Scribunt Acco mulierem fuisse quampiam notae stultitiae, quae solita sit ad speculum cum imagine sua perinde atque cum alia muliere confabulari, ut hinc vulgo, quae stultius aut ineptius aliquid agerent, accissare dicerentur et Accus nomine compellarentur. Apparet illud etiam moribus hujus mulieris adfuisse, ut recusaret, quae tamen cupiebat, unde et accismus ficta hujusmodi recusatio dicatur.

[18] Trad. cit. II, 141.

[19] *Adages* IV, II, 83 (éd. cit. 2088) ; **In armis accissat** / Ἐπὶ τοῖς ὅπλοις ἀκκίζεται, id est In armis accissat. In eum qui sibi placet seseque quasi contemplatur in armis suis. Conveniet in quemvis sibi stulte placentem. Refertur praeter alios Etymologico. De Accone muliere, quae ad speculum inepte sibi blandiebatur, dictum est alibi in proverbio Accisare.

[20] Trad. cit. IV, 97.

[21] L'histoire vient de Cicéron [*De Inventione*, II, 1] et est reprise, entre autres, par Pline l'Ancien [*Histoire Naturelle*, XXXV, 36] et Érasme dans la collection de similitudes des *Parabolæ sive Similia*, éd. par Margolin, p. 276, paraboles 68–71.

[22] *Adages* III, VII, 54 (éd. cit. p. 1846) : **Quid caeco cum speculo ?** / Σὶ τυφλῷ καὶ κατόπτρῳ;

Que ferait un aveugle d'un miroir ?
Très ressemblant : 'un choucas n'a rien à faire d'une lyre'[23]
Ce vers tiré d'Épicharme est cité par Stobée : 'Quand on n'a plus ses yeux, que faire d'un miroir ?'
Quel besoin de livres pour un illettré ? Pourquoi confier le royaume à un jeune sot ? Pourquoi des richesses pour qui ne saurait qu'en faire[24] ?

Outil de la connaissance par la similitude, le miroir ne sert de rien à ceux qui ignorent la médiation et, partant, ignorent ce qu'est le miroir. Trompés par l'apparence, incapables de regarder le reflet pour ce qu'il est, ces aveugles et sots ne savent ni regarder ni produire une image. Du coup, ils ne savent pas, non plus, utiliser le langage : comme les avis d'Akkô, leurs paroles sont vaines car elles ne sont pas définies comme ressemblance au sujet, comme représentation du moi. Dans l'immédiateté et la naïveté de leur profération hors de la structure en miroir (ou en écho) avec leur auteur, les mots n'ont pas de sens. Ainsi, l'adage consacré aux 'éclairs du bassin'[25] compare les vaines menaces, paroles sans effet, aux illusions spectaculaires produisant l'image d'un éclair. Si le spectacle n'est pas reconnu comme montage, comme représentation qui fait œuvre, alors la méprise fait passer l'illusion pour le réel. Seuls les enfants, ou les ignorants, confondent fiction de l'éclair et phénomène naturel. Et, ce faisant, ils ne peuvent voir ni le dispositif, ni l'éclair. Aveugles au miroir...

id est Quid caeco cum speculo ? Finitimum illi : Nihil cum fidibus graculo. Citatur a Stobaeo versiculus ex Epicharmo :/ Σἰς γὰρ κατόπτρῳ καὶ τυφλῷ κοινωνία ; id est / Luminibus orbo cum speclo quid est rei ? Quorsum opus est libris illitterato ? quo regnum juveni stulto ? quo divitias ei, qui nesciat uti ?

[23] Voir les adages 335 (I, IV, 35) *Asinus ad Lyram* et 337 (I, IV, 37) *Nihil graculo cum fidibus*, regroupés en un seul adage portant sur la même idée : la lyre ne sert de rien à un âne, ni à un choucas.

[24] Trad. cit. III, 338.

[25] *Adages* II, VII, 90 (éd. cit. p. 1312) : **Fulgur ex pelvi** / Ἀστραπὴ ἐκ πυέλου, id est Fulgur ex pelvi. De minis inanibus eorum, qui laedere non queant. Fulgor enim, qui fulguris imitatione quadam e vasculorum nitentium vibratione aut ex aquis redditur, nocendi vim nullam habet, pueros modo territat. Habentur et hodie specula quaedam, quae vibrata ad solem fulgur imitantur ; verum id fulmen non est, quod montes ac maria percutiat, ut inquit Plinius. Paroemiam recenset Diogenianus. Ad quam allusit Augustinus, aut si quis alius est auctor, in libro De quinque haeresibus : Non nos hostium turba, non bellantium revocet forma, non quasi vitrea fulgentia terreant arma ; quanquam legendum opinor vitra.

> **L'éclair du bassin**
> Cet adage désigne des menaces vaines, émanant de personnes qui n'on aucun pouvoir de nuisance. Un éclat qui imite l'éclair — qu'il soit provoqué par le déplacement d'un plat bien astiqué ou par des reflets d'eau — n'a, en réalité, aucun pouvoir de nuisance ; il effraie juste les enfants. De nos jours aussi, il existe des miroirs qui imitent l'éclat de la lumière lorsqu'on les fait bouger au soleil. Mais il ne s'agit pas d'un vrai éclair, qui, comme l'écrit Pline, frappe mers et montagnes. Cet adage est recensé par Diogénianos ; Saint Augustin — ou un autre auteur, quel qu'il soit — y fait allusion dans *Les Cinq Hérésies* : 'Ne soyons pas effrayés par l'éclat des armes de verre (*vitrea*) !' Toutefois, je pense qu'il faut lire *vitra* [imitant le verre].[26]

Au rebours, l'homme sage, qui pèse ses mots, mesure combien sa parole est le reflet de son âme. Une fois encore, le miroir est le comparant de cette sagesse qui se définit d'abord comme conscience de la médiation puis maîtrise de cet outil qu'est la projection de soi : l'adage 'Tel homme, tel langage'[27] explore la similitude entre l'orateur et son discours, entre le sujet et sa parole :

> **Tel homme, tel langage**
> [...] bref c'est toute la vie d'un homme, c'est son âme entière dont on voit l'image dans son langage, comme dans un miroir, et même les secrets les plus intimes de son cœur peuvent se révéler par les traces qu'ils laissent pour ainsi dire dans la parole.[28]

Voire, la connaissance par le miroir, ou par le langage, est plus sûre que l'appréhension directe. À lire le miroir, le regard décèle 'les secrets les plus intimes' tandis que le regard posé sur le visage n'en voit que la surface. Car le miroir, en faisant passer le regard du simple au semblable, est l'outil du discernement. Passant par la profondeur du dédoublement qu'est le reflet, le miroir ouvre en effet l'espace de l'invisible, du secret, de l'insu : le miroir révèle l'homme. Du coup, en une variation du positif et du négatif, le miroir qui montrait la sagesse montrera de la même façon la bêtise. Tout comme l'excellent orateur se reconnaît dans ses mots, 'Insensé est le discours de l'insensé'[29] et l'adage précise bien

[26] Trad. cit. II, 417.

[27] *Adages* II, VI, 50 (éd. cit. pp. 515–16) : **Qualis vir, talis oratio** / Breviter omne vitae simulacrum, omnis animi vis in oratione perinde ut in speculo, repraesentatur ac vel intima pectoris arcanis quibusdam vestigiis deprehenduntur.

[28] Trad. cit. I, 441–42.

[29] *Adages* I, I, 98 (éd. cit. pp. 166–67) : **Stultus stulta loquitur**/ Euripides in Bacchis, Μῶρα γὰρ μῶρος λέγει, id est Nam stulta stulti oratio est. Totidem verbis propheta noster Esaias eam sententiam extulit. Seneca ad Lucilium Apud Graecos, inquit, in proverbium cessit : Talis hominibus fuit oratio, qualis vita. Hoc cujusmodi fuerit parum liquet, nisi quod tale quoddam

que le discours reflète 'l'image de l'âme', plus vrai que 'l'aspect du corps'. 'Miroir de la vie', le discours renseigne non sur l'apparence mais sur l'être. Par rebond et retour, le miroir montre non pas la surface mais la structure de la chose.

> **Insensé est le discours de l'insensé** [Ma traduction de ce titre]
> Euripide écrit dans *Les Bacchantes* : 'Insane est le discours de l'insensé'.
> Notre prophète Isaïe exprime cette idée dans les mêmes termes.
> Sénèque écrit dans une lettre à Lucilius que, chez les Grecs, il est devenu proverbial que le discours de l'homme était comparable à sa vie. On ne sait pas exactement quelle était la formule si ce n'est qu'il reste un vers de ce genre très connu des Grecs :
> À un discours on reconnaît un caractère.
> Chez Diogène Laërce, Démocrite qualifie le discours de 'miroir de la vie', comme si c'était son ombre. On ne pouvait rien dire de plus exact. Car l'aspect du corps n'est en rien réfléchi dans un miroir de façon plus parfaite ou plus expressive que l'image de l'âme n'est reflétée dans le discours. Les hommes se distinguent en effet par leur discours tout comme les vases de bronze par le son qu'ils produisent.[30]

Pour le géomètre, sa médiation permet de saisir les abstractions autrement indiscernables : dans l'adage 'Que nul n'entre ici s'il n'est géomètre',[31] les choses 'intelligibles et empreintes du vrai', les lointaines et claires idées, sont visibles 'comme dans un miroir', c'est-à-dire dans la représentation et la dialectique.

> [...] Philon aurait ensuite appelé la géométrie [le point de départ et le pays d'origine] de toutes les disciplines parce qu'elle nous donne à voir les images de toutes choses intelligibles et les empreintes du vrai comme dans un miroir très finement poli, et qu'elle détache peu à peu des sens l'esprit purifié.[32]

Abstraction par la distance, par la conscience de la ressemblance et de la dualité, le miroir est alors l'outil de la connaissance pure, hors des naïvetés et cécités de

carmen extat, Graecis celebratum : / Ἀνδρὸς χαρακτὴρ ἐκ λόγου γνωρίζεται, id est /Hominis figurae oratione agnoscitur./ Democritus philosophus apud Laertium orationem εἴδωλον τοῦ βίου, id est vitae simulacrum, quandamque velut umbram esse dicebat. Qua quidem sententia nihil dici poterat verius. Nam nullo in speculo melius, expressiusque relucet figura corporis, quam in oratione pectoris imago repraesentatur. Neque secius homines ex sermone quam aerea vasa tinnitu dignoscuntur.

30 Trad. cit. I, 131.

31 *Adages* III, III, 60 (éd. cit. p. 1639–40) : Αγεωμέτρητος ούδεὶς εἰσίτω / [...] Tum Philo, qui geometriam ἀρχὴν καὶ μητρόπολιν appellarit omnium disciplinarum, quod haec rerum intelligibilium imagines verique vestigia tamquam in politissimo speculo demonstret mentemque purgatam paulatim abducat a sensu.

32 Trad. cit. III, 152.

la prime apparence. C'est bien une similitude, comme celles qui se disaient et se lisaient en série, que le miroir : double, dynamique, heuristique.

Comme dans un miroir, comme dans un adage

L'adage consacré au miroir lui-même est alors un adage en partie double, appuyé dans son titre par la conjonction polyptotique *tamquam* qui exprime comparaisons, similitudes, paraboles, ou encore métaphores : *Tanquam in speculo, tanquam in tabula* (Comme dans un miroir, comme sur un tableau).[33] Construit comme un miroir, l'adage compare miroir et tableau, les deux images à la ressemblance de l'original, et propose dans sa structure la dualité essentielle du reflet. Voire, le discours se développe également en deux genres, tandis qu'une sentence, centrale, tient lieu de miroir du miroir.

> **Comme dans un miroir. Comme sur un tableau.**
> Ce sont des hyperboles proverbiales, Comme dans un miroir, comme sur un tableau. Au sujet d'une chose évidente et limpide.
> Térence dans *Les Frères* [414–16]

[33] *Adages* II, III, 50 (éd. cit. pp. 1043–44) : '**Tanquam in speculo, tanquam in tabula.** / Hyperbolae proverbiales sunt et hae : καθάπερ ἐν κατόπτρῳ, καὶ ὥσωπερ ἐν γραγῇ id est, Tanquam in speculo, tanquam in tabula. De re vehementer euidente, perspicuaque. Terent. in Adel. Denique inspicere tanquam in speculum vitas omnium iubeo.

Plato lib. De legibus 10. ὡς ἐν κατόπτροις, id est, tanquam in speculis. Propterea quod inter multa, quae rerum referunt imagines, velut aqua, aes politum, gemmae quaedam, nihil aeque perspicue reddit simulacra, vt vitru plumbo sublitum. Nam vitrum quidem, vt rarum ac pellucidum maxime capax est imaginum ; verum acceptas non perinde continet ob pellucentiam : proinde sublitum plumbum, ob nigrorem aliquem, naturaeque densitatem, sistit impressam imaginem, neque dilabi sinit. Euanescunt enim simulacra, non secus, atque candelae lumen in sole, quoties excipiuntur in corpore nimium tenui, perspicuoque velut in aere, pura aqua, crystallo, vitroque non obscurato corporis alicuius oppositu. Rursum in nimium densis, et obscuris non haerent, velut in silice, lignove. Proinde natura, solertissima rerum omnium architectrix, oculum animantium ad speculorum imaginem finxisse videtur, quippe parti pellucenti nigrorem quendam adiunxit a tergo, quo sublato, protinus tollatur et videndi facultas.

Vnde scite dictum est illud, Vt speculum oculus est artis, ita oculum esse naturae speculum. Illud miraculo dignum, speculum non modo clarissime repraesentare simulacra rerum omnium, quaecunque obiecta sunt, verumetiam interualla, colores, motus, breuiter res ipsas pene dixerimus euidentius, quam sunt, hoc vno discrimine, quod aduersa refert. Cuius causa haec est, quod species, quae a corpore solido per aerem manat ad speculi superficiem, simplex, puraque est, vnde fit, vt non alia sit auersa, quam aduersa. Quibus de rebus acutissime multa disseruit Ambrosius Leo Nolanus in quaestionibus suis.

Iam ars picturae a rudibus initiis profecta, eo subtilitatis excreuit, vt non oculis hominum, sed etiam auibus, imposuerit, effeceritque vt veras res esse crederent, quae coloribus erant fucata.'

Enfin,
Je donne ordre d'examiner leur vie à tous,
Comme dans un miroir.

Platon, au livre X [905] des *Lois*, dit 'comme dans les miroirs', parce que parmi toutes les substances qui réfléchissent l'image des choses, comme l'eau, le bronze poli, certaines pierres précieuses, rien ne reproduit aussi fidèlement qu'une lame de verre au dos enduit de plomb. Le verre, en effet, corps rare et translucide, est le plus apte à rendre les images ; mais une fois qu'il les a reçues, il ne les garde pas, à cause de sa translucidité ; or le plomb dont on l'enduit arrête l'image imprimée et ne la laisse pas échapper grâce à sa couleur noire et à la densité de sa nature. Car les images reproduites s'évanouissent comme la lumière de la chandelle au soleil, toutes les fois qu'elles sont reçues sur un corps de matière ténue, transparent comme l'air, l'eau, le cristal, et le verre non assombri par l'obstacle d'un corps. À l'inverse, elles ne se fixent pas sur les corps trop denses et trop sombres, comme la pierre ou le bois. La nature, architecte suprêmement intelligente de l'univers, semble avoir conçu les yeux des êtres vivants sur le modèle du miroir, en appliquant au revers de la partie transparente une membrane noire, dont la disparition ôte aussitôt la faculté de voir. D'où la subtile sentence :

Tout comme l'œil est le miroir de l'art, le miroir est l'œil de la nature.

C'est vraiment merveille que le miroir reproduise si clairement les formes de tous les objets qui lui sont présentés, et qu'en outre, il rende les distances qui les séparent, leurs couleurs, leurs mouvements, en un mot, les choses elles-mêmes, j'irais jusqu'à dire plus vives qu'elles ne sont, avec cette seule différence qu'il les restitue inversées. La raison en est que la figure illusoire, émise par le corps solide et qui traverse l'air, est simple et pure, d'où advient qu'elle est identique dans un sens comme dans l'autre. Notre cher Ambrogio Leone de Nola a très savamment et longuement traité le sujet dans ses *Questions*.

Désormais l'art de peindre, après la maladresse des débuts, est parvenu à un tel degré de subtilité qu'il a pu abuser les yeux des hommes et même des oiseaux, et leur faire croire à la réalité de ce qui était peint de trompeuses couleurs.[34]

Premier développement : la définition, physique, du miroir. Eau, bronze poli, pierre précieuse, lame de verre, le miroir reproduit les images sans les conserver. Figures éphémères de l'original, les reflets introduisent une première comparaison : le miroir est un œil. Là se situe le miroir central de l'adage, dont l'axe de symétrie est constitué par une sentence, elle-même en miroir : 'Tout comme l'œil est le miroir de l'art, le miroir est l'œil de la nature'. Or la sentence commence par le reflet et termine par sa source. Elle donne en effet préséance aux deux médiations, le miroir et l'art, avant d'évoquer l'outil de la perception et le monde premier, l'œil et la nature. En mettant ainsi l'originel après sa repré-

[34] Trad. cit. II, 181–82.

sentation ou son image, la sentence escamote, tout simplement, la simplicité d'un regard naïf : il n'est d'œil que second, dans la section des comparés, dans la seconde partie de la phrase. Imitant le titre en partie double de l'adage, la sentence en fournit donc une variation en partie double, tout aussi énigmatique. Centrale et fondatrice, la citation relance l'adage, dans le reflet de lui-même. À l'éloge de la nature qui précédait la sentence, répond l'éloge du miroir qui la suit. Et aux citations classiques qui ouvraient l'adage, répond l'évocation d'Ambrogio Leone, savant ami et médecin d'Érasme dont les *Questions Naturelles* expliquaient la nature des miroirs. Voire, le dispositif des reflets se poursuit, car à l'évocation de l'ami, succède un court paragraphe sur les tableaux qui 'trompent l'œil', ici bouclant la boucle des maladresses de la répétition stérile, de l'ignorance qui refuse la médiation, et des illusions dont s'aveuglent les sots. Définition topique de l'image fidèle à son modèle et qui attire les oiseaux comme un vain reflet, les 'effets' (*effecerit*) de la peinture ressemblent au modèle, mais ne sauraient être le modèle. Une fois encore, c'est la dualité qui fait le tableau : seuls les oiseaux sans cervelle picorent les raisins de Zeuxis.

L'ami, cet autre moi

La convocation d'Ambrogio Leone dans l'adage sur le miroir introduit, comme dans un tableau, une ultime et essentielle dimension du reflet. Correcteur avec Érasme chez les Aldes[35] — au moment, donc, de la première rédaction des *Adages* –, médecin d'Alde Manuce, d'Érasme et du Cardinal Giovanni de' Medici (connu comme pape sous le nom de Léon X), Ambrosius de Nola (aussi appelé Ambrogio Leone da Nola, 1459–1525)[36] fut un ami cher et fidèle d'Érasme qui, à sa mort, fit son éloge.[37] Médecin, archéologue, philologue, prolifique auteur de traités scientifiques,[38] Ambrogio Leone da Nola apparaît nommément trois fois dans les *Adages*. L'adage de la troisième chiliade, 'Priusquam gallus cecinerit',[39] cite et commente l'explication 'naturelle' d'Ambrogio sur

[35] Voir Pastorello, *L'epistolario manuziano* , p. 281 et '*Aldina*', in Dazzi, *Aldo Manuzio e il dialogo veneziano de Erasmo*, p. 153–79.

[36] Voir Ammirati, *A.L. Nolano* et *'Salutate Messer Ambrogio'*, pp. 173–81.

[37] Voir Erasmus, *Opera omnia*, III, éd. par Jean le Clerc, pour ses *Epistolae Ad Baptistam Egnazium*, DCCLXIX (3 oct. 1525).

[38] On citera, entre autres, les titres *De Nola* ; *Castigationes adversus Averroem [...]* ; *De urinis* (trad. de Giovanni Attuario) ; *Novum opus questionum seu problematum*.

[39] *Adages* III, VII, 66 (éd. cit. p. 1853–54).

les particularités des œufs pondus avant le chant du coq. Symétriquement, en aval de l'adage sur les miroirs situé dans la deuxième chiliade, l'adage 'Δὶς διὰ πασῶν'[40] (*Bis per omnia.* Tout en deux.) est présenté comme un dialogue avec l'ami à propos des octaves et des harmoniques :

> De fait, alors que je noircissais mes cahiers de ces réflexions, je reçus la visite impromptue d'Ambrogio Leoni di Nola, le plus éminent philosophe de notre époque, doué d'un soin et d'une habileté incroyable pour pénétrer les mystères des sciences, et rompu à la lecture et l'examen précis des auteurs grecs et latins.[41]

Harmonique, octave, reprise, reflet, la conversation avec l'ami se fait au miroir des adages, pièces commentées servant de miroir de la parole vraie, celle qui montre l'âme. Dès lors, la définition de l'amitié, en exergue des *Adages*,[42] et son expression comme 'égale réflexion' du moi dans les adages qui suivent ces prémices, est celle de la reconnaissance, non pas dans l'identité mais dans la ressemblance, au miroir… Quand 'l'ami est un autre soi-même', c'est dans la médiation, du texte, qu'il faut lire son âme et donner la sienne à lire. Dès lors, les adages, pour la plupart en partie double entre latin et grec, et dans le dédoublement des échos et reprises, sont une conversation de part et d'autre du miroir des sentences : dialogue entre l'auteur et les auteurs classiques, qu'il cite et regarde 'comme dans un tableau', ils établissent également le dialogue du lecteur avec ces textes, donnés dans le cadre de la page 'comme dans un miroir'. Coquettes et aveugles ne verront dans la collection d'adages qu'un amas savant et divers : l'ami lecteur y verra la conversation d'un lecteur avec sa bibliothèque et, passant de l'autre côté de la lecture, entrera dans la danse des langues, réminiscences et commentaires. L'intertextualité sert alors de profondeur de champ et ouvre l'espace où le lecteur, à son tour, crée le texte. Citationnel et dialogique, le projet des *Adages* toujours en mouvement est ainsi celui d'un regard au miroir : reconnaissance de soi par la connaissance de la similarité. Au vertige des reprises et symétries, au miroir des médiations et de l'amitié, dans la forêt des *Adages*, le moi découvre son reflet en se faisant lecture.

[40] *Adages* I, II, 63 (éd. cit. p. 222–27).

[41] Trad. cit., I, 182–83.

[42] Le premier adage, sans cesse travaillé et augmenté, est en effet 'Amicorum communia omnia', suivi par 'Amicitia æqualitas : Amicus alter ipse', adage en partie double sur le reflet de soi que propose l'ami sincère. Sur la valeur et les implications de ce premier adage, voir Eden, *Friends Hold All Things in Common* et '"Between Friends All is Common"'. Sur la poétique de cette amitié, voir Cazes, 'Partage des *Adages*' et 'Démonstrations d'amitié et d'humanisme'.

Ouvrages cités

Sources primaires

Erasmus, Desiderius [Érasme], *Les Adages*, trad. et éd. par J.-C. Saladin, Collection 'Le Miroir des humanistes' (Paris : Belles Lettres, 2011)

——, *Adagia*, éd. par J. Le Clerc, in *Desiderii Erasmi Opera Omnia*, II (Olms : Hildesheim 1961) [Lugduni Batauorum, 1703]

——, *Adagiorum Chiliades*, éd. par M. L. van Poll-van de Lisdonk, M. Mann Phillips, Chr. Robinson (I), M. Szymanski (II), F. Heinimann, E. Kienzle (III), R. Hoven, et C. Lauvergnat-Gangnière (IV), in *Opera Omnia Desiderii Erasmi Roterodami* Ordinis secundi, tomus primus-octavus (Amsterdam : North-Holland, 1993–99)

——, *Adagiorum Collectanea*, éd. par Felix Heinimann et M. L.van Poll-van de Lisdonk, in *Opera Omnia Desiderii Erasmi Roterodami* Ordinis secundi, tomus nonus (Amsterdam : North-Holland, 2005)

——, *Collected Works of Erasmus. 31 : Adages, I Il 1 to I V 100*, trad. par Margaret Mann Phillips, annot. par R. A. B. Mynors (Toronto : University of Toronto Press, 1982)

——, *Epistolae*, in *Opera omnia III*, éd. par Jean le Clerc (Leiden : Vander, 1703)

——, *Parabolæ sive Similia*, éd. par Jean-Claude Margolin, in *Opera Omnia Desiderii Erasmi Roterodami* Ordinis Primi, tomus quintus (Amsterdam: North-Holland, 1975)

Nola, Ambrogio Leone da, *Castigationes adversus Averroem [...]* (Venice : B. e M. Vitali, 1517)

——, *De Nola* (Venice : G. Rosso, 1514)

——, *De urinis*, trad. par Giovanni Attuario (Venice : B. e M. Vitali, 1519)

——, *Novum opus questionum seu problematum* (Venice : B. e M. Vitali, 1517)

Sources secondaires

Ammirati, Luigi, *A. L. Nolano* (Marigliano : Anselmi, 1983)

——, *'Salutate Messer Ambrogio', Ambrogio Leone entre Venise et l'Europe* (= *Les Cahiers de l'humanisme*, 1 (2000)), pp. 173–81

Cazes, Hélène, 'Démonstrations d'amitié et d'humanisme : Alba, adages et emblèmes chez les petits-enfants d'Érasme', *Acta Conventus Neo-Latini Monasteriensis : Proceedings of the Fifteenth International Congress of Neo-Latin Studies (Münster 2012)*, éd. par Astrid Steiner-Weber et Karl A. E. Enenkel (Leiden : Brill, 2015), pp. 18–49

——, 'Partage des *Adages* : Un geste d'amitié humaniste en tête des *Adagiorum Chiliades* (1508–1533) d'Érasme', in *À la Recherche d'un sens : Littérature et vérité* : *Mélanges offerts à Monique Gosselin-Noat*, vol. I, éd. par Yves Baudelle et autres, Collection 'Actes' (= *Roman 20–50* (2014)), pp. 123–34

Eden, Kathy, '"Between Friends All is Common" : The Erasmian Adage and Tradition', *Journal of the History of Ideas*, 59.3 (1998), 405–19

——, *Friends Hold All Things in Common : Tradition, Intellectual Property, and the 'Adages' of Erasmus* (New Haven : Yale University Press, 2001)

Dazzi, Manlio T., éd., *Aldo Manuzio e il dialogo veneziano de Erasmo* (Vicenza : N. Pozza, 1969), pp. 153–79

Pastorello, Ester, *L'eistolario manuziano : Inventario cronologico-analitico 1483–1597* (Florence : Istituto per la Collaborazione Culturale, 1957)

Figure 8.1. 'Contre les blasonneurs des membres', from Gilles Corrozet, *Les Blasons domestiques contenantz la decoration d'une maison honneste* (Paris, 1539); Cambridge, MA, Harvard University, Houghton Library, Typ 515.39.299, p. 38r / E6r.

Mirror/Window, Reflection/ Deflection: Regulating the Gaze Inside and Outside the House in Gilles Corrozet's *Blasons domestiques* (1539)

Elizabeth Black

Gilles Corrozet, Parisian bookseller and author, published his *Blasons domestiques* with famed printer Denis Janot in 1539. The *Blasons* follow the model of the better-known anatomical blazons quite closely; each one is devoted to an individual room or piece of furniture found in a house, just as the blazons for which Clément Marot and Maurice Scève are known were written in praise of individual women's body parts. Corrozet's poems extol the virtues of the object or room in question, and they are also accompanied by woodcut images. Yet Corrozet writes his book of blazons as a response to his fellow poets, condemning their use of women's bodies for poetic material. At the end of the volume, he includes a poem entitled 'Contre les blasonneurs des membres' (Against Those Who Write Blazons of Body Parts), excoriating the writers of anatomical blazons for their licentious texts.

The poem is accompanied by an image of an artist painting a woman who is standing on the viewer's left (see Fig. 8.1). On the right is a group of three women who are other models for the painting. The image is adapted from one found in a series of editions of the *Roman de la Rose* produced by Clément Marot in France during the 1520s and 1530s. It illustrates the tale of Zeuxis, a

Elizabeth Black (eblack@odu.edu) is Assistant Professor of French at Old Dominion University. Her research focuses on domesticity and representations of the home in sixteenth-century French literature, visual culture, and architectural treatises. She has published on Marguerite de Navarre, Gilles Corrozet, Pierre Coustau, and the digitization of the emblem corpus.

The Mirror in Medieval and Early Modern Culture: Specular Reflections, ed. by Nancy M. Frelick, CURSOR 25 (Turnhout: Brepols, 2016), pp. 190–209 10.1484/M.CURSOR-EB.5.111048
BREPOLS PUBLISHERS

fourth-century Greek who painted Helen of Troy but could not find one model beautiful enough, so he painted parts of five different women to make a composite Helen with matchless beauty. Nancy Vickers has traced the use, reuse, and copying of the Zeuxis image across several volumes and by several printers and booksellers in Paris in the early sixteenth century, noting that Zeuxis was often invoked in discussions surrounding the poetic dismemberment of women's bodies in the blazons.[1]

Corrozet not only imitates the anatomical blazons' form, but also aims to replace them with his own poems. The gaze, invoked at every turn in the anatomical blazons, must therefore be diverted away from the female form, with the house proposed as an alternate source of viewing pleasure. Two glass surfaces — the windowpane and the mirror — negotiate and manipulate the ideal limits of the gaze both of outsiders and of the house's residents, attempting to establish clear boundaries between domestic and public space. The two concerns of privacy in the home and the writing of immoral poetry are linked: by allowing visitors into the home, they might gaze upon the women who live there, potentially turning them into fetishized, poetic objects. Corrozet is a bookseller; any business that takes place in his home includes the exchange of books containing poetic and engraved images of women's bodies. But the transparency of the window and the reflective nature of the mirror call these boundaries into question. Furthermore, the mirror obliges an interrogation of Corrozet's role as reactionary poet. At a time when relationship between subject and object is in flux,[2] it becomes increasingly unclear who is being gazed upon and who is actively gazing, and it becomes necessary to reconsider whether Corrozet's goal of ridding poetry of the spectacle of the body is genuine or even plausible.

The windows invoked are part of the description of the bedroom. The title of the blazon, 'Le Blason de la sasle et chambre' (The Blazon of the Reception Room and Bedroom), suggests that two rooms will be the focus of the poem — a bedroom ('chambre') and a 'sasle', which would be a reception room or meeting room.[3] A 'sasle' would often be included in suites of rooms in noble houses, but more modest abodes may not boast an entire suite, with the result that some rooms had to serve more than one purpose.[4] The bedroom often takes

[1] Vickers, 'The Unauthored 1539 Volume', pp. 176–79.

[2] De Grazia, Quilligan, and Stallybrass, 'Introduction', pp. 2–5.

[3] This is in contrast to the privy, whose blazon is entitled 'Blason de la chambre secrete, ou retraict', the 'ou' (or) indicating two titles for one room which is clearly stated to have one purpose.

[4] Architectural treatises printed in France during the sixteenth century confirm this building practice.

LES BLASONS

Le blaſon de

LA SASLE ET

CHAMBRE.

CHambre treſclere & bien quarrée,
Chambre au corps humain preparée,
Chambre baſtie d'ung maſſon

Figure 8.2. 'Le Blason de la sasle et chambre', from Gilles Corrozet, *Les Blasons domestiques contenantz la decoration d'une maison honneste* (Paris, 1539); Cambridge, MA, Harvard University, Houghton Library, Typ 515.39.299, p. 14v / B6v.

over the function of a meeting room and is precisely the place where the public intersects with the familial.

In Corrozet's blazon, the doubly named room seems to indicate that this is one room with a dual purpose. This is first apparent in the woodcut image, which clearly shows only one room (see Fig. 8.2). Moreover, the text of the poem also shows only one room by always making reference to the 'chambre'. The 'sasle' and its function are never mentioned, not even in passing. In the poem, therefore, the possibility of using this room as a reception room is erased from consideration. If the 'sasle et chambre' is poetically only a 'chambre' then it functions only as a bedroom and not as a reception room, and visitors have effectively been banished from the room.

The blazons' images up until this point in the volume have placed the reader inside the house, except for the initial 'Blason de la maison' (Blazon of the House), in which the reader seems to be approaching the house as if to pay it a visit. With the bedroom image, the reader's perspective is now placed outside the house, with a view into the room. If the reader is imagined as a visitor to the house taking a tour, then just as in the poem, visitors are again banished from the bedroom. Positioning readers outside and having them look in makes them interlopers and confirms the bedroom's status as a room that is not to be visited, that is, clearly not a 'sasle'. The readers' status as voyeurs is confirmed by the female figure on the left of the image who can be seen through an opening in the side of the wall of the house. This aperture follows the medieval artistic convention of cutting away part of a wall to see the interior of a building, but from the reader's perspective it functions as a window, allowing an illicit view of a personal bedroom scene. The female figure is framed by the opening, but this framing is also reminiscent of a painter's canvas on which she might be depicted. If we look more closely at her, we can see that she is similarly nude and holding the same drapery as the models of Zeuxis in the image which concludes Corrozet's volume in the poem 'Contre les blasonneurs des membres'. This visual echo to the ancient Greek painter's models, coupled with the figure's framing, casts her as a spectacle, a representation of her own nudity. Since she is equated with a painted figure, both through her framing and her pose, she comes to represent ideal beauty as expressed in Renaissance painting. She risks the same objectification as Zeuxis's models and all subjects of paintings and poetry that glorify the female form. She may be framed not only by the opening in the walls but also by poet-painters. As suggested above, a woman inside the home may be blazoned by visitors: viewed, imagined, dissected, and represented in parts, in words or images.

In addition to the woodcut's nude female figure on the left, on the right is another figure. Outside of the house, hovering next to the other aperture — this time, it is definitely a window — is a winged, bearded figure wearing a crown and carrying a sceptre or torch. He resembles the god of love, Amors, from various editions of the *Roman de la Rose*, having all the same features: wings, crown, and sceptre. But in none of the miniatures as yet examined is the god of love nude, nor bearded, nor is he depicted flying.[5] The figure here also shares attributes with a number of different mythological characters: the pudendum looks female, suggesting maybe Venus or Nike, but the face looks male and appears to be bearded. Hymenaeus, the Roman god of marriage, carries a torch. Thanatos, the Greek personification of death, is winged. Hades, Greek ruler of the underworld, is depicted as a bearded man with a crown, a sceptre, and a key. Venus can be depicted with sceptre, but she is never winged. Chantal Liaroutzos suggests that the figure is Cupid,[6] based on part of the blazon's text, which states:

> L'amoureuse deité
> De Cupido, à chascune heure
> Y vouldroit bien faire demeure[7]
>
> (The amorous god,
> Cupid, at every moment
> Would like to make his home there.)

Again, the attributes seen here do not correspond to depictions of Cupid common in the first half of the sixteenth century.[8] I propose that the figure, like Zeuxis's Helen or the book of blazons, is a composite one — part Cupid, part Venus, part Thanatos, part Hymenaeus, part Hades. Cupid, Venus, and Hymenaeus represent sexuality and its regulation, while Thanatos and Hades suggest damnation. The composite figure would therefore personify all the pos-

[5] My principal research into *Rose* imagery has been carried out using the 203 images depicting the god of love from the image database 'The *Roman de la Rose* Digital Library' <romandelarose.org> and within the sixteenth-century illustrated versions of Marot's translation available on the Gallica's website for the Bibliothèque nationale de France <gallica.fr> [both sites accessed 29 December 2015].

[6] For Chantal Liaroutzos's analysis, see her article: 'De pièces et de morceaux', p. 51.

[7] Gilles Corrozet, *Blasons domestiques*, fols Bviir–Bviiv. Note: All translations are my own.

[8] My conclusion is based on images of Cupid found in the canonical French emblem books which can be consulted at 'French Emblems at Glasgow' <http://www.emblems.arts.gla.ac.uk/french/> [accessed 29 December 2015].

sibilities of the bedroom; site of conjugal union, but also site of lust and sin. In emblematic fashion, allusions to the god of love must be taken into consideration, especially if we bear in mind Corrozet's other borrowings from the *Roman de la Rose* such as the Zeuxis image. The god of love threatens the chastity of the rose, resulting in a final assault on the tower where she is guarded, and a breach of the walls that hold her. The intertext provides a commentary on the ability of walls — here specifically the walls of the family home — to provide the kind of protection needed for female members of the household. The woodcut places both the reader and the winged figure in the same space. The figure's presence therefore threatens the home's stability and the female figure's purity, just as much as the reader's does.

Visitors to the home are interlopers who should not have access to the site of conjugal harmony and sanctioned sexual mores. In order to turn the reader's gaze away from the female body, the poem of the blazon focuses, as so many do, on the physical attributes of the room itself: the walls, the floor, the decoration. When the windows are mentioned, their function is ignored, while their beauty is praised: 'Chambre dont les vitres sont telles | Qu'on n'en vidt jamais de plus belles' (Bedroom whose windowpanes are such | That more beautiful ones have never been seen).[9] One is not expected to look *through* these windows, one is expected to look *at* them in order to admire them. The reader's gaze is to be arrested at the limit of the walls of the house, and not trespass inside. The text draws attention to the different depths of field within the image and instructs the reader's eye to focus on the plane on which the walls and window sit. If the reader is looking beyond the walls and through the window, rather than looking at the window, as the poet helpfully instructs, the reader may see something forbidden. The text demands an immense act of self-control on the part of the readers, asking them not to look at a naked body placed directly in front of their eyes. It seems to be the reader's responsibility to avoid looking at the omnipresent representations of women's bodies.

After presenting the bedroom from outside the building, the reader-visitor is taken back inside, into the bedroom to inspect its contents. The first object encountered is the bed. The female figure who was visible through the window from outside the house has disappeared. Having suggested the potential dangers of allowing visitors into the bedroom, the next blazon removes those suggestions, showing a sanitized space that the eye can contemplate without temptation. The bedroom ceases to be a place where women's bodies are available or

[9] Corrozet, *Blasons domestiques*, p. Bviir.

LES BLASONS

Le blaſon de

L'ESTVY DE CHAMBRE.

ESTuy de fin veloux couuert
De cramõyſi, de bleu ou vert
Eſtuy de marroquin paré
Eſtuy tant bien faict & doré
Eſtuy ou pignes ſont dedans,
A groſſes & menues dentz

Figure 8.3. 'Le Blason de l'estuy de chambre', from Gilles Corrozet, *Les Blasons domestiques contenantz la decoration d'une maison honneste* (Paris, 1539); Cambridge, MA, Harvard University, Houghton Library, Typ 515.39.299, p. 27v / D3v.

on display. If there is to be a visitor in the bedroom, which is inevitable if the room doubles as a reception room, then the visit will not coincide with any time when a woman might be present. The exchange of representations of women's bodies or their body parts will not be allowed in this most intimate of spaces.

The window is one of two glass objects in the volume, both of which manipulate the gaze of the reader. The other is the mirror, which can also be considered to be located in the bedroom if we group together the suite of objects immediately following the blazon of the bedroom — bed, chair, bench, table, dresser, chest, stool, broom, comb-case, mirror, cabinet — until we clearly change rooms and move into the stable.[10] In Corrozet's sequence of blazons, the mirror follows directly after the comb-case. This object is clearly coded as a man's since it contains items for tending the beard. It is exaggerated in size, depicted in a landscape, and seemingly independent of any human interaction (see Fig. 8.3). The original French title given to the blazon — 'Blason de l'estuy de chambre' — identifies the object with the bedroom even though the comb-case's woodcut depicts it not only outside of the room in question but away from the house altogether.

In contrast, the blazon of the mirror once again takes the reader back inside the house. The mirror is also drawn in relation to a landscape, one that recalls the landscape in which the comb-case is situated, but the mirror is placed inside the house, in an archway that leads to the outside (see Fig. 8.4). It stands in a liminal space between the interior and exterior of the house, with the reader placed in front of it. It also blocks most of the view of the landscape outside. Another game of vision and viewing is set up. The mirror seems placed to play again with the reader's consciousness of depth of field. Just as the reader was instructed to look at the window's panes of glass to admire them instead of looking through them to the bedroom inside, the mirror invites the reader's gaze towards its glass as opposed to beyond it and through the arch to the world outside. A second time, the gaze should not go beyond the wall of the house; instead, it is to focus on the plane which is aligned with the house's walls, and on which the mirror sits. Taken together, the 'Blason de la sasle et chambre' and the 'Blason du miroir' give a clear indication that the apertures in a house's walls are not for looking through, either from outside the house or from inside, marking a strict dividing line between public and familial.

[10] This grouping is justified based on some references in these blazons to the idea that they belong to the bedroom. The cabinet, although a small room and not an object, would be accessible through the bedroom and would not necessarily have any other access point; it would be considered even more intimate than the bedroom.

LES BLASONS

Le blaſon du

MIROIR.

Miroir cler & reſplendiſ-
ſant,
Miroir plaiſant, reſiouyſ-
ſant,
Miroir ardant de grand ſplendeur,

Figure 8.4. ‘Le Blason du miroir’, from Gilles Corrozet, *Les Blasons domestiques contenantz la decoration d’une maison honneste* (Paris, 1539); Cambridge, MA, Harvard University, Houghton Library, Typ 515.39.299, p. 28v / D4v.

Although this division of space is constructed mostly visually in the *Blasons domestiques*, Corrozet's subsequent collections of emblems, *L'Hecatomgraphie* (1540–44) and the emblems in *Cebes* (1543) reprise these discussions of the moral dimensions of the family house, particularly with respect to gender, and more clearly state his positions regarding the circulation of male and female inhabitants, the status of visitors in the home, and the problem of women's visibility.[11] Across both the blazons and the emblems there is a consistency of message guiding all family members' attention inwards to the home, disallowing women's appearance in public space, and warning heads of households about the dangers of inappropriate visitors. The only figure to appear in the house is the naked woman in the woodcut from the 'Blason de la sasle et chambre'. It is presumably her body and her gaze that should stay inside.

The mirror does not only stop the gaze at the surface of the house. It also throws the gaze back into the room, deflecting it away from the wall's plane, away from the view through the arch, and back at the gazing subject. The poem states:

> [A] chascun tu monstre sa forme
> S'elle est belle, laide ou difforme,
> Et ne refuse en ta clarté
> D'aulcun la laidure ou beaulté.[12]
>
> (To all people you show their form
> Whether beautiful, ugly or deformed
> And do not exclude from your light
> Anyone's ugliness or beauty.)

This time the mirror is indiscriminate, with the indefinite pronoun 'chascun' indicating that anyone can look in the mirror and see him or herself, and indeed there is nobody depicted looking in the mirror; the gazing subject is absent from the woodcut image. At first glance, and before alighting on the accompanying verse, it is the reader who finds him or herself looking into the

[11] Corrozet's 1540 *Hecatomgraphie* contains a cautionary note to women, which fits in with the general message of the *Blasons domestiques*. Entitled 'La Statue de Caia Cecilia' (fol. N8v), it warns that women should not leave the house. If a woman is 'pudique' she will be 'domestique', literally of the home. The danger of leaving the house is that the woman might 'mielx montrer son corps' (better show off her body). In the same volume, the emblem 'Nature foeminine' (p. L7v) equates women with flighty birds who care little for public reputation and whose freedom to leave the house should be restricted. For more in-depth analysis of the notion of domesticity in Corrozet's emblem works, see my article: Black, 'Gilles Corrozet's "Domestic" Emblems'.

[12] Corrozet, *Blasons domestiques*, p. Dvr.

mirror. Yet the poem does specify a particular viewer: a beautiful, unnamed woman. The mirror is:

Ou la belle, plaisante, & clere
Se void, se mire, & considere
En regardant sa contenance
Et de son gent corps l'ordonnance.[13]

(Where she who is beautiful, pleasing and radiant
Sees herself, looks at her own reflection, and reflects
As she looks at her face
And at the arrangement of her noble body.)

Linked by the adjective 'clere' to the mirror's own 'clarté', the woman is set up as the counterpart to the mirror's reflective surface, its brilliance matched by her own. The gaze here is again female, reinforcing the idea that not even the woman's gaze, let alone her physical body, should be allowed outside of the house, as opposed to the man who owns the comb case who can roam freely.[14] The blazon's poem continues with a detailed description of the woman's body:

Ses yeulx scintillans & sa face
Son fronc poly, sa bonne grace,
Sa doulce bouche vermeillette.
Son menton qui faict la fossette
Son dur tetin, ses bras gentilz
Ses blanches mains, ses doigs traitifs
Et tout le reste de son corps,
Dont les membres sont bien concordz.[15]

(Her sparkling eyes and her face
Her shining forehead, her good grace,
Her sweet red mouth.
Her dimpled chin
Her firm breast, her graceful arms
Her white hands, her slender fingers
And all the rest of her body,
Whose parts are in harmony with each other.)

[13] Corrozet, *Blasons domestiques*, p. Dvr.

[14] As will be explained in the *Hecatomgraphie*, Corrozet's emblem book published a year after the *Blasons domestiques*, women should remain inside, even though — and especially because — they desire nothing but to be as free as the birds.

[15] Corrozet, *Blasons domestiques*, p. Dvr–v.

Her prolonged contemplation of herself leads naturally to a comparison with Narcissus. The exhortation to avoid mirrors is often stated in Renaissance texts, with the implied risk of self-obsession, commonly held to be a vice. Mirrors were considered particularly dangerous for women, whose minds were thought to be weaker than men's. Narcissus could not tear his gaze away from the stream in which he saw himself mirrored, to his own demise and that of poor Echo. Yet the fountain of Narcissus as described in the *Roman de la Rose*, from which Corrozet borrows so many references and images, does not reflect the wandering lover's own face but shows him the location where his love object is to be found. Here, too, Corrozet's mirror not only reflects the woman's body back at her, but also projects her reflection so that the reader can see it by reading the itemized description of her body parts. What was a clear gender identification with the female occupant of the bedroom becomes less stable as the sight-lines shift. While self-obsession is Narcissus's vice, the mirror offers another possibility: self-contemplation that morphs into contemplation of the beloved. As remarked above, the image shows neither the woman nor her reflection.

The reader therefore finds him or herself in front of the mirror and sees not his or her own reflection but the view of the beautiful woman described in textual form. The reader is placed in her position inside the bedroom. The juxtaposition of text and image in this case compounds the woman's presence in the bedroom in order for her reflection to be seen and the reader's presence in order to see both the mirror and the woman's reflection. Visitors, despite the conclusions we could take from the blazons of the bedroom and bed, do find their way into the bedroom at the same time as its naked female occupant.

The reader is not the only other character who might look in the mirror — the poet-narrator also places himself in front of it, but fears what he might see. The sin of lasciviousness that Corrozet associates with gazing upon the female form takes on a personal note following the lengthy description of the female body in the 'Blason du miroir'. The poem ends with a direct address to the mirror by the poet, in which he enlists its help in becoming a model of virtue:

> O Miroir je te prie cache
> De mon corps la laidure ou tache,
> Et de l'ornement de vertu
> Me feray beau & bien vestu.[16]

[16] Corrozet, *Blasons domestiques*, p. Dvv.

(O Mirror, I pray you hide
The ugliness or stain of my body
And with the ornament of virtue
I will make myself beautiful and well-dressed.)

The Fountain of Narcissus from the *Roman de la Rose* is again called to mind; the narrator looks into the mirror, sees his beloved, but recognizes in what he sees his own lustful nature. His fear is that the mirror will show not her body but his, and with it the truth of his character. Mirrors were thought to reveal truth — an idea that dates back to Plato — and it was believed that bodily appearance reflected the condition of the soul. The woman reflected is naked but perfectly beautiful, demonstrating her own purity, while the poet knows his own lack of virtue will be shown in the mirror by his ugly reflection.[17] Woman's naked body again finds its place in the bedroom, while the poet's act of looking upon it is marked with the language of lust and sin.

Returning to the description of the woman, it is made up of a list of individual body parts.[18] Although the vast majority of anatomical blazons were dedicated to a single part of the body, there was at least one, possibly written by Francis I, which praised a female body divided up into such a list as included here.[19] In addition, Charles de la Hueterie's first *contreblason* vilifies an entire body made up of a list of individual parts.[20] Thus, as observed by Chantal Liaroutzos,[21] the volume which ostensibly opposes the practice of blazoning the body contains an anatomical blazon, and one which conforms to the original project outlined by Marot — to praise beautiful bodies. Furthermore, the image of the mirror contains a visual echo of the original blazon in praise of one part of a woman's body. The mirror's rounded glass, shaded with lines to

[17] Suzanne Conklin Akbari also addresses issues of Narcissus and seeing the self in the other or the other in the self. Unlike her example, however, Corrozet's poet-narrator will not find his redemption in the contemplation of the other; rather, he is afflicted with his own corruption when he beholds his beloved.

[18] As Sabine Melchior-Bonnet explains in her *Histoire du miroir*, the technology to produce glass mirrors larger than one metre did not exist at this time, and the vast majority of households would have a mirror no bigger than forty centimetres, and most would be made of metal. Most mirrored images of a woman in reality would be necessarily fragmentary; no mirror would be large enough to contain the whole woman and would therefore only be able to show one part at a time unless from a great distance (pp. 31–33).

[19] Vickers, 'Members Only', p. 4.

[20] Vickers, 'Members Only', p. 13.

[21] Liaroutzos, 'De pièces et de morceaux', p. 50.

show its convexity, would be a close likeness of the woodcut image accompanying Marot's 'Blason du beau tetin' (Blazon of the Beautiful Breast) if it had the nipple added to its centre.[22]

The woman who stands in the bedroom and looks in Corrozet's mirror is framed twice — once by the mirror frame, and again by the frame of the archway. This double-framing has already been observed in the blazon of the bedroom, in which the naked woman, already framed by Zeuxis as a painted nude, is framed anew by the aperture in the wall. The woman's reflection in the mirror therefore resembles a painted image — a fitting analogy for what is a poetic representation of the body, in an era when poets in the garden of poetry were equated with painters such as Zeuxis and Apelles.[23] But Corrozet still claims to object to this type of objectification. While the female figure starts out as the gazing subject, the sight-lines shift from the woman's eyes, to the reader's, to the narrator's. Her status as subject — as the person standing in front of the mirror and viewing herself — casts her equally as an object to be viewed by others — by both the reader who reads and the narrator who composes the blazon.

As David Hillman and Carla Mazzio explain in their introduction to *The Body in Parts*, the collected volume in which Vickers's article about the anatomical blazons appears, parts of the body, when isolated, become 'objects of libidinal cathexes',[24] and Chantal Liaroutzos argues that replacing the body with a house does not dislodge the desire that is activated by the gaze, it simply displaces that desire on to a set of replacement objects, with the net result that the journey through the fictional house becomes an extended metaphor for the conquered female body.[25] The mirror, in the same manner as all other objects in the house, is the recipient of the libidinal cathexis that parts of the body receive in the anatomical blazons, since the house as a whole replaces the presence of the female body according to Liaroutzos's reading.

I propose that the mirror occupies a special juncture between the anatomical and the domestic blazon projects because of its unique ability to both contain and reflect an image of the body. It is both the object to be observed and

[22] This resemblance to Marot's 'tetin' is complementary to Tom Conley's observation in his keynote lecture that the shading lines on Corrozet's mirror resemble both a half-moon and a fingerprint. In the spirit of the best emblematists, the semiotics of Corrozet's imagery is richly multiple.

[23] Vickers, 'The 1539 Unauthored Volume', p. 170. See, more generally, Lecercle, *La Chimère de Zeuxis*.

[24] Hillman and Mazzio, 'Introduction', p. xii.

[25] Liaroutzos, 'De pièces et de morceaux', pp. 51–53.

simultaneously presents and contains the body it replaces. Cathexis is deflected away from the body physical to the body reflected, but it is still directed to an image of the body. Consequently the mirror is both a marker of the body's absence but also an enabler of its presence, albeit in virtual form.

I observed above that the female figure becomes both subject and object through her gaze and the projection of her image to be gazed upon. Margreta de Grazia, Maureen Quilligan, and Peter Stallybrass, in their introduction to *Subject and Object in Renaissance Culture*, propose the Renaissance as an exceptional period when the status of subjecthood and objecthood are in flux, countering the Burckhardtian model of the subject coming instinctively into existence prior to its awareness of the objects it then dominates: 'If the subject (or author or painter) is no longer assumed to be prior to and independent of objects, criticism can attend to a dialectic in which subjects and objects reciprocally take and make each other over.'[26] The mirror as object 'makes over' the woman's subjecthood, untying it from her and bestowing it on anyone who sees her reflection in the blazon's text. The reader's subjecthood is accidental to the logic of textual production and distribution, but the narrator's is the result of shifting perspective and agency. When he looks into the mirror, the mirror itself is given agency in the lines 'O Miroir je te prie cache | De mon corps la laidure ou tache' (O Mirror, I pray you hide | The ugliness or stain of my body), following the proposition that the mirror holds the ability to choose whether or not to show something in its reflective surface. It is only when the (subjective) poet-narrator stands in front of the object (the mirror recast as acting subject) and becomes the reader of an image that he emerges at all in the poem as something other than an omniscient narrator. Ironically, his visibility as desiring subject — indicated by the ugliness and stain on his sinful body — is constituted through the mirror's ability to turn him into an object to be gazed upon as it did to the female figure.[27]

It is only in gazing upon the other that the poet sees himself, in a reversal of the Fountain of Narcissus, in which the poet gazes upon himself and sees the other. The other becomes the mirror for the self. But in this case it is not her person but her reflection that provides the locus for self-contemplation. If in the mirror the self is seen as other, then the narrator sees himself in the other othered, for he is not gazing directly at the woman as other but at the mirror image that she also sees as other. Nonetheless, the gaze deflected away

[26] De Grazia, Quilligan, and Stallybrass, 'Introduction', p. 8.

[27] Suzanne Conklin Akbari also discussed the concept of the construction of the self through the gaze into the mirror and the contemplation of the other.

from the physical body to the virtual body does not solve the problem of lustful desire. Poetic and reflected bodies are as dangerous as real ones, echoing the idea found in Corrozet's 'Contre les blasonneurs des membres' at the end of the *Blasons domestiques* that while the names bestowed on body parts may be harmless, the verbal painting of them should be avoided. Such contemplation of textual objects reveals a self that is to be feared, for it is in this image that the sinful truth is revealed. Unlike De la Hueterie who blames the bodies themselves for the sins they elicit, Corrozet's text reveals a self-reflexive fear concerning his own blameworthiness.[28]

It should not be forgotten that Corrozet's volume of domestic blazons aims to replace the collection of anatomical blazons that he apparently found so abominable. However, writing within the tradition, even in order to oppose it, will not allow him to escape it. Nancy Vickers observes the same phenomenon in the case of De la Hueterie. Vickers has argued that the blazon competition established by Marot was less about the establishment of a hierarchy of body parts, or the adoption of a position in praise or in blame of a body part, since epideixis, the rhetorical procedure to which blazoning belongs, is the act of both praising and blaming:

> This significant casting of the function of epideixis not as rhetorical praise *or* blame [...] but as poetic praise *and* blame [...] is at the heart of the blazon enterprise. [...] Clearly, the part described is postulated as secondary to the artful making of rhetorical moves.[29]

Rather it is a process in which to exhibit poetic prowess and through which the poetic self can make itself manifest:

> The fact that the visual and rhetorical fragmentation and bifurcation of the female body was explicitly constitutive not only of the competition between the *blasonneurs* but, ultimately, of the collective product titled by 1543 the *Blasons anatomiques du corps fémenin* almost literalizes the logic whereby the self is collected, defined, and displayed through the *process* of splitting and exhibiting the other.[30]

Corrozet, in adopting the same rhetorical procedures as Marot and others enters into the blazoning game, and his overt references to the anatomical *blasonneurs*, even to condemn them, inscribe him further in that tradition. Furthermore, the emergence of the poetic persona as a desiring subject, in an

[28] Vickers, 'Members Only', p. 13.

[29] Vickers, 'Members Only', p. 12.

[30] Vickers, 'Members Only', pp. 13–14.

example of a praising blazon found in the midst of a volume designed to blame the whole genre, grounds Corrozet firmly in the epideictic tradition like Marot who sought both praising and blaming blazons from his contemporaries.[31] The overt manifestation of the self 'through the *process* of splitting and exhibiting the other' is literally written out on the page in the figure of the mirror which first describes the female body reflected in pieces and then threatens to show the poetic narrator his own body, revealing the poetic desire that he seems so desperate to hide behind vitriolic rhetoric and what can be considered banal, everyday subject matter.[32]

That subject matter turns out not to be so banal after all given the capacity to read it as an extended metaphor for the body,[33] and the mirror's bridging of the gap between body and material object. In addition, the very choice of the house as the replacement for a collection of body parts does not move him away from discussions of the body but takes him to the heart of the debate about women's visibility and circulation in both public and domestic space, since the house is the very setting in which these debates are played out, as emerges more strongly in Corrozet's emblematic works.

In conclusion, the texts and images in the *Blasons domestiques* stage complex relations between outside and inside, between housing and being housed, between body and spectacle, between writing within a tradition and attempting to escape from it. The visitor's gaze from the outside must be arrested at the exterior limit of the house. As Corrozet will state later in his emblem books, the woman's body must remain inside the home, but here the mirror also keeps the gaze within the confines of the house — it must not go beyond the window frame, and is actually cast back inside.

Corrozet's carefully constructed notions of inside and outside, dividing worldly lasciviousness from chaste home life, is undermined by the lack of division between public space and family space — *sasle* and *chambre* are stubbornly one room. The bedroom, and the bed itself, cannot help but suggest the subject he claims to be trying to avoid. The need for some kind of space where other people will not intrude upon the family is suggested by the difficulty of negotiating the dual function of the bedroom. The notion of privacy — time and space to oneself — is far from being established, but will come in the following century.

[31] Vickers, 'Members Only', p. 17.

[32] For a full discussion of the various subjects treated by blazon writers throughout the sixteenth century, see Saunders, *The Sixteenth-Century 'Blason Poétique'*.

[33] Liaroutzos, 'De pièces et de morceaux', pp. 51–53.

Yet Corrozet also thwarts his own stated project by the very act of including images of the female form in the blazons. The exhortation to look at the window instead of the female body behind it feels like a taunt — a trap set by the text. Corrozet's book containing these images will enter into people's houses; he perpetuates the cycle of representation and circulation that he claims to abhor. This discrepancy suggests that the slippage between family life and the business world, both hosted in the bedroom, is unavoidable, with Corrozet admitting the impossibility of the task he has set himself. In order to combat poetry, which induces lust, he is obliged to mention it, caught between a desire to eliminate references to the body and the inability to escape fully from them. As long as domestic space remains both public and familial, Corrozet's task to do away with representations of the female form there will be an impossible one, and even a disingenuous one. His approach must therefore be pragmatic: admit that the unmentionable is unavoidable, and develop diversionary tactics.

Works Cited

Primary Sources

Corrozet, Gilles, *Les Blasons domestiques contenantz la decoration d'une maison honneste, & du mesnage estant en icelles: Invention ioyeuses, & moderne* (Paris: [E]n la boutique de Gilles Corrozet Libraire, 1539)
——, *Hecatomgraphie* (Paris: Denis Janot, 1540)
——, *Tableau de Cebes* (Paris: Denis Janot, 1543)

Secondary Studies

Black, Elizabeth, 'Gilles Corrozet's "Domestic" Emblems: Gender and Ethics at Home in the *Hecatomgraphie* and the Emblems in *Cebes*', *Emblematica*, 20 (2013), 111–40
Grazia, Margreta de, Maureen Quilligan, and Peter Stallybrass, eds, 'Introduction', in *Subject and Object in Renaissance Culture*, ed. by Margreta de Grazia, Maureen Quilligan, and Peter Stallybrass (Cambridge: Cambridge University Press, 1996), pp. 1–13
Hillman, David, and Carla Mazzio, eds, 'Introduction', in *The Body in Parts: Fantasies of Corporeality in Early Modern Europe* (New York: Routledge, 1997), pp. xi–xxix
Lecercle, François, *La Chimère de Zeuxis: Portrait poétique et portrait peint en France et en Italie à la Renaissance* (Tübingen: Narr, 1987)
Liaroutzos, Chantal, 'De pièces et de morceaux: *Les Blasons domestiques* de Gilles Corrozet', *Littérature*, 78 (1990), 46–53
Melchior-Bonnet, Sabine, *Histoire du miroir* (Paris: Imago, 1994)
Saunders, Alison, *The Sixteenth-Century 'Blason Poétique'* (Bern: Lang, 1981)
Vickers, Nancy, 'Members Only', in *The Body in Parts: Fantasies of Corporeality in Early Modern Europe*, ed. by David Hillman and Carla Mazzio (New York: Routledge, 1997), pp. 3–22
——, 'The Unauthored 1539 Volume in Which is Printed the *Hecatomphile*, *The Flowers of French Poetry*, and *Other Soothing Things*', in *Subject and Object in Renaissance Culture*, ed. by Margreta de Grazia, Maureen Quilligan, and Peter Stallybrass (Cambridge: Cambridge University Press, 1996), pp. 166–88

A Mosaic Mirror: Fracture and Fault Line of Ronsard's *Amours* of 1552

Tom Conley

Diligent readers of the first edition of Ronsard's *Amours*, a slim volume published in 1552, intuit quickly what must have been the fuss that came with its reception.[1] The poems lead us everywhere, all about and around the classical and contemporary world, from the Aegean archipelago to the author's homeland in the Touraine. Some seduce, others resist, a few scintillate, others baffle. As he notes in the final sonnet, the author himself, of a proudly grandiose ego, launches a 'French' Petrarch and a new art and science of verse that will assure him a chapel in the pantheon in which are inhumed the greatest poets of all time, to which the visual rhetoric of the renowned frontispiece (Figure 9.1) attests: following the title page, a frontispiece, on the verso and recto of two folios, offers a double portrait of the poet and his beloved, seen in profile, who gaze upon each other across the frames in which their heads and busts are confined. Ronsard, wearing laurels in the style of Petrarch and set in a profile reminiscent of a Grecian demigod, casts his gaze toward the beloved, whose bare virtues, we soon discover, will be the subject of his praise.[2]

[1] For reasons that bear on the visual presentation of the verse, reference is made to the first edition, *Les Amours de P. de Ronsard Vandomoys: Ensemble le cinquiesme de ses odes.*

[2] The oval surround, whose ellipse would distort the images they contain, reminds the reader that it could be a distorting glass in which Ronsard and Cassandre replace Petrarch and

Tom Conley (tconley@fas.harvard.edu), Lowell Professor of French and Visual Studies at Harvard, has recently authored *An Errant Eye* (2011) and *À fleur de page: Voir et lire le texte de la Renaissance* (2015). With T. Jefferson Kline he has co-edited Wylie-Blackwell's *Companion to Jean-Luc Godard* (2014).

The Mirror in Medieval and Early Modern Culture: Specular Reflections, ed. by Nancy M. Frelick, CURSOR 25 (Turnhout: Brepols, 2016), pp. 211–233 10.1484/M.CURSOR-EB.5.111049
BREPOLS PUBLISHERS

Figure 9.1. Frontispiece, *Les Amours de P. de Ronsard Vandomoys : Ensemble le cinquiesme de ses odes* (Paris: Chez la Veufve de Maurice de la Porte, au clos Bruneau, à l'enseigne S. Claude, 1552); Cambridge, MA, Harvard University, Houghton Library, *FC R6697.552a.

Ὡς ἀπὸ ῥωνσάρδου
εἰς τὴν κάσσανδραν.

Φοιβάδα τὴν Κάσσανδραν, ἔρως τὸν ἑταῖρον ἐκείνης,
Φοιβομανῆ τεῦξεν φοῖβος ἐρωμανέων.
Ἡ δ' ἄλλη Κασσάνδρη Κελτίδος, οὐκέτι φοιβάς,
Νῦν ἔμ' ἐρωμανέα ῥέξ' ἰδὲ φοιβομανῆ.

Ἰα. Ἀντω. Βαΐφιου.

The reader is in several places at once — in his native land, the *terroir* of the garden of France, his *Vendomois*, but also in Italy, in the classical world, and yet less immediately in Paris, where the book is published. And so too, drawn from a variety of classical and Italian sources, the poems that follow are set in a broad geographical and historical frame.

A collection that would be the *speculum* on which an *amorous universe* is reflected, at once a cosmogony and a cosmography of love, can be appreciated as a composite topography, a series of parcelled reflections that in the book where they are printed comprise a paradoxically cohering scatter. That design seems protean because a reader cannot move through the sum of 185 poems in the first edition without having to consult myriad works of reference in the classical and Italian canons. It is implied that readers seeking to obtain the sense of a sum are either deficient or in need of the assistance that Marc-Antoine de Muret provides with ample commentary in the edition published the following year. In 1552, however, Ronsard suggests that his readers ought to 'educate' themselves enough to appreciate the fabulously recondite iconography, the tortuous turns and 'mannered' mode of the poems. The reader who reads the verse from beginning to end discovers that poems of different sort and circumstance fill out the final pages. It becomes clear that the collection is open-ended and composed so as to be subject to change. And notably for reason of the force of the collection's form: contrary to modern editions, not set under Roman or Arabic numerals, the poems are arranged in a quadrature. Given that the sonnets are set over another on each folio, when the book is opened wide we see an arrangement of four sonnets, each in its integral form (never do the lines flow from one page onto the next as they do in most anthologies), conveying a sense of what could be an architecture, a garden-plot, or even a disposition of island-like units juxtaposed to one another.

For the purpose of the collection of essays in which these words are placed, I would like to suggest that the *Amours*, first in 1552 and soon after in the edition of 1553, especially in respect to the final poems in the last pages, resemble something that might be called *a mosaic mirror*: a sum of reflective fragments conveying affective shatter, a work that illuminates and sparkles with the light and fire of wit while dealing with the tribulations at work in the world at the time of its writing. The poems can be imagined as surfaces that focalize, reflect,

Laura, who had been shown apposite each other in a heart-shaped frame in recent editions of the *Canzoniere*, most evidently *Il Petrarca* (Lyon, 1547). Studying the way the poems are shaped in the collections, Cécile Alduy saliently develops the connection in *Politique des 'Amours'*, pp. 320–25.

and refract the intellectual energies — invention, creative drive, force of desire — on the part of readers who invest their physical and intellectual energies in them. Given that the mirror, an object that figures in the poetry, can be implied to be emblematic of the collection, wherever it happens to be found, it would be analogous to an *imago mundi*, an image of the greater world that would have strong geographic latency. Because they mix the space and place of their origin with the Medieval and Classical worlds of their sources and avatars, the poems belong to a tradition in which poetry is infused with geography and topography, and where the concept of a *speculum mundi* gives way to reflection on the world as a loose aggregate of singularities.

The *Amours* remind us that Ronsard's relation with geography runs through all of the writings, from praise of the landscape in his homeland in the early odes to developed reflections on shape and substance of the world. A cursory overview shows how and why. Most famously, albeit later in his career, Ronsard's elegy to Nicolas de Nicolaï, a preface to the author of the *Quatre premiers livres des Navigations et Peregrinations orientales* (Lyon: Guillaume Roville, 1567–68), elegantly betrays grounding in the geography of Ptolemy. Praise is cast on Nicolaï, for whom the globe, of infinite variety, whets our curiosity and a desire to 'see and visit' the world at large. Ronsard writes that his friend, an engineer, cartographer and diplomat, brings the fruits of travel to his diligent readers:

> Ains les divers portraitz tu nous monstres au vif,
> Des temples, des chasteaux, des regions entieres,
> Des palais, des citez, des ports et des rivieres,
> Par tout où tu passois ne laissant rien de beau
> Sans le representer en ton docte tableau

(these diverse portraits you show us so vividly, temples, castles, entire regions of palaces, of cities, of ports, and of rivers: wherever you went leaving nothing of beauty without displaying it in your learned picture.)[3]

Earlier, in 1558, he had praised Nicolaï's copperplate map of Boulogne in conjunction with Henri II's recent wresting of the region from the English, in which the poet emphasizes how the map allows the reader to see all the parts and pieces of the area:

[3] 'Elegie à N. de Nicolay', in Ronsard, *Œuvres complètes*, ed. by Céard, Ménager, and Simonin, II, 425–27, ll. 83–88. Here and elsewhere all translations from the French are my own. See Louisa Mackenzie's treatment not only of Ronsard and Nicolaï but also of the greater relation of cartography and poetry in her magnificent *Poetry of Place*, pp. 41–50 and elsewhere.

Vous verrez la grandeur, les places, et les forts
Du Boulongnois, et d'Oye, et les ports,
Monts, fleuves, et forests, qui s'ejouissent d'estre
Reduits de sous la main de leur ancien maistre.[4]

(You'll see the grandeur, the places, the fortresses in Boulogne, and in Oye both the ports, mountains, rivers and forests which take joy in being brought back under their former master.)

In the same year, although published in the second volume of his *Meslanges* (1559), as if in dialogue with Du Bellay's famous thirty-first sonnet in the *Regrets* of 1558 on exile ('Heureux qui, comme Ulysse [...]'), Ronsard writes that he has had enough of Ulysses' island-hopping in the Aegean:

Qu'on ne me vante plus d'Ulysse le voyage,
Qui ne vit en dix ans que Circe ou Calypson,
Le Cyclope, et Sylla qui fut demy poisson [...][5]

(May I be spared of praise for Ulysses whose voyage in ten years witnessed only Circe or Calypso, the Cyclops, and Sylla who was a half-fish [...])

In 1560, in the first edition of his *Œuvres complètes*, Ronsard includes a sonnet dedicated to cosmographer André Thevet that varies on the same line:

Si du nom d'Ulysses l'*Odyssée* est nommée,
De ton nom, mon Thevet, un livre on deust nommer,
Qui n'as veu nostre terre, ou sa prochaine mer.

(If with the name of Ulysses the *Odyssey* is named, with your name, my dear Thevet, a book ought to be titled, for it is you who have seen our earth and our neighbouring seas.)

He adds that the author of *Les Singularitez de la France antarctique* (1557) has travelled about many places and written of them in his own hand.[6]

In this respect, because it is an agglomeration of poems — or even due to its proximity to Ronsard and Thevet, author of books of islands — for heuristic purposes the *Amours* can be likened to an *isolario*, a specific cartographic genre that in the history of cartography is a precursor of the atlas, an avatar

[4] 'En faveur de M. Nicolaï, à Monseigneur le Connestable', in Ronsard, *Œuvres complètes*, ed. by Céard, Ménager, and Simonin, I, 1208, ll. 9–12.

[5] 'À Monsieur d'Avanson,' in Ronsard, *Œuvres complètes*, ed. by Céard, Ménager, and Simonin, I, 509, ll. 1–3.

[6] Ronsard, *Œuvres complètes*, ed. by Céard, Ménager, and Simonin, I, 488, ll. 1–3.

of an *imago mundi*. Peculiar to the fifteenth and sixteenth centuries, the *isolario* develops so as to classify and describe an expanding world whose lands take the shape of islands. Islands unto themselves, like their smaller counterparts, continental masses float in the midst of an ocean sea, and as a result each unit is seen as a self-contained and autonomous whole. The conception owes to Cristoforo Buondelmonti (1386–*c.* 1430), whose *Liber insularum archipelagi* (1430), a collection of maps of the islands of the Aegean archipelago, witnessed the production of many manuscript copies up to 1470. Bartolomeo dalla Sonetti's *Isolario* (1486), an incunabulum in which are collected woodcut images of many of the same islands, features poems — many of which are sonnets — describing the notable traits of the masses adjacent to them.[7] The jagged shoreline of each of the islands recalls the stylization of coasts that marks the fourteenth- and fifteenth-century portolan charts of the Mediterranean. Author Bartolomeo Zamberti renames himself fittingly for the collection, as if to suggest that a strong force of attraction exists between his pseudonym, the island-images, and the poems — especially the sonnets — that float near them. The typographical mass of each unit of printed writing becomes a textual analogue to the surface area of the island it describes. The reader of the map notices that the adjacent poem is located within a wind rose, the compass and radiating lines of which designate the directions of the prevailing Mediterranean winds. An implied spatial narrative nestles in the description of the commanding traits of the island, set adjacent to cardinal points that indicate how and where it can be approached, where shoals must be avoided (as marked by agglomerations of dots), and how peninsulas or inlets may serve as anchoring points. By way of analogy, a prevailing mode of reason in the Renaissance, each of the poems becomes both a totality and a fragment in what might be a scatter of land in the aftermath of a cosmic deluge.[8] The reader becomes a navigator who encounters the poems from angles at once lexical and perspectival. They can be discerned

[7] A copy (of which forty-nine are extant, some coloured, others not) in the Boston Public Library is available for easy reading online at: <http://archive.org/details/isolario00bart> [accessed 4 January 2016].

[8] In the opening pages of *Le Livre d'îles*, Frank Lestringant studies how, from the standpoint of theology, the *isolario* is born of a deluge that floods and pulverizes continental masses, scattering them as island fragments anchored in the substrate of a great ocean sea, just 'as the continent-ocean sea (*oekoumène*) of times past had to break apart and be opening before the caravels of Columbus and Magellan' (p. 52). George Tolias offers a history of the medieval *isolario* in Harley and Woodward, *The History of Cartography*, I: *The Medieval World*.

in view of their sources, surely, but also diagonally, and transversally, across the surface on which their italic characters are printed.

Sonetti's modest *isolario* is rich for what it implies about its relation to the spatiality of its verse. Each poem can be construed as an island in the collection in which it figures. At once an assemblage of singularities (hence an archipelago) and a mirror of the world (a totality), the poems of the *Amours* appear to anticipate what poet René Char brilliantly calls 'la parole en archipel'.[9] The poems are *both* what they utter and the mass of islands to which their words are addressed. No doubt Maurice Blanchot had a keen grasp of the island effect when noting that Char's art was one of juxtaposition rather than composition. In sum, it presents an infinity of fragments:

> *Parole en archipel*: découpée en la diversité de ses îles et ainsi faisant surgir la haute mer principale, cette immensité très ancienne et cet inconnu toujours à venir que seule nous désigne l'émergence des terres profondes, infiniment partagées.[10]
>
> (*Speech in archipelago*: divided into the diversity of its islands and thus causing the principal high sea to surge upward, this very ancient immensity and this unknown forever to come that for us only the emergence of deep, infinitely scattered lands can designate.)

Or earlier on, in what Blanchot remarks of Char's *Poème pulvérisé* (Pulverized Poem):

> [é]crire, lire ce poème, c'est accepter de ployer l'entente du langage à une certaine expérience morcellaire, c'est-à-dire de séparation et de discontinuité. Pensons au dépaysement. Le dépaysement ne signifie pas seulement la perte du pays, mais une manière plus authentique de résider, d'habiter sans habitude; l'exil, c'est l'affirmation d'une nouvelle relation avec le Dehors. Ainsi le poème fragmenté est un poème non pas inaccompli, mais ouvrant un autre mode d'accomplissement, celui qui est en jeu dans l'attente, dans le questionnement ou dans quelque affirmation irréductible à l'unité.[11]
>
> (to write, to read this poem is to agree to bend the harmony of language to a certain fragmentary experience, in other words, an experience of separation and discontinuity. Let's think about getting uprooted. Uprooting means not only the loss of

[9] Included are poems written 1952–60, first published in 1962 (Paris: Gallimard) and included in *Oeuvres complètes*, pp. 337–415; in English as *The Word in Archipelago*, trans. by Baker.

[10] Blanchot, 'Parole de fragment', p. 454.

[11] Blanchot, 'Parole de fragment', p. 452.

one's ground but a more authentic way of residing, of inhabiting without habitude; exile is the affirmation of a new relation with the Outside. Thus the fragmented poem is not a poem unfinished, but a poem opening another mode of accomplishment, what is at stake in waiting, in questioning, or in any kind of affirmation [or statement that is] irreducible to unity.)

In view of collections of poems that take the form of the archipelago, Blanchot's point applies as much to Ronsard as to René Char. Buondelmonti notes that he fashioned his book of islands to include accounts of various encounters and to offer the pleasure of 'an errant writing and reading' in which the chance and happenstance of travel come forward in the mix of text and image.[12] In turn it can be said that each island — in other words, each text in its relation to other texts — can be seen as both a mirror and an island that 'repeats' itself through its own variation and fragmentation. The poem that varies on its own formulation, that through repetition and difference turns its clichés into creative forms, itself belongs to an insular geography. In its iteration it confirms that its own beginnings — from its *incipit* and throughout its performance — are re-beginnings, its commencements being *recommencements*.[13]

The fissured poem is paradoxically finished and unfinished, one to itself but part of a mass that extends to an 'outside', to what is real, to what is unknown and without limit. The fragment reflects its own incompletion and, no less, the site and 'situation' in which it is found, juxtaposed to other poems, units, and fragments of a kaleidoscopic entity that acquires the appearance of an archipelago of verse. The poems are set on the page to suggest that each is one of the corners of an implicit quadrangle (Figure 9.3). Two sonnets, one above the other, are printed on the recto and verso of each folio, a design enabling the reader to *behold* each poem in its integrity (without breakage or visual interruption, as in modern editions where the poems are cut haphazardly and where Roman numerals, not found in the original text, are innocuously set above the poems in order to keep track of them). Each poem is of a piece, an island unto itself, but its fragments are seen making different sense in different places. Discursive

[12] In a keen treatment of the material aspect of the manuscripts of the *Liber insularum archipelagi*, Simone Pinet notes that 'the space of the island seems to be especially apt for conveying new forms of interrogating the real through the incorporation or assemblage of a variety of discourses', and that the maps 'are there to illustrate his text, to mirror the information, to visualize the stories in a collage and in enumerations'; *Archipelagoes*, pp. 47–53.

[13] Deleuze, 'Causes et raisons des îles désertes', especially p. 17. See also Benítez-Rojo, *La isla que se repite* (translated into English by James Maraniss as *The Repeating Island*).

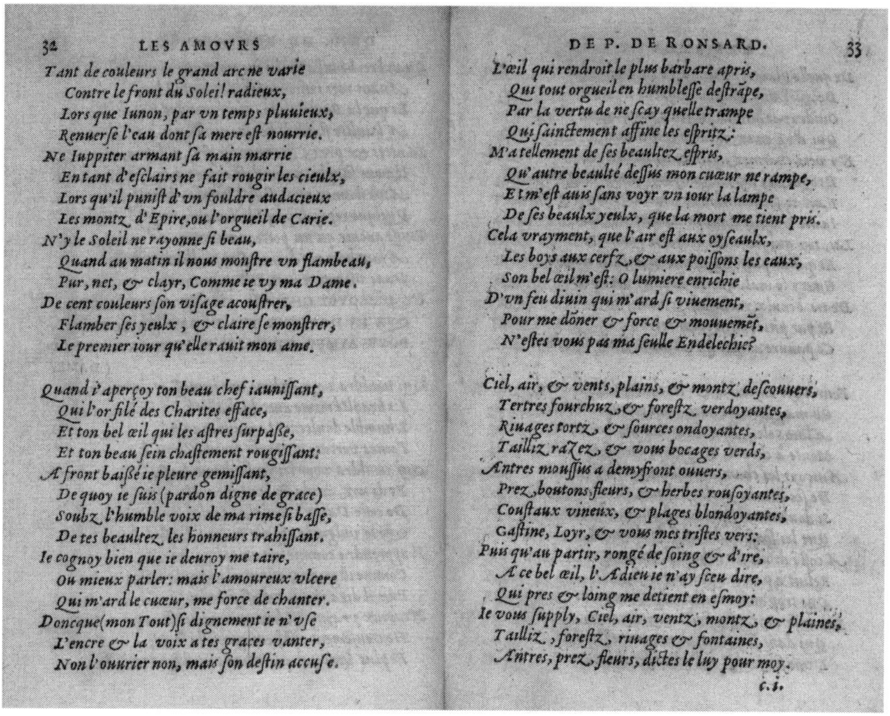
32 LES AMOVRS

Tant de couleurs le grand arc ne varie
Contre le front du Soleil radieux,
Lors que Iunon, par vn temps pluuieux,
Renuerse l'eau dont sa mere est nourrie.
Ne Iuppiter armant sa main marrie
En tant d'esclairs ne fait rougir les cieulx,
Lors qu'il punist d'vn fouldre audacieux
Les montz d'Epire, ou l'orgueil de Carie.
N'y le Soleil ne rayonne si beau,
Quand au matin il nous monstre vn flambeau,
Pur, net, & clayr, Comme ie vy ma Dame.
De cent couleurs son visage acoustrer,
Flamber ses yeulx, & claire se monstrer,
Le premier iour qu'elle rauit mon ame.

Quand i'aperçoy ton beau chef iaunissant,
Qui l'or filé des Charites efface,
Et ton bel œil qui les astres surpaße,
Et ton beau sein chastement rougissant:
A front baißé ie pleure gemissant,
De quoy ie suis (pardon digne de grace)
Soubz l'humble voix de ma rime si basse,
De tes beaultez les honneurs trahissant.
Ie cognoy bien que ie deuroy me taire,
Ou mieux parler: mais l'amoureux vlcere
Qui m'ard le cuœur, me force de chanter.
Doncque (mon Tout) si dignement ie n'vse
L'encre & la voix a tes graces vanter,
Non l'ouurier non, mais son destin accuse.

DE P. DE RONSARD. 33

L'œil qui rendroit le plus barbare apris,
Qui tout orgueil en humblesse destrãpe,
Par la vertu de ne sçay quelle trampe
Qui sainctement affine les espritz:
M'a tellement de ses beaultez espris,
Qu'autre beaulté dessus mon cuœur ne rampe,
Et m'est auis sans voyr vn iour la lampe
De ses beaulx yeulx, que la mort me tient pris.
Cela vrayment, que l'air est aux oyseaulx,
Les boys aux cerfz, & aux poissons les eaux,
Son bel œil m'est: O lumiere enrichie
D'vn feu diuin qui m'ard si viuement,
Pour me dõner & force & mouuemẽt,
N'estes vous pas ma seulle Endelechie?

Ciel, air, & vents, plains, & montz descouuers,
Tertres fourchuz, & forestz verdoyantes,
Riuages tortz, & sources ondoyantes,
Tailliz razez, & vous bocages verds,
Antres moussus a demyfront ouuers,
Prez, boutons, fleurs, & herbes rousoyantes,
Coustaux vineux, & plages blondoyantes,
Gastine, Loyr, & vous mes tristes vers:
Puis qu'au partir, rongé de soing & d'ire,
A ce bel œil, l'Adieu ie n'ay sceu dire,
Qui pres & loing me detient en esmoy:
Ie vous supply, Ciel, air, ventz, montz, & plaines,
Tailliz, forestz, riuages & fontaines,
Antres, prez, fleurs, dictes le luy pour moy.

c. i.

Figure 9.2. *Les Amours de P. de Ronsard* (1552), pp. 32–33.

pieces and iterations are shown as if refracted in the two or three others in their immediate context, above, below, to the left, or to the right.

A sonnet evoking the landscapes of Ronsard's homeland, the topographical gist of *Ciel, air, & vens, plains & montz descouverts* implies that it is as much an island as a description of the author's countryside. Set at the bottom of page 33, the sonnet seems to figure in an archipelago of verse. In what it says the sonnet portrays the lover, seeking to attenuate the anguish he feels over the departure of the beloved, asking the landscape to say to her in its words — words that he relates to himself, in an inner speech — what he cannot express. There emerges from the sum an erotic topography in which his own locale becomes confused with the parts of the beloved's body that good taste names through euphemism, notably: *montz descouverts* (l. 1, 'unveiled mountains' — but also, in a pun that the spacing of the letters reveals as *mondes couverts*, 'veiled worlds') *tertres fourchuz* (l. 2, 'forked hillocks'), *tailliz rasez* (l. 4, 'sheared hedgerows'), and, stunningly, *Antres moussus à demyfront ouvers* (l. 5, 'mossy lairs halfway

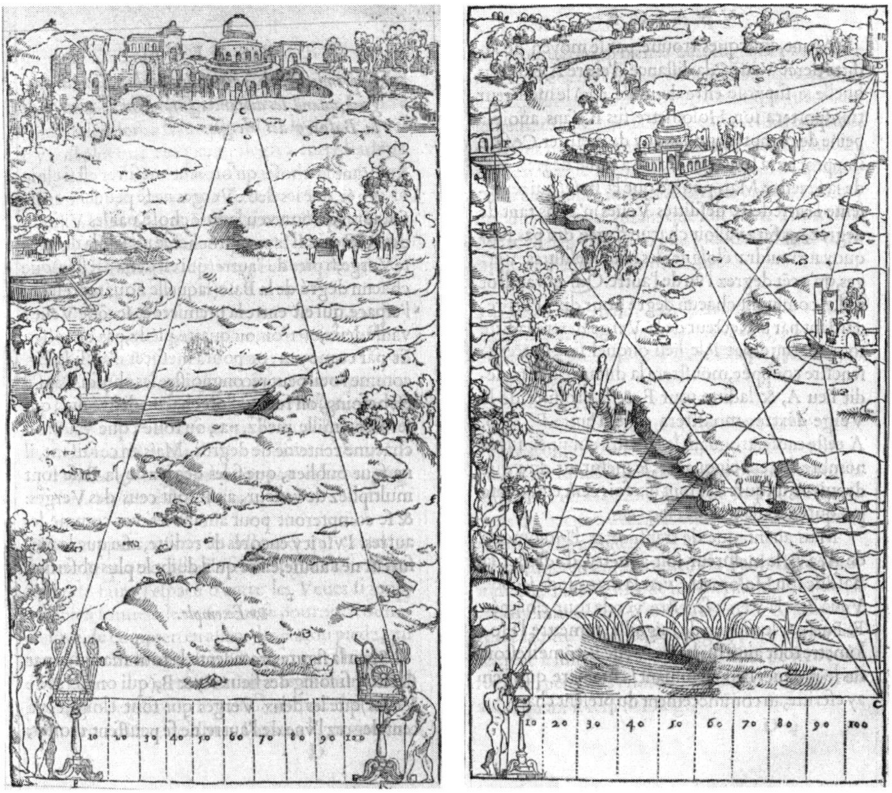

Figure 9.3. Abel Foullon, *Vsaige et description de l'holometre. Pour scavoir mesvrer toutes choses qui sont soubs l'estenduë de l'œil : Tant en longueur & largeur, qu'en hauteur & profondité. Inuenté par Abel Foullon Vallet de chambre du Roy* (Paris: Pierre Beguin 1561), fols D1r° and D3r°.

opened'), terms implying that the undulating hills and valleys are those of his lady's nude body.[14] Some of the same and similar figures and turns of speech are scattered about the poems in the immediate vicinity. The poem can thus be read 'unto itself' as a whole and also in the midst of others that, like islands or garden plots or erotic *points de capiton*, repeat and vary upon the general form. Everywhere visual and verbal correspondences can be read crosswise with the three adjacent sonnets. *Les montz d'Epire* (mountains of the Epire) on the

[14] The reading is in dialogue with Weber, *La Création poétique au XVIe siècle*, pp. 320–21, in which the poem is appreciated for the 'presence of the image tied to the Vendômois countryside and the impetuous vivacity of rhythm, which expresses the upturn of an entirely mobile sensibility' (p. 321).

upper left (l. 8) are set on a diagonal with *montz descouvers* on the lower right (disengaging the hidden verse or *vers* in the latter); *Et ton bel oeil* (And your lovely eye) in the sonnet below (l. 3) is on a sightline at an angle inclining to *A ce bel oeil* (To this lovely eye). The same pupil that inaugurates the sonnet above (l. 1) becomes *Son bel oeil* (Her lovely eye) in the first tercet set above the sonnet in question. The experience of reading along different trajectories, which seems all but lost in modern editions of the *Amours*, suggests that the poems invite the eye to move inventively across the geography of the page. In this way, in fashioning various itineraries, the reader might become a sort of surveyor or engineer, if a far-fetched comparison is allowed, like the classical nude in Abel Foullon's treatise on the *holomètre* who establishes sightlines that cut across a lush fluvial landscape (Figure 9.3).[15]

Sightlines and mirrors become cardinal points of the compass. In this way the most celebrated of the mirror sonnets in the *Amours* can be seen as the very object it emblazons: a mirror capturing and scattering the rays of spiritual light — the wit — that its reader brings to it.[16] Initially a description of the mirror on which the face of the beloved is reflected, its very words, their characters too, become luminous particles:

> Je parangonne à voz yeulz ce crystal,
> Qui va mirer le meurtrier de mon ame:
> Vive par l'air il esclate une flamme
> Vos yeulx un feu qui m'est sainct et fatal.
> Heureux miroer, tout ainsi que mon mal

[15] In 'The Engineer-Poet and his Garden Poem', I have used the same illustration to show how the same author's poetry relates to landscape design.

[16] Catherine Weber and Henri Weber note cogently: 'Ce sonnet semble accompagner le présent d'un miroir que le poète envie de pouvoir contempler à loisir sa Dame. Les pétrarquistes précieux comme Serafino et les néo-Latins comme Angeriano ont exécuté d'innombrables variations sur ce thème, suivis par Scève, Saint-Gelais et les auteurs de Blasons' (This sonnet seems to accompany the present of a mirror that the poet envies for being able to contemplate his Lady at leisure. Petrarchan poets tending to preciosity, such as Serafino and neo-Latin Angeriano executed countless variations on the theme, followed by Scève, Saint-Gelais, and the authors of the *blasons*), in their critical edition, Ronsard, *Les Amours*, pp. 538–39. In a reading of Scève's use of the mirror, in his critical edition of *Délie* Gérard Defaux notes the presence of Petrarch's *Rime disperse*, sonnet 4, and Lucain's *De Bello Civili*, IX, 669–70; in *Délie: Object de plus haulte vertu*, II, 226–27. Basing research on the Lacanian mirror-stage, in a keen psychoanalytical (and forcibly psycho-geographical) reading of Scève's *dizain* (*Délie*, II, 186) of the basilisk and the mirror, Nancy M. Frelick aptly studies the alterity of reflection in *Délie as Other*.

Vient de trop voyr la beaulté qui m'enflamme:
Comme je fay, de trop mirer ma Dame
Tu languiras d'un sentiment egal.
Et toutesfoys, envieux, je t'admire,
D'aller mirer le miroer où se mire
Tout l'univers dedans luy remiré.
Va donc miroer, va donq, & pren bien garde,
Qu'en mirant ainsi que moy ne t'arde
Pour avoir trop ses beaulx yeulx admiré.

(I compare to your eyes this crystal, that will gaze upon the murderer of my soul: alive in the air explodes a flame, your eyes a fire for me sacred and fatal. Happy mirror, just as my ill comes from seeing too much the beauty enflaming me: as I do, of gazing too much upon my Lady you will languish from the same feeling. And yet, envious, I admire you for beholding the mirror where the entire universe is reflected within its view. Go then, mirror, go then, and take good care, that in gazing as do I you won't burn for having her handsome eyes too much admired.)

The incipit plays on *parangonner*, to compare or to make one object equal to that which is juxtaposed to it, linked with the substantive *parangon*, or flint-like touchstone, which, when rubbed against a metal object, is employed to test the purity of gold or silver contained within.[17] Through paronomasia Ronsard sets the metallurgy of the mirror into the mix of printed language. The comparative object that makes visible the presence of a precious metal is likened to the poem's movement of a metaphor, itself a mirror, extending across — and thus inventing — the reflective surface of the poem. Reflection and scintillation flicker in the comparison that makes the beloved's eyes the 'parangon' of the mirror (or 'crystal'). Between the one and the other is found the 'humeur crystalline', the visual complexion of the sonnet, indeed what makes it a partial object reflecting or implying the presence of a greater sum, an open whole, that could be composed of equally partial pieces.[18] The sonnet is a mosaic mirror, its words becoming

[17] In their critical edition Henri Weber and Catherine Weber signal archaic terms by using asterisks (*) that turn many of the poems into sparkling constellations of enigmas that for thumbnail explanation require reference to a glossary-gazetteer in the appendix. In their printing of this sonnet (LXXVI, 48), starred words include *parangonne* (l. 1), *mirer* (l. 2), *esclate* (l. 3), *mirer* (l. 7), *remirer* (l. 11), and *arde* (l. 13), the meanings of which are fabulously protean.

[18] Randle Cotgrave in his *Dictionarie of the French and English Tongues*: 'Humeur cristalin. *The crystaline humour; seated in the middest of the eye, and of a round figure, somewhat flattened before and behind; it is the first instrument of sight; and a glasse wherin the spirit examines, Judges of, the formes represented unto it*' <http://www.pbm.com/~lindahl/cotgrave/search/260r.html> [accessed 4 January 2016].

reflective surfaces, and their characters signifying particles.[19] The poem organizes elements that resemble, repeat, and reshape each other within the very drift of the conceit. From the outset *mirer* (that Randle Cotgrave translates as 'to aim, level at; to looke, view, regard, observe, prie into, behold through, or thoroughly') anticipates what is seen incised into *meurtrier* (l. 2, 'murderer'), before the fragment is iterated in *miroer* (l. 5), prior to settling near the centre of the poem (l. 7), where the poet regrets what happens when 'de trop *mirer* ma Dame' (having gazed too much upon my lady), he is burned, where the words reflect the flame of an inventive desire: 'je t'*admire*, | D'aller *mirer* le *miroer* où se *remire* | Tout l'univers dedans luy *remiré*' (I admire you, for going to gaze upon the mirror where the entire universe is viewed reflected in it). And it further refracts in 'Va donc *miroer*' (l. 12), returns 'en le *mirant* ainsi' (l. 13), and reveals itself as the keystone to the poem, the *pierre angulaire*, in the final word in the formula 'ses beaulx yeulx *admiré*' (l. 14), which echoes the end of the first line.

Redundancy mobilizes variation and lends creative license to the reading. Words are shown bending and twisting as if on the very surface of the mirror on which the universe is reflected. Reflection and refraction are of the same order: in the first tercet the poet begs his beloved to contemplate where she might be in the image of the greater world — at the same time the latter is reflected on the convex surface of her pupil. The optical faculty that the poem is celebrating is underscored in the surface tension of the iteration of the fragment — the letter *m* — that seems to be mediating the soul (*mon ame*, roughly 'my *m*' that a productive misprision in an English reading yields as 'my name') and the physical world reflected on the beloved eye. As in the geographical latency of the *blason*, the bodily part becomes a topography detached from the world-body to which it ought to belong. In the geographical tradition on which the sonnet is based, the *blason* fragments the body, which is given to be a world, into isolated elements constitutive of a topography (Figure 9.4).[20] Hence the letter *m*,

[19] In a post-semiotic idiolect, in this sonnet it can be said that what signifies is what *s'ignifie* (what means and what catches fire) or what exudes a blazing effect when its meaning takes spark and is no sooner extinguished. It is here, if Gilles Deleuze's definition of the term is appropriate, where the poem becomes an *event*. He notes that images must be inserted into language, especially into names and the voices that utter them, in order to *spatialize* them: 'What counts in the image is not the impoverished content, but the mad energy contained within on the verge of explosion, which means that images don't last for long. They are confused with the detonation, combustion, and dissipation of their condensed energy. Like ultimate particles, they are of very short duration'; 'L'Épuisé', pp. 71 and 72.

[20] In an illustrated gloss of Ptolemy, Pieter Apian makes clear the presence of the tradition

printed with greater frequency here than any sonnet in the *Amours*, tends to bind the verse while, in capturing the gaze of the reader, it acquires more visual force as a graphic unit whose iteration implies that as a fragment it has far greater presence than the whole (the sonnet) or even the words (twenty-five) in which it is found. Further even, given the mix in which the practice of the *Grands Rhétoriqueurs* informs the paradoxical encomium, in which a play of letters causes the words that convey them to recede in a background where the letters or ciphers become visually prominent, the poem becomes a partial reflection both on and of the autonomous letter that is 'm'.[21] It acquires hilarious form in early *Menschenalphabets*, where it is seen to be, from his feet, a supine male, his raised legs, and the juncture of the two traits of 'V' at its centre, his genitalia, but it is more in line with glosses with which Ronsard had no doubt been familiar, where it is something *m*edian, at the *m*iddle of the alphabet, it becomes a *m*eans to distinguish between one side of things and another, at the very same time (at least where it is written sans serif), its two equal sides assure that when reflected in a mirror or reversed, *m* remains *m*.[22]

of the *blason* in the context of geography in the famous similitude where the globe is to a portrait of a person as the image of a piece or part is to a topographical representation, in *Cosmographia* (notably in the French edition: Paris: Vivant Gaultherot, 1551, fol. iiir).

[21] For example, Jean Molinet's 'Ad laudem irundinis' (In praise of the swallow), a tour de force of 144 lines that begins in honour of the 'divine genre, the sweet angelic species' (l. 1) and ends with the author's decrying (hence eternizing) the poetic seed he feeds to the bird, the words that are the grist of the little mill indicated by his name, 'le gros grain d'ung rude molinet' (l. 144, 'the fat seed of a rough little mill'): available in Gray, *Anthologie de la poésie française*, pp. 11–15. Molinet treats words and their spacings as ciphers in his 'Dictier joyeux', a poem whose format implies the presence of a cartographic latency taken up in Conley, *The Self-Made Map*, pp. 48–54.

[22] Bodily treatment of the letter *m* is shown in early modern alphabets in Massin, *La Lettre et l'image*, especially in variants of Peter Flötner's *Menschenalphabet* (*c.* 1534). In his gloss of the letter Geoffroy Tory notes how *m* is given to murmur wherever it is placed in the words in which it figures: 'Priscian au lieu ou il traicte De literarum commutatione, dit que M. a trois manieres de son: Obscur, Apert, & Moyen. Ses mots sont telz quil s'ensuyt. M. obscurum in extremitate dictionum sonat, vt Templum. Apertum in pricipio, vt Magnus. Mediocre in mediis, vt Vmbra. M. dit-il, en lextremite & a la fin des dictions sonne obscurement, comme en ceste diction Templum. Elle a son aussi au commencement qui est apert, comme en cette diction Magnus. Pareillement elle a son moyen au mylieu, comme en cette diction Vmbra' (Where Priscian writes in *De literarum commutatione* he notes that M has three manners of sound: Obscure, Open, and Middle. His words are as follows: [...] M, he says, in the extremity and at the end of uttered words is sounded obscurely, as in this word *Templum*. It is also voiced at the beginning, which is open, as in the word *Magnus*. Likewise, it has middle in the midst, as in the word *Vmbra*); *Champ fleury*, fol. Kiir°.

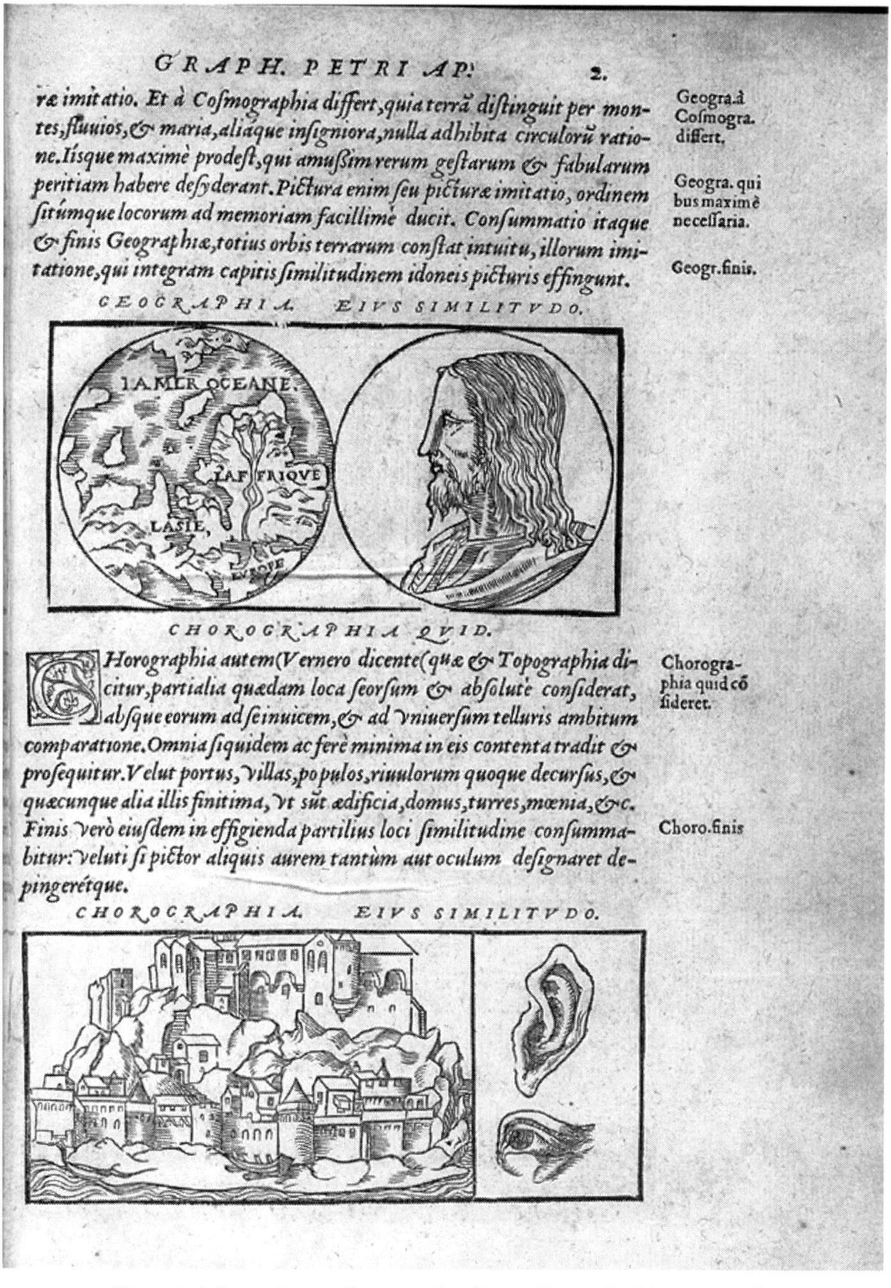

GRAPH. PETRI AP. 2.

ræ imitatio. Et à Cosmographia differt, quia terrã distinguit per montes, fluuios, & maria, aliaque insigniora, nulla adhibita circulorũ ratione. Iísque maximè prodest, qui amußim rerum gestarum & fabularum peritiam habere desyderant. Pictura enim seu picturæ imitatio, ordinem sitúmque locorum ad memoriam facillimè ducit. Consummatio itaque & finis Geographiæ, totius orbis terrarum constat intuitu, illorum imitatione, qui integram capitis similitudinem idoneis picturis effingunt.

Geogra. à Cosmogra. differt.

Geogra. quibus maximè necessaria.

Geogr. finis.

GEOGRAPHIA. EIVS SIMILITVDO.

CHOROGRAPHIA QVID.

Chorographia autem (Vernero dicente (quæ & Topographia dicitur, partialia quædam loca seorsum & absolutè considerat, absque eorum ad seinuicem, & ad vniuersum telluris ambitum comparatione. Omnia siquidem ac ferè minima in eis contenta tradit & prosequitur. Velut portus, villas, populos, riuulorum quoque decursus, & quæcunque alia illis finitima, vt sũt ædificia, domus, turres, mœnia, &c. Finis verò eiusdem in effigienda partilius loci similitudine consummabitur: veluti si pictor aliquis aurem tantùm aut oculum designaret depingerétque.

Chorographia quid cõsideret.

Choro. finis

CHOROGRAPHIA. EIVS SIMILITVDO.

Figure 9.4. Pieter Apian, *Cosmographia* (Paris: Vivant Gaultherot, 1551). Cambridge, MA, Harvard University, Houghton Library, Typ 515.53.149, fol. iiir°. Courtesy of the Houghton Library, Harvard University.

In this area of the *Amours* (*roughly speaking,* sonnets 50–80 or so), in part for reason of the mirror effect of this poem, Ronsard beckons us to look about and around the work. He invites us to treat mythology and typography on the same level, as if thus to show that the latter does not disappear when conveying the images of famous scenes from the former. In this poem, in French as in English, the I and the eyes, *Ie* and *yeux*, resume the mirrored gaze that marks so many of the immediately contextual pieces where eyes are described with obsessive delight: 'Quand ces beaulx yeux jugeront que je meure' (sonnet 63, 'When these comely eyes will decide that I must die'); 'Qui vouldra voyr les yeulx d'une deesse' (sonnet 64, 'Whoever might wish to see the eyes of a goddess'); 'Voïant les yeux de toi, Maitresse elüe' (sonnet 68, 'Seeing your eyes, my chosen Mistress'); 'L'oeil qui rendroit le plus barbare apris' (sonnet 69, 'The eye that would tame the most barbarous soul'); 'Pour voyr ensemble & les champ & le bord' (sonnet 81,'To see together the fields and the border'), and so on.[23] To relegate allusions or thematize them in respect to a Neoplatonic, Petrarchan, or Ficinian worldview would risk mediating the visual beauty of the verse and, at the same time, minimize the play of topography and of fractured mirror seen on the printed page.

In sum, the *amoureux univers* to which the words and letters of 'Je parangonne à vos yeux ce crystal' appeal, the whole of wholes (*Tout l'univers*) can only be reflected in the verbal fragments that are the touchstones of the poem.[24] If the poem is building on something of the order of Gilles Corrozet's blazon of the mirror of some years before, it projects a domestic object, found earlier in a poem describing a settled world of domesticity, into an erotic topography. Corrozet writes:

> Miroir de verre bien bruni
> D'une riche chasse garny
> Ou la belle, plaisante, & clere
> Se void, se mire, & considere
> En regardant sa contenance

[23] Numbering follows the Weber edition noted above.

[24] 'Dedans mon coeur l'amoureux univers' (inside my heart the amorous universe): the eighth line that caps the second quatrain of the thirty-seventh sonnet, 'Les petitz corps, culbutans de travers' (The little bodies, bouncing all about), makes clear that his heart takes heed of and strives to contain the accident and chance of proto-Brownian movement of atomic particles — but only before he finds that 'little whole' that he is (l. 12) is cast in doubt before the more dubious form of 'le grand Tout' (l. 14, 'the great Whole'), allowing him to discern chaos.

Et de son gent corps l'ordonnance,
Ses yeulx scintillans & sa face
Son front poly, sa bonne grace,
Sa doulce bouche vermeillette.[25]

(Mirror of well burnished glass set within a rich shrine where the pretty, pleasant, beauteous lady sees, reflects herself, and considers in beholding her appearance and the order of her lithe body, her scintillating eyes, and her face, her polished forehead, her good grace, her sweet little red mouth.)

The *blason* of 1539 isolates the mirror in a static place, much as does the woodcut image that heralds it, on a windowsill that opens onto a world at large, where the mirror reflects the inner space whence the spectator is placed and not the greater landscape on the horizon.[26] By contrast, in 'Je parangonne à voz yeux', a sonnet of greater ambition, Ronsard relates the mirror to the ambiguous relation that Ptolemy had described at the outset of his *Geographia*, in which the study of parts of the world is engaged irrespective of their exact place in the greater whole.[27]

In a first conclusion it can be surmised that throughout *Les Amours* the world at large is seen as a dynamic configuration of reflective pieces and parts. Although many of the places named in the work are situated in the Touraine, the 'garden of France' of Ronsard's origin, the composition bears more than striking resemblance to the vision of a book of islands, an *isolario* born of the Aegean, in the way that a reader who navigates through and about the many pieces constantly gets lost, retrieves a sense of compass, is dazzled, and then

[25] Corrozet, *Les Blasons domestiques*, fol. 29r°.

[26] See the illustration (on fol. 28v°) accompanying Elizabeth Black's contribution to this volume.

[27] For the Alexandrian scientist, cosmography (what he calls geography) 'is a representation in picture of the whole known world together with the phenomena which are contained within. It differs from Chorography [or topography] in that Chorography, selecting certain places from the whole, treats more fully the particulars of each by themselves — even dealing with the smallest conceivable localities, such as harbors, farms, villages, river sources, and such'. He soon infers that those who inquire of the nature of the world are artists, since '[t]he end of Chorography is to deal separately with a part of the whole, as if one were to paint only the eye or the ear by itself', while the 'task of Geography is to survey the whole in its just proportions, as one would the entire head. For as in an entire painting we must first put in the larger features, and afterward those detailed features which portraits and pictures may require, giving them proportion in relation to one another so that their correct measure apart can be seen by examining them, to note whether they form the whole or a part of the picture'; from Claudius Ptolemy, *The Geography*, trans. and ed. by Stevenson, pp. 25–26.

lost, and found over and again. The sonnets belong to an ambiguous, open-ended form where the reader, once cognizant of the way he or she chooses to read through the work, navigates from one port of call to the next. In a second and, it is hoped, projective conclusion we can speculated that the latency of the idea of a collection in the form of a mosaic of islands is made manifest when Ronsard publishes an expanded edition of *Les Amours* in 1553. Thirty-six new sonnets are added to the volume, two are suppressed and, in the final pages, are appended: a sonnet by Nicolas Denisot to which Ronsard responds with a 'Sonet sur *Les Erreurs amoureuses* de Pontus de Tyard' (published in 1549); an ode of 174 lines, 'A Melin de Saint Gelais', in which he seems — but only seems — to make peace with his rival in the French court; a longer ode of 268 lines, 'Les Isles fortunées', that he dedicates to Marc-Antoine de Muret; the short 'Ode sur les misères des hommes' (seventy-two lines) for 'Ambroise de la Porte Parisien'; last and famously not least, the short ode of eighteen lines, that every student of French literature knows by heart, 'Mignonne, allons voir si la rose', bequeathed to Cassandre, the love-object of *Les Amours*.[28]

On cursory view the added items appear to be *copia* that 'fill out' the volume, attesting to an aesthetic of addition, of open-endedness or, like the *Essais* of Montaigne, of a work that evolves with the writer who lives consubstantially with it.[29] Close inspection reveals the presence of a political aesthetic that the mosaic aspect of *Les Amours* at once veils and, in the end, ultimately brings forward. Nicolas Denisot's sonnet stages Venus, her fingers idly fashioning a *mol chapeau* (soft headdress) with which no sooner, as she kisses the poet, she crowns his head to assure him that he is her eternal equal. The following 'Sonet sur les *Erreurs amoureuses* de Ponthus de Tiard màconnois', no doubt intended to accompany a re-edition in Lyon (1554) of the work first published in 1549, is a literal tour de force about errant movement about the globe, beginning with the island of Thule, the northernmost point of the known world and descending to the unknown reaches of Scythia. The verbal magic of the poem — 'Puis retombant par les spheres à bas, | Pour contre'errer tu fais errer mes pas | Après l'erreur de ton erreur si sainte (Then falling back among the lower spheres, to counter-wander you cause my feet to move errantly after the error of your very saintly wander), showing how Tiard (or Tyard) genially leads the poet to wan-

[28] See Ronsard and Muret, *Les Amours, leurs Commentaires, texte de 1553*, ed. by de Buzon and Martin, pp. 257–73; the digitized copy in the Gordon Collection at the University of Virginia: *Les Amours de P. de Ronsard Vandomois, nouvellement augmentées par lui, et commentées par Marc-Antoine de Muret*.

[29] The idea runs through Cave, *The Cornucopian Text*, as well as Jeanneret, *Perpetuum mobile*.

der and to and fro and back again.[30] The ode to Mellin de Saint-Gelais, an olive branch Ronsard tenders to his dedicatee, seeks to settle quarrels with his enemy while displaying flourish, erudition, and wit that Saint-Gelais would be hard put to equal. At the outset their stormy relation is figured as a stormy seascape in the Aegean archipelago, close in visual effect to an angry windhead blowing icy air from the boreal corner of a world map:

> Tousjours ne tempeste enragée
> Contre ses bords la mer Egée,
> Et toujours l'orage cruel
> Des vens, comme un foudre ne gronde,
> Elochant la voute du monde
> D'un soufflement continuel. (ll. 1–6)

(As enraged blasts may always thrust the Aegean Sea against its shores, so also the cruel windstorm, as a great thunderclap, with continual thrust shakes the world's vault.)

The poem reaches into the classical past to show how calm and friendship can follow the violence of storms that since time immemorial sailors have endured in the Grecian archipelago. Dedicated to the commentator, referring to the Canary Islands, 'Les Iles fortunées' first reads as a precursor to an 'Invitation au voyage', as a romantic eclogue, somewhere between Baudelaire and Rimbaud's 'Bateau ivre', in which the poet invites Muret to accompany him and his friends — members of the 'Brigade', albeit *minus* Mellin de Saint-Gelais, who is never mentioned — to weigh anchor and sail, far from the tempest and ruin of Europe at the time of its writing (when civil strife and an economic downturn were manifest, and the Wars of Religion were beginning to rumble):

> Ramons la nef dans les chams bienheureus,
> Au port heureus des Iles bienheurées,
> Que l'Océan de ses eaus asseurées,
> Loin de l'Europe, & loin de ses combas,
> Pour nous, pour nous emmure de ses bras. (ll. 88–92)

(Let's row the ship to the prosperous fields in the fortunate port of the ever-fortunate Islands, that the Ocean with is assured waters, far from Europe, and far from its battles for us, for us emprisons in its arms.)

[30] On this theme, see Rigolot, *L'Erreur de la Renaissance*, as well as Alduy, *Politique des 'Amours'*, p. 126. On Tyard as geographer, see Céard, 'Pontus de Tyard cartographe', pp. 67–76.

It could be said that the idealized *Iles bienheurées* are poems among which we as ideal readers would navigate in the warm breeze of a bookish sea. Ronsard envisages Muret's exile to Rome where his friend would spend the rest of his days. At the same time the great world-mirror he had fashioned in the poems to which the 'Iles fortunées' are appended is implied to be fissured and cracked.[31] And thus the effect of counterpoint in the last two poems, the ode on the misery of humankind adjacent to the poet's final whisper to Cassandre, leaves the reader uneasy: uneasy for reason of an overriding sense of isolation and of the presence of an unsettled world whose fault lines indicate that the mirror on which it is reflected is shattering.

[31] See Daniel Ménager's decisive 'Conflits et évasion dans "Les Isles fortunées"', pp. 21–36, in which the context and the writing of the poem are studied through a Freudian lens.

Works Cited

Archival Sources

Ronsard, Pierre de, *Les Amours de P. de Ronsard Vandomoys: Ensemble le cinquiesme de ses odes* (Paris: Chez la Veufve de Maurice de la Porte, au clos Bruneau, à l'enseigne S. Claude, 1552); Cambridge, MA, Harvard University, Houghton Library, *FC R6697.552a

Primary Sources

Apian, Pieter, *Cosmographia* (Paris: Vivant Gaultherot, 1551)

Char, René, *Oeuvres complètes* (Paris: Éditions Gallimard/Pléiade, 1983)

——, *The Word in Archipelago*, trans. by Robert Baker (Richmond, CA: Omnidawn, 2012)

Cotgrave, Randle, *Dictionarie of the French and English Tongues* (London: Adam Inslip, 1611); <http://www.pbm.com/~lindahl/cotgrave/search/260r.html> [accessed 30 December 2015]

Corrozet, Gilles, *Les Blasons domestiques contenantz la decoration d'une maison honneste, & du mesnage estant en icelles: Invention ioyeuses, & moderne* (Paris: [E]n la boutique de Gilles Corrozet Libraire, 1539)

Foullon, Abel, *Vsaige et description de l'holometre. Pour scavoir mesvrer toutes choses qui sont soubs l'estenduë de l'œil: Tant en longueur & largeur, qu'en hauteur & profondité. Inuenté par Abel Foullon Vallet de chambre du Roy* (Paris: Pierre Beguin, 1561)

Ptolemy, Claudius, *The Geography*, trans. and ed. by Edward Luther Stevenson (New York: Dover, 1952/91)

Ronsard, Pierre de, *Les Amours*, ed. by Catherine Weber and Henri Weber (Paris: Garnier, 1964/98)

——, *Les Amours de P. de Ronsard Vandomois, nouvellement augmentées par lui, et commentées par Marc-Antoine de Muret, Plus quelques Odes de l'auteur, non encor imprimées* (Paris: Chez la veuve Maurice de la Porte); <http://static.lib.virginia.edu/speccol/gordon/gordonimages/Gordon1554_R65_No_4/>

——, *Œuvres complètes*, ed. by Jean Céard, Daniel Ménager and Michel Simonin, 2 vols (Paris: Gallimard/Pléiade, 1994)

——, and Marc-Antoine de Muret, *Les Amours, leurs Commentaires, texte de 1553*, ed. by Christine de Buzon and Pierre Martin (Paris: Classiques Didier Érudition, 1999)

Scève, Maurice, *Délie: Object de plus haulte vertu*, ed. by Gérard Defaux, 2 vols (Geneva: Droz, 2004)

Sonetti, Bartolomeo dalla, *Isolario* (1486), available online at Boston Public Library, <http://archive.org/details/isolario00bart> [accessed 30 December 2015]

Tory, Geoffroy, *Champ fleury* (Paris: Geoffroy Tory, 1529)

Secondary Studies

Alduy, Cécile, *Politique des 'Amours': Poétique et genèse d'un genre français nouveau* (Geneva: Droz, 2007)

Benítez-Rojo, Antonio, *La isla que se repite* (Barcelona: Casiopea, 1998)

——, *The Repeating Island: The Caribbean and the Postmodern Perspective*, trans. by James Maraniss (Durham: Duke University Press, 1996)

Blanchot, Maurice, 'Parole de fragment', in *L'Entretien infini* (Paris: Gallimard, 1973), pp. 438–58

Cave, Terence, *The Cornucopian Text* (Oxford: Clarendon Press, 1979)

Céard, Jean, "Pontus de Tyard cartographe," in *Illustrations inconscientes: Écritures de la Renaissance: Mélanges offerts à Tom Conley*, ed. by Bernd Renner and Philip John Usher (Paris: Garnier, 2014), pp. 67–76

Conley, Tom, 'The Engineer-Poet and his Garden Poem: Ronsard's *bocages*', in *Technology and the Garden*, ed. by Michael G. Lee and Kenneth I. Helphand, Dumbarton Oaks Colloquium on the History of Landscape Architecture, 25 (Washington, DC: Dumbarton Oaks Research Library and Collection, 2014), pp. 11–23

——, *The Self-Made Map: Cartographic Writing in Early Modern France* (Minneapolis: University of Minnesota Press, 1997/2011)

Deleuze, Gilles, 'Causes et raisons des îles désertes', in *L'Île déserte et autres textes*, ed. by David Lapoujade (Paris: Minuit, 2002), pp. 11–17

——, 'L'Épuisé', postface to Samuel Beckett, *Quad et autres pièces pour la télévision, suivi de 'L'Épuisé'*, trans. by Edith Fournier (Paris: Minuit, 1992), pp. 72–75

Frelick, Nancy, *Délie as Other: Toward a Poetics of Desire in Scève's 'Délie'* (Lexington: French Forum Monographs, 1994)

Gray, Floyd, ed., *Anthologie de la poésie française du XVIe siècle* (New York: Appleton-Century-Crofts/Meredith, 1967)

Harley, J. Brian, and David Woodward, eds, *The History of Cartography*, I: *The Medieval World* (Chicago: University of Chicago Press, 1987)

Jeanneret, Michel, *Perpetuum mobile* (Paris: Macula, 1997)

Lestringant, Frank, *Le Livre d'îles: Atlas et récits insulaires de la Genèse à Jules Verne* (Geneva: Droz, 2002)

Mackenzie, Louisa, *The Poetry of Place: Lyric, Landscape and Ideology in Renaissance France* (Toronto: University of Toronto Press, 2011)

Massin, *La Lettre et l'image: La Figuration dans l'alphabet latin du huitième siècle à nos jours* (Partis: Gallimard, 1973)

Ménager, Daniel, 'Conflits et évasion dans "Les Isles fortunées"', in *Writers in Conflict in Sixteenth-Century France: Essays in Honour of Malcolm Quainton*, ed. by Elizabeth Vinestock and David Foster (Durham: Durham University Press, 2008), pp. 21–26

Pinet, Simone, *Archipelagoes: Insular Fictions from Chivalric Romance to the Novel* (Minneapolis: University of Minnesota Press, 2010)

Rigolot, François, *L'Erreur de la Renaissance: Perspectives littéraires* (Paris: Champion, 2002)

Weber, Henri, *La Création poétique au XVIe siècle* (Paris: Nizet, 1958)

Between Stage-Prop and Metaphor: Mirrors in Giovan Battista Della Porta and Giordano Bruno

Sergius Kodera

Introduction

This essay is about real and metaphorical mirrors in the late Renaissance. It relates some texts by Giovan Battista Della Porta (1535–1615) and by Giordano Bruno (1548–1600), which describe mirrors as real objects (for instance, as a tools for visualization) and as metaphors to explain abstract concepts. Mediating between 'inside' and 'outside', between human beings and the cosmos, between matter and form, mirrors function in these texts as tools and at the same time as literary metaphors that have the capacity to render visible what normally cannot be seen or what cannot be grasped by reasoned discourse. I will argue that Della Porta's inventions of new optical tools, such as an improved form of the *camera obscura* or certain cabinets of mirrors, did not only lend themselves to metaphorical uses. With all their playfulness, these material objects were also fascinating to contemporary audiences because they were referring to well-known discourses on reflection.[1] In Bruno the *speculum*

[1] On the intertwining between humanist literary culture and natural objects, especially with regards to collectionism, see Findlen, *Possessing Nature*, pp. 53, 56, and especially pp. 299–301. On these relationships, see also the contributions by Anna Dysert, Hélène Cazes, and Ulrike Feist in this volume.

Sergius Kodera (sergius.kodera@univie.ac.at) is Dean of the Faculty of Design at New Design University, St. Pölten, Austria. He teaches Renaissance philosophy at the Department of Philosophy at the University of Vienna. He is the author of *Disreputable Bodies: Magic, Medicine, and Gender in Renaissance Natural Philosophy*.

The Mirror in Medieval and Early Modern Culture: Specular Reflections, ed. by Nancy M. Frelick, CURSOR 25 (Turnhout: Brepols, 2016), pp. 235–255 10.1484/M.CURSOR-EB.5.111050
BREPOLS PUBLISHERS

videns, for instance, figures not only as a crucial, but rather a fantastic metaphor to explain human consciousness, and Della Porta's *camera obscura* gave this old idea about magic mirrors a kind a material basis. The following study will discuss a few examples in which the discourses on real and imaginary mirrors become intertwined.

Della Porta and Bruno: Two Scholars from the Italian South with Divergent Intellectual Approaches

In his time, Della Porta was certainly much more famous than Bruno, whom he outlived by nearly two decades. Born in 1535 into a family of lower Neapolitan nobility, Della Porta had the means and the leisure to conduct experiments and to publish books which were quickly translated into many vernacular languages. He died in 1615, not only as Europe's most renowned natural magician; Della Porta had also belonged to one of the most advanced circles for the development of what was later termed the scientific revolution. Together with Galileo Galilei and Federico Cesi, he was a founding member of the *Lincei*, the famous academy of the Lynx.[2] A true Renaissance polymath, Della Porta did not only publish on mirrors, optical instruments, and all kinds of magical tricks, distillation, the art of memory, and cyphers, to mention but a few of his subjects, but he was also a prolific and successful writer for the stage, who left more than a dozen comedies and three tragedies.[3] What characterizes his entire literary and scientific work is a deep interest in the production of the marvellous. Della Porta's experimental and scientific endeavour aimed at a kind of histrionic *scienza*, because this erudite magus sought to produce marvels that would leave his audiences stunned.[4] Book XVII of the second edition of

[2] For Della Porta's biography, see Romei, 'Della Porta, Giovambattista'. For a succinct introduction to the intellectual background, and Della Porta's forerunners, see Copenhaver, 'The Occultist Tradition and its Critics', pp. 454–57; Zambelli, *White Magic, Black Magic in the European Renaissance*, chap. 1. On Della Porta's relationship to the *Lincei*, see Freedberg, *The Eye of the Lynx*, pp. 72–75, with many references. On the relationship between Bruno and Della Porta, see Védrine, 'Della Porta e Bruno'.

[3] Still by far the best introduction to Della Porta's theatre is Clubb, *Giambattista della Porta, Dramatist*, and Giovan Battista della Porta, *Teatro*, ed. by Sirri.

[4] Like Marsilio Ficino, Pietro Pomponazzi, Agrippa von Nettesheim, or Girolamo Cardano, Della Porta belonged to an influential group of what Lorraine Daston and Katharine Park have labelled as 'preternatural philosophers' in *Wonders and the Order of Nature, 1150–1750*, pp. 159–64. In this tradition, Della Porta developed a 'sublime science' that was catering to

his enormously popular *Magia naturalis* (1589) describes the shape and use of many different mirrors and lenses: here one encounters semi-translucent mirrors as well as burning glasses, telescopes, and devices by which you can see in the dark; Della Porta also describes lenses and mirrors which he intended for use in what we would now call telecommunication.[5]

At first sight, Giordano Bruno's life and career could not have been more different: born of humble origins in Nola, he attended the elite university of the south, the *Convento di San Domenico Maggiore* in Naples, in 1565. Bruno studied there for more than a decade to become a priest and start a career in the ranks of the Dominican order.[6] This entailed a much more formal training than Della Porta had ever experienced (indeed, in order to graduate, the students at San Domenico had to know the writings of their most famous and now canonical teacher, Thomas Aquinas, virtually by heart).[7] This intellectual backdrop accounts for much of Bruno's theoretical and speculative orientation. More than Della Porta, he regarded himself as a general philosopher-prophet who was not interested in conducting more or less innocuous *sperimenti*. Bruno, rather, and dangerously for his personal safety, insisted on the idea that the Copernican heliocentric cosmology would entail a general revision of the outdated natural philosophy, and, in particular, of Aristotelian geocentrism. Hilary Gatti has aptly characterized Bruno's texts on the finite universe as the first documents

a courtly environment: 'Wonder became a reflection not of ignorance but of virtuosity and connoisseurship; the product not only of great experience and erudition, but also of impeccable taste' (p. 170).

[5] On the impact of the two editions of the *Magia*, see: Balbiani, 'La ricezione della *Magia naturalis* di Giovan Battista della Porta'; and Balbiani, *La Magia naturalis di Giovan Battista Della Porta*, pp. 20–23. On Della Porta's optics, see: Thielemann, 'Lenti e specchi nella scienza e nella pittura del primo Seicento'; Eamon, *The Professor of Secrets*, pp. 197–202, on Della Porta's indebtedness to Ettore Ausonio; on this topic, see also Reeves, *Galileo's Glassworks*, pp. 70–80, for a lucid account of Della Porta's claim to have invented the telescope. For many references to literature on magic mirrors, which allow for telecommunication, see Grabes, *Speculum, Mirror und Looking-Glass*, pp. 140–47. See also the contribution by Berthold Hub in this volume.

[6] For a general and succinct introduction to Bruno, see: Copenhaver and Schmitt, *Renaissance Philosophy*, pp. 290–303 and pp. 315–17; a fine intellectual biography by Rowland, *Giordano Bruno: Philosopher/Heretic*; cf. Ciliberto; *Giordano Bruno*, pp. 7–28, for Bruno's early years as well as on Bruno's monastic experience. For a concise recent summary on Bruno research, see Saiber, *Giordano Bruno and the Geometry of Language*, pp. 9–12. For Bruno's cosmology, see Michel, *The Cosmology of Giordano Bruno*, trans. by Maddison; for an excellent exposition of Bruno's science, see Gatti, *Giordano Bruno and Renaissance Science*.

[7] Miele, 'L'organizzazione degli studi dei Domenicani di Napoli al tempo di Giordano Bruno'.

of what would later be called 'philosophy of science'.[8] From 1576 on, when he fled the Convent in Naples, Bruno had to lead the life of an itinerant scholar, which took him across some of Europe's most important capitals and university towns, always in fierce opposition to the peripatetic academic culture of his day. Eventually, and through a series of serious mishaps, he was captured in Venice in 1592 and handed over to the Roman Inquisition, which had him burned alive as a heretic after a series of trials that lasted eight years.

Setting aside these differences between the two authors, one has to remember that Della Porta's literary production did not escape the attention of the inquisitors either: from 1574 on, he was under the lifelong surveillance of the Roman Holy Office. The ecclesiastical censors prohibited (or severely delayed) the publication of many of his works on magic, physiognomy, and astrology: even so, Della Porta's elevated social position and his extreme prudence, as well as his refusal to write on the larger philosophical (let alone theological) implications of his *scienza*, protected him from suffering more serious personal consequences.[9] Another trait that is common to both authors is that, in order to convey their ideas, they used very different literary registers: as Della Porta with his comedies, so Bruno propagated the intricacies of his new infinitist metaphysics by means of half a dozen (sometimes very!) hilarious dialogues, written in Italian. Moreover, the *Nolanus*, as he called himself, authored one facetious *commedia*, the *Candelaio*.[10] It is in these contexts that the mirror — in its double function as metaphor as well as optical instrument — becomes an important means to propagate some of the most salient ideas of these two Neapolitan scholars.

Bruno's Seeing Mirrors

More frequently than Della Porta, Bruno writes about mirrors in metaphysical and in psychological contexts. The metaphor of the *speculum videns*, a mirror

[8] Gatti, *Giordano Bruno and Renaissance Science*, p. 51.

[9] See Valente, 'Della Porta e l'inquisizione', for the best summary of the proceedings; for a comparison of the court cases of Girolamo Cardano and Della Porta, see Ricci, *Inquisitori, censori, filosofi sullo scenario della Controriforma*, pp. 145–59; on Della Porta's (self-)censorship of the *Physiognomia*, see Trabucco, 'Riscrittura, censura, autocensura' and 'Il corpus fisiognomico dellaportiano tra censura e autocensura'.

[10] For an interesting performative reading of these texts, see: Hufnagel, *Ein Stück von jeder Wissenschaft*; and also Yates, *Giordano Bruno and the Hermetic Tradition*. On the *Candelaio*, see Moliterno, 'Introduction'.

that is not only reflecting but also capable of perceiving and even of producing forms, is a case in point.[11] The *De imaginum, signorum et idearum compositione* (1591), one of Bruno's most sophisticated treatises on mnemonics,[12] describes the mirror of internal vision as different from external vision. Bruno says that a 'seeing mirror' is different from one that does not see (*non videns*).[13] The former is a mirror that is

> [...] quodque simul lux est et speculum, et in quo obiectum sensibile cum subiecto sensibili sunt unum. Hic est mundus quidam et sinus quodammodo inexplebilis formarum et specierum.'[14]
>
> ([...] at the same time illuminated and endowed with forms, which is light and mirror at the same time, and in which the sensible object is identical (*unum*) with the sensible subject. This is a kind of world and in a certain way the insatiable bosom of the forms and species.)

In this mirror of consciousness, subject and object collapse into a unity. At the same time, the *speculum videns* is capable of constantly producing all forms and species. This mirror has, therefore, paradoxical qualities: Bruno says that it functions as a source of unity *and* of diversity, and that this is the way in which nature itself assembles the myriad different species out of a few elements.[15] The

[11] The crucial difference between Bruno's and Cusanus's versions is that the latter is talking not about *specula videntes* but about a mirror that is alive (*speculum vivum*), and this is a metaphor for God, not for individual conscience, as in Bruno; see von Kues, *Philosophisch-theologische Schriften*, ed. by Gabriel, III, 126, 142, 160 (*De visione Dei* 4.12; 8.30–31; 12, 48, 15.63). On the differences between Cusanus and Bruno, see Copenhaver and Schmitt, *Renaissance Philosophy*, pp. 300–03.

[12] For a concise introduction to the topic in Bruno, see Clucas, 'Giordano Bruno's *De imaginum, signorum et idearum compositione*'; Clucas, '*Simulacra et Signacula*'.

[13] Bruno, *Opera latine conscripta*, ed. by Fiorentino and others, II.3, p. 119 (*De imaginum compositione*): 'Tandem differt oculi visus a visu interni spiritus, quem admodum speculum videns a speculo non vidente, sed tantum repraesentante speculum se ipso illuminatum et informatum'; Bruno, *On the Composition of Images*, ed. by Higgins, p. 38: 'Finally, eyesight differs from the seeing power of the internal spirit as a mirror that sees should be distinguished from a mirror that does not see, but is only distinguished by him who represents it as a mirror illuminated and informed at the same time.'

[14] Bruno, *Opera latine conscripta*, ed. by Fiorentino and others, II.3, p. 119 (*De imaginum compositione*); on this passage and the connections between the plasmatic function of phantasy and matter see also Tirinnanzi, *Umbra naturae*, pp. 249–50.

[15] Bruno, *Opera latine conscripta*, ed. by Fiorentino and others, II.3, p. 119 (*De imaginum compositione*): 'Rursumque sicut ex paucis elementis natura innumerabiles species componuntur

mirror is therefore a source of unity and of multiplicity. Bruno thus maintains that the metaphor of the *speculum videns* allows us to understand the basic unity of all things as well as the synthetic nature of individual conscience.[16] In the quote above, Bruno uses an optical metaphor (seeing mirror) and a somatic metaphor (bosom of forms) to explain a highly abstract set of metaphysical concepts, that is, the relationship between unity and multiplicity as well as the analogy between individual consciousness, and the creative acts of nature.

This malleable and plasmatic *speculum videns* is not the only metaphorical mirror in Bruno, who also talks about shattered mirrors, again in a context in which the mirror functions as medium between unity and multiplicity. Bruno argues that every individual mind (as mirror) reflects the infinite creation at least in part, just as the pieces of a broken mirror are still capable of making visible fragments of the whole. In *Lampas triginta statuarum* (1591), another of his mnemonic treatises, Bruno uses the metaphor of the broken mirror to demonstrate this central theorem of his philosophy;[17] here again the malleabil-

et coalescunt, ita et opere istius intrinseci efficientis non solum specierum naturalium formae in isto amplissimo sinu reservantur, verum quoque ad innumerabilium imaginum multiplicationem improportionabiliter concipiendarum'; Bruno, *On the Composition of Images*, ed. by Higgins, p. 39: 'And, on the other hand, just as in nature innumerable species are composed of and coalesce out of small elements, so too by the action of this intrinsic cause not only are the forms of natural species preserved within this most ample inlet, but also will they be able to be multiplied for the multiplication of the innumerable images conceivable beyond compare.'

[16] See for instance, Bruno, *Opera latine conscripta*, ed. by Fiorentino and others, II.3, p. 305 (*De imaginum compositione*). On this, see the remarks by Suzanne Conklin Akbari in this volume on Christine de Pizan and her search for self-reflection that is not identical with narcissism.

[17] For this and the following, see Bruno, *Opera latine conscripta*, ed. by Fiorentino and others, III, 59 (*Lampas triginta statuarum*): 'Cum materia sit causa multitudinis et divisionis, forma vero unitatis, dicimus fulgorem divinitatis spiritum esse per se unum et facere unum (ab uno enim secundum quod unum non procedit nisi unum), tamen quia est, operatur in universo extento et materiali, quo quidem divisionem recipiente et in partium multiplicationem materiam distribuente accidit multitudo, ut ea anima quae in toto tota et in uno una videbatur, iam, in multa veluti fragmenta distracto corpore et in diversas hypostases numerales multiplicato, multae fiunt animae, sicut multa sunt subiecta, et totidem producuntur animalia, vel saltem nihilominus animata corpora, quamvis non ubique anima speciem viresque suas exerat, unde quaedam sine anima a quibusdam iudicantur. Quod ita ferme est, quemadmodum, si unus sit sol et unum continuum speculum, in toto illo unum solem licebit contemplari; quod si accidat speculum illud perfringi et <in> innumerabiles portiones multiplicari, in omnibus portionibus totam repraesentari videbimus et integram solis effigiem, in quibusdam vero fragmentis vel propter exiguitatem vel propter infigurationis indispositionem aliquid confusum vel prope nihil de illa forma universali apparebit, cum tamen

ity (despite the obvious fragility of the looking glass) is the most salient quality of the mirror. Due to its plastic quality, the mirror also becomes a metaphor for matter, for Bruno maintains that matter (like the mirror of consciousness mentioned above) produces the myriad forms from its *sinus*, just as the mirror is not only capable of absorbing any form that is put in front of it, but also of reflecting that image. The mirror thus recreates a form it has received into a new shape or mode of being. The relationship between form and matter can thus be explained by means of an optical metaphor: as optical fraction, as a distortion of an image in a mirror. In this process the mirror has to be imagined not merely as hard and brittle: the looking glass is also malleable. Bruno's mirror metaphors are thus evocative of objects which are hard and soft at the same time: the qualities of the metaphorical looking glasses are as paradoxical as the more or less elusive abstract concepts (consciousness, matter) that this optical tool is supposed to elucidate. Such paradoxical qualities are of course to be found in water, the surface of which is the oldest mirror, and this leads us to

nihilominus insit, inexplicata tamen. Itaque si quemadmodum uno perfracto speculo propter partium multiplicationem animalium animarum multiplicata sunt supposita, si accidat iterum partes omnes in unam massam coalescere, unum erit speculum, una forma, una anima, sicut si omnes fontes, flumina, lacus et maria in unum concurrant Oceanum, unus erit Amphitrites' (As matter is the cause of multitude and of division, and form [is the cause of] unity, we say that the splendour of the divinity is a spirit which is one in itself and which brings forth one (because from the one as one only the one appears), yet, because this is so, it operates in the extended and material universe, in which, by means of a certain division of the recipient and of the multiple parts of dividing matter occurs the multitude: in such a way that the soul which seemed to be entirely and wholly one, already, as if torn apart into many pieces in the body and multiplied into different hypostases, of which as many pertain to soul just as many are subject to it and as many bring forth animals or in any event at least animated bodies, yet not everywhere, but according to the extent to which soul may stretch forth its species and forces (which is why it was judged to be without soul by some). Which is almost as if the soul was, as it were, but one continuous mirror, in which all this sun in which one may contemplate the one sun in its entirety; and [the division of soul is] what happens when this mirror breaks and multiplies into innumerable parts, then we see in all these parts an untouched representation and image of the whole sun, yet in certain fragments there appears really nothing (*propre nihil*) or this universal form only confusedly, even though it is inherent to it, but not in its unfolded state (*inexplicatus*): and this either because of the [mirror's] smallness or because of its wrong arrangement which is confused in some way. Therefore, just as in this one broken mirror the parts are multiplied, so the animals are to be supposed to be under multiple souls: yet, if they would happen to become unified again in one mass there would be one mirror, one form, one soul just as all fountains, rivers, lakes and seas would concur in one Ocean, there would be only one Amphitiries) (author's translation). On the analogy of the broken mirror, see also Marsilio Ficino, *The Philebus Commentary*, ed. and trans. by Allen, pp. 211–13 (*In Philebum* I, 22).

non-metaphorical uses of mirrors in Bruno's cosmology: he says that the oceans reflect sunlight, and thus are functioning like gigantic mirrors. Bruno concludes that this is the reason why the planets in our solar system become visible to us during the night. The infinitely numerous suns in the universe outshine the dim light that is emitted by their satellites, because the reflected light is visible only at much smaller distances than the *vera lux* (true light) of the suns. Bruno adapts these observations to his cosmology in concluding that most of the stars we see in the night sky are actually suns, centres of other planetary systems, while the rest of the celestial bodies — the planets in a solar system — are visible only at comparatively short distances. In a Platonic vein of thought, Bruno argues that the mirror is incapable of producing the objects themselves (*res ipsae*); it brings forth only their reflections or shadows (*umbrae*).[18] Mirrors, therefore, are tools that have the capacity to render things visible; but this capacity is limited. Bruno departs from the Platonic tradition in that, for him, this is not entirely negative, because the shadowy images are making visible to limited human minds what cannot be seen with the eye and what can be experienced only at the peril of the individual.[19]

Bruno's observation that mirror images are of inferior quality to the originals they represent is quite accurate, given the poor quality of early mirrors (even after the invention of Venetian *cristallo*, in the mid-fifteenth century).[20]

[18] Bruno, *Centoventi articoli*, ed. by Canone, p. 56: 'Ipsorum tellures seu aquae non sunt visibiles propter distantiae momentum, et quia sunt minora corpora, utpote ita ad suos soles proportionabiles, ut istae ad unum istum, et quoniam quae speculariter lucent, non adeo longius visibilitatis diametrum servant, atque principaliter splendentia. [...] E speculo enim propinquo lumen simile apparet atque e propinqua candelae luce, sed ambobus distantiam nactis, longe, incomparabiliterque citius visio specularis quam verae lucis evanescit'; Bruno, *Opera latine conscripta*, ed. by Fiorentino and others, I.1, pp. 179–80 (*Camoeracensis Acrotismus seu rationes articulorum physicorum*, art. LXVII): 'These earths or waters are invisible because of the importance of distance, and because they are smaller bodies, that is in proportion to their [respective] sun, just as [the proportion of the planets] of this solar system to their [sun] and because [the bodies] which emit light by reflection (*speculariter lucent*) are less visible in their diameter than those which shine without intermediary (*principaliter splendentia*). For a nearby light which is reflected in a mirror (*lumen*) looks similar to the light which is emitted by a candle nearby, but if it lights both from a greater distance the vision in the mirror fades away incomparably faster from the candle than from the true light (*vera lux*)' (author's translation).

[19] On these, which Bruno associates with the myth of Actaeon in his dialogue the *Eroici furori*, see, for instance, Tirinnanzi, *Umbra naturae*, pp. 217–28.

[20] See Schechner, 'Between Knowing and Doing', pp. 145–60, on the poor and distorting quality of early mirrors, well into the eighteenth century; on *cristallo*, an almost colourless flat glass, that was covered with lead, and later tin foil, see pp. 148, 153–54.

Viewed from this perspective, we have so far encountered four different mirrors in contexts which were well known to medieval and early modern readers: the surface of water (in which the beholder may drown); the polished metal mirror (which reflects only dark images); the looking glass (which may break); and the *speculum videns* (a set of magical tools, closely related to the human imagination, because these mirrors have the alleged capacity to visualize what cannot be seen).[21] The last kind of looking glass finds its material representation especially in mirrors, which render distorting images. In a remarkable passage from his didactic poem *De monade, numero et figura* (1591), Bruno says that the different species of animals are produced by distorting the one single original intelligible species, just as the shapes of an object are altered in curved mirrors.

> Nam varie in variis eadem est impressa figura, | Ut speculi varia est sors, vis, positura tomorum,| Materies, magis atque minus per imaginis actum | Unius, illustris. [...] | At species eadem, pro appulsu materiei, | Hanc unam oblongat formam, sinuatque recurvo | A speculo, heic vultum canis et bovis atque elephantis | Efficit, in longos artus trahit inde lacertas, | Spargitur in ramos eadem hic substantia plantae, | Omneis telluris spectas in corpore partes. | Nilque adeo exiguum est, quod sensibus obiiciatur, | Quod non et speciem hanc referat pro viribus unam.[22]

> (The same figure is impressed differently into different things, because the fate, force, the disposition of the swellings, and the material varies in mirrors. Consequently, they are more or less capable of delivering that single image brightly. [...] While through the impact of matter, this one species becomes oblong in its form, and bends into a curve, [just as] in this mirror the faces of dogs, of oxen, and of elephants are transformed into long arms and legs, and drawn into the shape of lizards, and likewise the substance of the plant is unfolded into the branches, and all parts of the earth are to be seen in [one] body. And nothing can be so exiguous which is present to the senses that it would not refer (to the best of its ability) to this single species.)

The creation, development, and shape of all things out of a single substance, is, therefore, the result of a veritable *ars deformationium* (art of disfigurement), such is the title of one of Bruno's last mathematical manuscripts (1591).[23]

[21] On such magic mirrors, see Grabes, *Speculum, Mirror und Looking-Glass*, pp. 140–47; Kodera, *Disreputable Bodies*, pp. 100–05.

[22] Bruno, *Opera latine conscripta*, ed. by Fiorentino and others, I.2, pp. 328–29 (*De monade*, author's translation).

[23] The translators and editors of *De monade* — Bruno, *Über die Monas, die Zahl und die Figur als Elemente*, ed. by von Samsonow and Mulsow, pp. 253 ff., have pointed to Della Porta's works on physiognomy in that context.

Bruno's analogies with contorting creative mirrors are of course related to the many anamorphic experiments in sixteenth-century visual arts.[24] In this respect Bruno's and Della Porta's works are indirectly related to each other, because the construction and practical use of such mirrors to create marvellous apparitions is an important topic in Della Porta's *Magia naturalis*.

Della Porta's Mirror Cabinets

Della Porta describes the use of coloured or distorted mirrors that make people appear as animals (for instance, asses, dogs, pigs) as well as different forms or projections, or mirrors that produce contorted images.[25] Andreas Thielemann has recently published a detailed and fascinating account of one of Della Porta's more spectacular optical devices, a long glass cylinder that was reflective on the inside. This mirror not only produces highly complex anamorphotic images: when the beholder is changing his or her position, the magic cylinder renders optical transformations of the objects that are laid out at its bottom.[26] I would argue that Della Porta was thus trying to construct looking glasses that embodied the qualities of some of mirrors, which Bruno and others had used as metaphors in their metaphysics.[27] This of course does not mean that either Bruno or

[24] I am of course referring to Hans Holbein's *Ambassadors* (1533), the most renowned example. For an excellent introduction to the topic, with many examples, see Stafford and Terpak, *Devices of Wonder*, pp. 239–47. The term *anamorphosis* is actually slightly anachronistic, as the Jesuit Gaspar Schott is credited for having used it first in the seventeenth century in order to describe the distorted representations of bodies produced by using irregularly shaped mirrors: Findlen, 'Jokes of Nature and Jokes of Knowledge', pp. 322–23, at 323.

[25] Cf. Della Porta, *Magia*, pp. 547–48 (bk XVII, chap. 1) This topic is also present in Della Porta's theatre, for instance, in the complicated plot of the *Sorella*, in which 'Central to that idea [of the play] were the signature events, namely the discovery that a person feigning to be someone really is that person — a play upon the baroque commonplace regarding the convergence and confusion of appearances and reality'; Giovan Battista della Porta, *The Sister*, trans. by Beecher and Ferraro, p. 20.

[26] See Thielemann, 'Lenti e specchi nella scienza e nella pittura del primo Seicento', pp. 143–49, who says on p. 147: 'Gli oggetti nel cilindro non sono soltanto alzati, come dicono gli autori del seicento, ma anche deformati secondo la loro forma ed il punto di vista dello spettatore, e si vede nascere un uovo' (The objects in the cylinder appear not only elevated as the authors from the sixteenth century say, but also deformed in relation to their form and from the point of view of the beholder, so that one can see an egg cracking open) (author's translation).

[27] As Schechner, who is a museum curator and specialist in historical mirrors, argues from a slightly different perspective: 'There were many books printed in the sixteenth and seventeenth centuries that depicted marvellous instruments that never existed or if constructed, would not

Della Porta were directly referring to each other: in fact we have seen that the context of Bruno's infinitist metaphysics is quite different from Della Porta's spectacular science; yet from the last decades of the sixteenth century onwards, both fields of interest increasingly tended to overlap. In the seventeenth and eighteenth centuries this tendency resulted in the creation of outright anamorphotic games.[28]

In this ludic context, Della Porta also seems to have been working on what Bruno would have called *specula videntes*, because he invented a new and improved form of *camera obscura*.[29] Della Porta was among the first to experiment with sets of convex mirrors and concave lenses.[30] He observed that certain combinations of these glasses could be used for an improved form of the *camera obscura*, for they would render images standing on their feet right side up on the opposite wall instead of rendering the image upside down.[31] Della Porta

have performed as well as described nor been widely used. These books read like science fiction, and their primary goal was not to instruct readers in the use of instruments but to impress patrons of the author' ('Between Knowing and Doing', p. 141).

[28] Findlen, 'Jokes of Nature and Jokes of Knowledge', p. 323: 'In part, we can view the creation of anamorphic games as a social message, an intellectual metaphor for the age. Within the narrower parameters of scientific play, however, it exemplifies the desire to traverse the boundaries between the natural and the unnatural world.'

[29] On this topic, see Hammond, *The Camera Obscura*. One of Della Porta's very important magical sources describes the *camera obscura* as well as certain mirrors that create apparitions: Agrippa von Nettesheim, *De occulta philosophia libri tres*, pp. 97–98 (bk i, chap. 6).

[30] Della Porta explicitly states that these glasses are a contemporary invention, *Magia naturalis*, p. 552 (bk xvii, chap. 3): 'Speculum autem è planis compactum, cui si unum spectabile demonstrabitur, plura illius rei simulacra demonstrabit, prudens invenit vetustas: ut ex quibusdam Ptolemaei scriptis, quae circumferentur, percipitur, demum quod nostra adinvenit aetas adiungemus, longe mirabilius & jucundius'; Della Porta, *Natural Magick*, p. 359: 'Prudent antiquity found out a Looking-glass made of plain glasses, wherein if one object might be seen, it would represent more images of the same thing, as we may perceive by some writings, that go in Ptolemy's name. Lastly, I shall add to this what our age has invented, that is far more admirable and pleasant.' See also Reeves, *Galileo's Glassworks*, p. 71 and pp. 90–93; Eamon, *The Professor of Secrets*, p. 200.

[31] Della Porta, *Magia naturalis*, p. 562 (bk xvii, chap. 6): 'Oppositio foramini specillium è convexis fabricatum, inde in speculum concavum resiliat. Distet speculum concavum à centro, nam imagines, quas obversas recepit, rectas reddit, ob centri distantiam. Sic supra foramen, & papyrum albam jaculabit imagines rerum objectarum, tam clare, & perspicue, un non satis laetari, non satis mirari possis'; Della Porta, *Natural Magick*, p. 364: 'Put against a hole a convex glass; from thence the Image reflect on a Concave glass. Let the Concave glass be distant from the Centre, for it will those images right that it receives turned, by reason of the distance of the Centre. So upon the hole and the white paper, it will cast the Images of the objects so clearly and

recommends this invention not only to anyone who wishes to draw persons or landscapes:[32] with the help of this optical tool, he also seems to have imagined the development of a form of theatrical projection, a kind of proto-cinema. The spectators were sitting in the dark chamber and would watch the projections of carefully orchestrated shows that were performed outside in the open, or as Della Porta boasts: 'Those that are in the chamber shall see trees, animals, hunters' faces, and all the rest so plainly, that they cannot tell whether they be true or delusions.'[33] It was a new kind of improved Platonic cave, or more appropriate to Della Porta's cultural context, his *camera obscura* can be inserted into the old tradition of more or less fantastic shows with magic lamps that were performed in darkened banquet rooms.[34] But this is not all: by dint of its capacity to connect 'interiors' and 'exteriors', Della Porta recommends his tool for communicating with people who are being held 'in the remotest prisons'.[35]

plainly, that you will not wonder a little.' For diagrams that show the different combinations of concave lens and convex mirror, to achieve these effects, see Reeves, *Galileo's Glassworks*, p. 12.

[32] Della Porta, *Magia naturalis*, p. 561 (bk XVII, chap. 6): 'Ut quisque picturae ignarus rei alicuius, vel hominis effigiem delineare possit. Dum modo solum colores assimilare discat'; Della Porta, *Natural Magick*, p. 364: 'If you cannot draw a picture of a man or anything else, draw it by this means; If you can but only make the colours.' It is doubtful whether Della Porta was actually capable of obtaining concave mirrors (let alone to have them produced in sufficient quality). On this, as well as the context of what has become known as the Hockney hypothesis, see Schechner, 'Between Knowing and Doing', pp. 137–40 and pp. 154–55; Reeves, *Galileo's Glassworks*, p. 78.

[33] Della Porta, *Natural Magick*, pp. 364–65; Della Porta, *Magia naturalis*, pp. 562–63 (bk XVII, chap. 6): 'qui enim in cubiculo adsunt, arbores, animalia, venatorum vultus, & reliquia conspicent, ut nesciant an vera, an praestigia sunt.'

[34] For just one instance, see Thomas Cantimpratensis, *Liber de natura rerum*, ed. by Boese, p. 108: 'Lacrima eius [sc. Asini] cum oleo concussa et mixta in lucernam et accensa, omnes in convivio videbunt habere capita asinina' (When you shake the tears of an ass together with oil and mix it in a lantern, and light it, all the banqueters will see themselves with asses' heads) (author's translation). Della Porta has many similar recipes, and Dupré, 'Images in the Air', accordingly says that 'the jocular and the demonic were mutually complementary forces in Renaissance culture.' (84). On the continuity of magical thinking in visual media of the twentieth century, see Stafford and Terpak, *Devices of Wonder*, pp. 1–142.

[35] Della Porta, *Magia naturalis*, p. 563 (bk XVII, chap. 6): 'Hinc rei conscio quispiam occulte narrandi auspicari poterit principia, quae voluerit, ut remote carceribus occluso'; Della Porta, *Natural Magick*, p. 365: 'From hence may one this principle of declaring any thing to one that is confederate with him, that is secret, though the party be far off, shut up in prison.' On the topic of insides and outsides, especially on the connection between mirrors and windows, see the contribution by Elizabeth Black in this volume.

He also claims that his *camera obscura* functions like an artificial eye, and that the apparatus thus explains the principle of vision.[36]

Moreover, the exact same combination of convex and concave lenses and mirrors is also identical to the principle of the very telescope (this time a material and not a metaphorical *speculum astronomiae*)[37] used by Galileo in 1609 for his ground-breaking astronomical observations. We have mentioned that in Bruno's metaphysics a metaphorical *speculum videns* acted as bridge and leveller between 'insides' and 'outsides'. Now, I would contend, a real mirror (again a *speculum videns* of some kind in the hands of Della Porta) is used by Galileo to argue for the essential material unity of the universe. For, by viewing the surface of the moon through his telescope, Galileo could demonstrate that the surface of celestial bodies is as irregular as the terrestrial geography. He was thereby eroding the peripatetic doctrine that the stars beyond the sphere of the moon are ideal geometric bodies and that they, therefore (and unlike the earth) moved in regular orbits. It is well known that Della Porta and Galileo competed with each other as to who had made the invention first; clearly Della Porta preceded Galileo, but he intended the tool for very different uses.[38] Della Porta's telescope was a 'device of wonder'.[39] Yet, I think it is noteworthy to observe that Della Porta uses devices other than the telescope for visualizing the cosmos. It is in this vein of spectacular cosmography that Della Porta describes a mirror-

[36] Della Porta, *Magia naturalis*, p. 562 (bk XVII, chap. 6): 'Hic Philosophis & opticis patet, quo nam fiat visio loco, ac intromittendi dirimitur quaestio, sic antiquitus exagitata, nec alio utrumque artificio demonstrare poterit. Intromittitur idolum per pupillam, fenestrae foraminis instar, vicemque obtinet tabulae crystalline sphaerae portio in medio oculi locata, quod scio ingeniosis maxime placitum'; Della Porta, *Natural Magick*, p. 365: 'Hence it appears to Philosophers and those that study Opticke, how vision is made; and the question of intromissions taken away, that was anciently so discussed; nor can there be any better way to demonstrate both, than this. The image is let in by the pupil, as by the hole of a window; and that part of the Sphere, that is set in the middle of the eye, stands in stead of a crystal Table. I know ingenious people will be much delighted by this.' Vanagt, 'Early Modern Medical Thinking on Vision and the Camera Obscura', at pp. 569–71, discusses Della Porta's thesis in light of later medical literature on the subject.

[37] On the classification of astronomy as a 'science of mirrors' from the twelfth century on, see Grabes, *Speculum, Mirror und Looking-Glass*, p. 147.

[38] See Reeves, *Galileo's Glassworks*, pp. 70–78.

[39] For a concise introduction to this topic as well as on Cesare Cremonini's famous refusal to look through Galileo's telescope, see Drake, *Galileo*, as well as the account in Biagioli, *Galileo, Courtier*, pp. 95–97 and 238 on the central role of the telescope as an instrument in Galileo's physics.

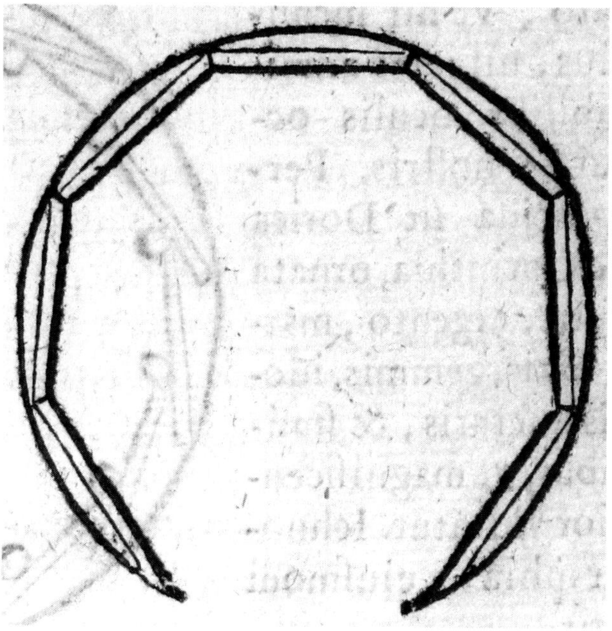

Figure 10.1.
Della Porta, *Magia*, p. 553.
Courtesy Universitäts-bibliothek, Wien.

cabinet in which one can see an infinitely large number of ordered stars. What he calls 'amphitheatrical' glasses are concentrically arranged sets of plain glass mirrors facing each other inside a box. One mirror is to be left out so that one can peep into the box.[40] The illustration from the *Magia* shows the ground plan of Della Porta's catoptric device (Fig. 10.1).

This is one of Della Porta's instructions on how to use this catoptric device:

> Si in medio candelam statuas, adeo ex resilientibus imaginibus, multiplicari videbitur, ut in coelo tot sydera non conspicias, ut non satis admirari queas, ordinem, symmetriam, et prospectum.
>
> (If you set a candle in the middle, it will seem so to multiply by the images rebounding, that you shall not see so many stars in the sky, that you can never wonder enough at the order, symmetry, and prospect.)[41]

[40] Della Porta, *Magia naturalis*, pp. 550–52 (bk XVII, chap. 2); for a reproduction of the illustration from the editio princeps of the *Magia naturalis*, see Grabes, *Speculum, Mirror und Looking-Glass*, pl. 15 (facing p. 121).

[41] Della Porta, *Magia naturalis*, p. 555 (bk XVII, chap. 3) (this is an anonymous translation from the seventeenth century, see *Natural Magick*, p. 359).

Della Porta's mirror-cabinet thus renders visions of a cosmos that is more perfect, and more regularly ordered, than the real sky. Unlike Galileo's and Bruno's universe, the candle in the centre of this catoptric device will create an immobile, geometrically ordered image of countless dots of light spots. This image of a noetic, utopian cosmos is reminiscent of the Neoplatonic metaphysics of light, according to which the entire cosmos is illuminated, and in that process shaped, and created by a single source of light (the ineffable one) in a process that does not happen in time.[42] Della Porta's marvellous image of a static and ordered infinite cosmos would have been completely unacceptable for Bruno: neither can you see the infinite universe (simply because its extension precludes any optical manifestation), nor do the celestial bodies move according to the strict rules of geometry (because they are animated and therefore have irregular shapes, just as the bodies we encounter here down on earth). I would like to suggest that Bruno must have been aware of such amphitheatrical glasses and that he repeatedly used this (perhaps only vague) knowledge in the context of one of his acrimonious caricatures of Peripatetic cosmology. In a famous passage of the *Cena delle ceneri*, he describes the Peripatetic geocentric world as a form of narcissism, because in this cosmology human beings perceive themselves as the centre and pinnacle of Creation: an error from which his *Nolana filosofia* was to free mankind. In a metaphorically rampant, prophetical mood Bruno speaks of himself as the one who 'ha donato gli occhi a le talpe, illuminati i ciechi che non possean fissar gli occhi e mirar l'imagin sua in tanti specchi che da ogni lato gli s'opponeno' (gave eyes to the moles and light to the blind, who could not fix their gaze and see their image reflected in the many mirrors which surround them on every side).'[43] This quote already bears a remote echo of Della Porta's catoptric device. In the *De l'infinito, universo e modi* (1584) Bruno gets more specific, because here he seems to have found it expedient to further elaborate on this metaphor[44] (and he therefore does not mention, this time, blind people who see themselves in mirrors):

[42] On this set of ideas, as developed by Marsilio Ficino, see Kodera, *Disreputable Bodies*, pp. 73–84.

[43] Bruno, *Dialoghi italiani*, ed. by Gentile and others, I, 33; Bruno, *The Ash Wednesday Supper*, trans. by Jaki, p. 55.

[44] On the textual history of these dialogues, see the remarks made in the first dialogue of *Della Causa*, in Bruno, *Dialoghi italiani*, ed. by Gentile and others, I, 191–224; Hufnagel, *Ein Stück von jeder Wissenschaft*, pp. 117–28.

> Quello dunque che ha fatto imaginar diversi cieli, son stati gli diversi moti astrali, con questo, che si vedeva un cielo colmo di stelle svoltarsi circa la terra, senza che di que' lumi in modo alcuno si vedesse l'uno allontanarsi da l'altro, ma, serbando sempre la medesima distanza e relazione, insieme con certo ordine, si versavano circa la terra non altrimenti che una ruota, in cui sono inchiodati specchi innumerabili, si rivolge circa il proprio asse. Là onde è stimato evidentissimo, come al senso de gli occhi, che a que' luminosi corpi non si conviene moto proprio, come essi discorrer possano, qual uccelli per l'aria; ma per la revoluzion de gli orbi, ne'quali sono affissi, fatta dal divino polso di qualche intelligenza.[45]

> (It was the different kinds of motions of the stars that gave rise to the fancy of the diverse heavens, together with one's view of the sky, [which appears to be] full of stars that rotate around the earth; because we cannot see that these lights are moving farther away from each other; in a fixed order, they rather [seemed to be] keeping always the same distance and ratio to each other and to turn around the earth like no other than a wheel which is turning on its own axis, and onto which innumerable mirrors are nailed. And from this assumption it is estimated most evidently, as though from ocular proof, that these luminous bodies have no motion of their own, by means of which they could fly through the air as birds do, [for they are rather believed] to be moving through orbs, onto which they are fixed [and their movement being caused] by the divine impetus of a certain intelligence.)

Bruno's shabby cabinet of mirrors is a caricature of an Aristotelian cosmology, which he perceived as anthropomorphic, and as fundamentally narcissistic, in the face of an infinitely large universe.[46] Moreover, I think it is obvious that Bruno is here describing an optical device that Della Porta would have recognized as a malicious caricature of his 'amphitheatrical glasses'. Of course, there are important differences: here the spectator is not outside the catoptric device, as in Della Porta. Bruno's merry-go-round with shattered mirrors is moving around its axis; in the centre, there is not a candle (or another source of light) but a human being whose image is infinitely reflected in the mirrors that encircle him entirely — thus allowing for a kind of grotesque narcissistic speculation in endless mirrors. Yet, we also have to remember that with the metaphorical use of this rotating mirror-cabinet Bruno is actually trying to come to grips with a problem that plagued his infinitism. For, because there is no image of

[45] Bruno, *Dialoghi italiani*, ed. by Gentile and others, I, 433–34 (author's translation). On the idea of crystalline spheres and solid orbs in Scholastic cosmology, see the excellent discussion in Grant, *Planets, Stars, and Orbs*, pp. 324–70.

[46] Bruno will come back to this topic once again, in his Latin epic, the *De immenso* (1591); see Bruno, *Opera latine conscripta*, ed. by Fiorentino and others, I.2, p. 125.

the infinite universe, it is neither visible, nor can it be fully grasped by means of advanced intellectual operations: therefore, any ocular proof against peripatetic geocentric cosmology is begging the question. Indeed, large parts of the *Cena delle Ceneri* are devoted to the task of replacing the evidence generated by eyesight with an awareness that, in order to reflect the immensity of the one infinite material universe, any discourse must needs be fragmented, distorted like the image in an uneven mirror, and dependent on time.[47] The infinity of the universe can only be inadequately represented by a process of inner reflection, a process that happens in what we come to know as Bruno's *speculum videns*.

Conclusion

I hope to have brought evidence to the claim of reciprocal relationships between real and metaphorical mirrors in two authors with quite different approaches to the topic, Della Porta and Bruno. In this study, I have looked at material mirrors with amazing representational capacities, which, in tandem with their invention, influenced metaphorical explanations of theoretical problems, for instance, in astronomy. (This was the case of Bruno's indirect reference to Della Porta's amphiteatrical glasses.) And I have also shown how one may relate Della Porta's improved version of the *camera obscura* to Bruno's concept of a *speculum videns*. In this case, the old (magical) dream of the seeing mirror finally had materialized in a spectacular optical device ('spectacular' in the true sense of the word) that mediated between insides and outsides, that could be used as a party gag, on a social and a playful level, as well as an instrument with which to explore the cosmos and to see images no human eye had seen before. These examples from the history of objects actually show how their use is codified by

[47] See for instance, Bruno, *Dialoghi italiani*, ed. by Gentile and others, I, 29 (*Cena delle ceneri)* where Teofilo says of Copernicus that '[…] quelli abietti e rugginosi fragmenti ch'ha possuto haver le mani da la antiquità, le ha ripoliti, accozzati e risaldati in tanto […] ch'ha resa la Causa, già ridicola, abietta e vilipresa, onorata, preggiata, piú verisimile che la contraria'; Bruno, *The Ash Wednesday Supper*, trans. by Jaki, p. 57: 'he had put together those rejected and rusty fragments which he could have from the hands of antiquity, and repolished them to such an extent that the argument once ridiculed, rejected and vilified, but now respected, appreciated and possessed of greater likelihood than its contrary.' See also, with similar cosmological undertones, Bruno, *Dialoghi italiani*, ed. by Gentile and others, II, 835–36 (*Cabala del cavallo pegaseo*). In deliberately highlighting the fragmented and provisional character of human knowledge, Bruno neglects the rules set up by Counter Reformation theologians who prohibited ambiguity in all literary production. See Cox, *The Renaissance Dialogue*, chap. 1, for a discussion of the different genres of the dialogue and their assessment by Renaissance censors.

ancient literary use, for instance, as analogy in metaphysics, and how these uses are formative in the understanding of their practical applications and vice versa. Both Della Porta and Bruno are, therefore, using the mirror to visualize things that under normal circumstances cannot be seen; the former in order to promote his personal ingenuity as a courtly administrator of marvels; the latter in order to use them as non-iconic, contradictory representations of the infinite cosmos. In using mirrors in different contexts, both Bruno and Della Porta had an acute awareness of the enormous power that could be exerted by dint of such media on the human imagination, and therefore on our perception of the world, and our choices.

Works Cited

Primary Sources

Agrippa von Nettesheim, Heinrich Cornelius, *De occulta philosophia libri tres*, ed. by V. Perrone Compagni (Leiden: Brill, 1992)

Bruno, Giordano, *The Ash Wednesday Supper*, trans. with intro. and notes by Stanley L. Jaki (The Hague: Mouton, 1975)

——, *Centoventi articoli sulla natura e sull'universo contro i peripatetici/Centum et viginti articuli de natura et mundo adversus Peripateticos*, ed. by Eugenio Canone, trans. by Carlo Monti (Pisa: Fabrizio Serra, 2007)

——, *Dialoghi italiani*, ed. by Giovanni Gentile and others, 2 vols (Florence: Sansoni, 1985)

——, *On the Composition of Images, Signs & Ideas*, ed. by Dick Higgins, trans. by Charles Doria (New York: Willis, Locker & Owens, 1991)

——, *Opera latine conscripta*, ed. by Francesco Fiorentino and others, 3 vols in 8 pts (Naples: Domenico Morano 1879–91)

——, *Über die Monas, die Zahl und die Figur als Elemente einer sehr geheimen Physik, Mathematik und Metaphysik*, ed. and trans. by Elisabeth von Samsonow and Martin Mulsow (Hamburg: Meiner 1991)

Della Porta, Giovan Battista, *Magia naturalis libri* XX (Rouen: Johannes Berthelin, 1650)

——, *Natural Magick*, facs. edn (New York: Basic Books, 1957) (orig. publ. London: Young and Speed, 1658)

——, *The Sister*, trans. by Donald Beecher and Bruno Ferraro (Ottawa: Dovehouse, 2000)

——, *Teatro*, ed. by Raffaele Sirri, 4 vols (Naples: Edizione scientifica Italiana, 2000–03)

Ficino, Marsilio, *The Philebus Commentary*, ed. and trans. by Michael J. B. Allen, (Berkeley: University of California Press, 1979)

Kues, Nikolaus von, *Philosophisch-theologische Schriften* [Latin and German], ed. by Leo Gabriel, trans. by Dietlind and Wilhelm Dupré, 3 vols (Wien: Herder, 1964–67)

Thomas Cantimpratensis, *Liber de natura rerum: Editio princeps secundum codices manuscriptos*, ed. by Helmut Boese (Berlin: de Gruyter, 1973)

Secondary Works

Balbiani, Laura, *La Magia naturalis di Giovan Battista Della Porta: Lingua, cultura e scienza in Europa all'inizio dell'età moderna* (Bern: Lang 2001)
——, 'La ricezione della *Magia naturalis* di Giovan Battista della Porta nella Cultura e scienza dall'Italia all'Europa', in *Bruniana et Campanelliana*, 5 (1999), 277–303
Biagioli, Mario, *Galileo, Courtier: The Practice of Science in the Culture of Absolutism* (Chicago: University of Chicago Press, 1991)
Ciliberto, Michele, *Giordano Bruno* (Rome: Laterza, 1990)
Clubb, Louise George, *Giambattista della Porta, Dramatist* (Princeton: Princeton University Press, 1964).
Clucas, Stephen, 'Giordano Bruno's *De imaginum, signorum et idearum compositione*: Art, Magic and Mnemotechnics', *Physis: Rivista internazionale di storia della scienza*, 38 (2001), 75–98
——, '*Simulacra et Signacula*: Memory, Magic and Metaphysics in Brunian Mnemonics', in *Giordano Bruno: Philosopher of the Renaissance*, ed. by Hilary Gatti (Aldershot: Ashgate, 2002), pp. 251–72
Copenhaver, Brian P., 'The Occultist Tradition and its Critics', in *The Cambridge History of Seventeenth Century Philosophy*, ed. by Daniel Garber and Michael Ayers, 2 vols (Cambridge: Cambridge University Press, 1998), II, 454–512
——, and Charles B. Schmitt, *Renaissance Philosophy* (Oxford: Oxford University Press, 1992)
Cox, Virginia, *The Renaissance Dialogue: Literary Dialogue in its Social and Political Contexts. Castiglione to Galileo* (Cambridge: Cambridge University Press, 1992)
Daston, Lorraine, and Katharine Park, *Wonders and the Order of Nature, 1150–1750* (New York: Zone Books, 1998)
Drake, Stillman, *Galileo* (Oxford: Oxford University Press, 2001)
Dupré, Sven, 'Images in the Air: Optical Games, Magic and Imagination', in *Spirits Unseen: The Representation of Subtle Bodies in Early Modern European Culture*, ed. by Christine Göttler and Wolfgang Neuber (Leiden: Brill, 2007), pp. 71–91
Eamon, William, *The Professor of Secrets: Mystery, Medicine, and Alchemy in Renaissance Italy* (Washington, DC: National Geographic, 2010)
Findlen, Paula, 'Jokes of Nature and Jokes of Knowledge: The Playfulness of Scientific Discourse in Early Modern Europe', *Renaissance Quarterly*, 43 (1990), 292–331
——, *Possessing Nature: Museums, Collecting, and Scientific Culture in Early Modern Italy* (Berkeley: University of California Press, 1994)
Freedberg, David, *The Eye of the Lynx: Galileo, his Friends, and the Beginning of Modern Natural History* (Chicago: University of Chicago Press, 2002)
Gatti, Hilary, *Giordano Bruno and Renaissance Science* (Ithaca: Cornell University Press, 1999)
Grabes, Herbert, *Speculum, Mirror und Looking-Glass: Kontinuität und Originalität der Spiegelmetapher in den Buchtiteln des Mittelalters und der englischen Literatur des 13. bis 17. Jahrhunderts* (Tübingen: Niemeyer, 1973)

Grant, Edward, *Planets, Stars, and Orbs: The Medieval Cosmos* (Cambridge: Cambridge University Press, 1994)

Hammond, J. H., *The Camera Obscura: A Chronicle* (Bristol: Hilger, 1981)

Hufnagel, Henning, *Ein Stück von jeder Wissenschaft: Gattungshybridisierung, Argumentation und Erkenntnis in Giordano Brunos italienischen Dialogen* (Stuttgart: Franz Steiner, 2009)

Kodera, Sergius, *Disreputable Bodies: Magic, Medicine, and Gender in Renaissance Natural Philosophy* (Toronto: Centre for Reformation and Renaissance Studies, 2010)

Michel, Paul Henri, *The Cosmology of Giordano Bruno*, trans. by R. E. W. Maddison (London: Methuen, 1973)

Miele, Michele, 'L'organizzazione degli studi dei Domenicani di Napoli al tempo di Giordano Bruno', in *Giordano Bruno: Gli anni napoletani e la 'peregrinatio' Europea*, ed. by Eugenio Canone (Cassino: Università degli Studi, 1992), pp. 29–50

Moliterno, Gino, 'Introduction', in *Giordano Bruno, Candlebearer*, ed. and trans. by Gino Moliterno (Ottawa: Dovehouse 2000), pp. 9–54

Reeves, Eileen, *Galileo's Glassworks: The Telescope and the Mirror* (Cambridge, MA: Harvard University Press, 2008)

Ricci, Saverio, *Inquisitori, censori, filosofi sullo scenario della Controriforma* (Rome: Salerno editore, 2008)

Romei, Giovanna, 'Della Porta, Giovambattista', in *Dizionario biografico degi Italiani*, xxxvii, ed. by Vincenzo Capelletti (Rome: Istituto della enciclopedia italiana, 1989), pp. 170–82

Rowland, Ingrid, *Giordano Bruno: Philosopher/Heretic* (New York: Farrar, Strauss and Giroux, 2008)

Saiber, Arielle, *Giordano Bruno and the Geometry of Language* (Aldershot: Ashgate, 2005)

Schechner, Sara J., 'Between Knowing and Doing: Mirrors and their Imperfections in the Renaissance', *Early Science and Medicine*, 10.2 (2005), 137–62

Stafford, Barbara, and Frances Terpak, *Devices of Wonder: From the World in a Box to Images on the Screen* (Los Angeles: Getty Research Institute, 2001)

Thielemann, Andreas, 'Lenti e specchi nella scienza e nella pittura del primo Seicento', in *L'arte della matematica nella prospettiva, Atti del Convegno Internazionale di Studi, Roma / Urbino, 8–11 Ottobre 2006*, ed. by Rocco Sinisgalli (Foligno: CB Edizioni, 2009), pp. 127–49 and pp. 369–82

Tirinnanzi, Nicoletta, *Umbra naturae: L'immaginazione da Ficino a Bruno* (Rome: Edizioni di storia e letteratura, 2000)

Trabucco, Oreste, 'Il corpus fisiognomico dellaportiano tra censura e autocensura', in *I primi Lincei e il Sant'Ufficio: Questioni di scienze de di fede* (Rome: Academia nazionale dei Lincei, 2005), pp. 235–72

——, 'Riscrittura, censura, autocensura: itinerari redazionali di Giovan battista della Porta', *Giornale critico della filosofia italiana*, 22 (2002), 41–57

Valente, Michaela, 'Della Porta e l'inquisizione: Nuovi documenti dell'Archivio del Sant'Uffizio', in *Bruniana et Campanelliana*, 3 (1997), 415–45

Vanagt, Katrien, 'Early Modern Medical Thinking on Vision and the Camera Obscura: V. F. Plempius' *Ophthalmographia*', in *Blood, Sweat and Tears: The Changing Concepts of Physiology from Antiquity into Early Modern Europe*, ed. by Manfred Horstmanshof, Helen King, and Claus Zittel (Leiden: Brill, 2012), pp. 569–93

Védrine, Hélène, 'Della Porta e Bruno: Natura e magia', *Giornale critico della filosofia italiana*, 65 (1986), 217–306

Yates, Frances A., *Giordano Bruno and the Hermetic Tradition* (London: Routledge & Kegan Paul, 1964)

Zambelli, Paola, *White Magic, Black Magic in the European Renaissance* (Leiden: Brill, 2007)

Portraits et miroirs dans les éloges collectifs de femmes au XVIIe siècle

Jean-Philippe Beaulieu

L'une des contributions des XVIe et XVIIe siècles à la célèbre *Querelle des femmes* est le développement d'ouvrages qui font l'apologie du sexe féminin par la réunion, au sein d'un même livre, d'éloges de figures méritoires. Au XVIIe siècle en particulier, les auteurs de tels éloges collectifs[1] attribuent une valeur spéculaire à ces textes où l'on cherche à établir la valeur du sexe féminin au moyen de portraits de femmes exemplaires — antiques ou chrétiennes — auxquelles on invite les lectrices à s'identifier. L'usage du terme 'galerie' pour désigner certains de ces recueils de femmes illustres attire notre attention sur le rôle qu'y jouent les métaphores picturales et les jeux de regards associés à une exemplarité qui repose en grande partie sur les effets de miroir. Bien qu'elle soit plus métaphorique que ce que l'on trouve dans les textes de l'époque qui, puisant à la tradition médiévale, se désignent comme des 'miroirs',[2]

[1] Comme le précise Renée-Claude Breitenstein ('Présentation' du dossier 'Publics et publications dans les éloges collectifs de femmes à la fin du Moyen Âge et sous l'Ancien Régime', p. 5), si, pour désigner un *compendium* de figures exemplaires, 'l'expression "éloge collectif" n'appartient pas au vocabulaire de l'époque', elle n'en exprime pas moins l'intérêt de la sensibilité prémoderne pour le regroupement de discours épidictiques.

[2] Dans le seul domaine des ouvrages moraux concernant le sexe féminin, signalons *La Vie de Mademoiselle de Neuvillars, miroir de perfection pour les femmes mariées et pour les âmes dévotes*, de Nicolas Du Salut, ou le *Miroir de la vanité des femmes mondaines* de Louis de Bouvignes.

Jean-Philippe Beaulieu (jean-philippe.beaulieu@umontreal.ca) est professeur au Département des littératures de langue française de l'Université de Montréal. Spécialiste de la Renaissance, il s'intéresse depuis bon nombre d'années aux discours des XVIe et XVIIe siècles sur les femmes. Sa plus récente édition critique : le florilège *Remontrances, prophéties et confessions de femmes (1575–1650)*.

The Mirror in Medieval and Early Modern Culture: Specular Reflections, ed. by Nancy M. Frelick, CURSOR 25 (Turnhout: Brepols, 2016), pp. 257–269 10.1484/M.CURSOR-EB.5.111051
BREPOLS PUBLISHERS

la nature spéculaire de ces recueils épidictiques mérite d'être dégagée, de façon à souligner l'infléchissement que connaît le genre lorsque des femmes, telles Jacquette Guillaume et Marguerite Buffet, se l'approprient au cours des années 1660. L'usage exemplaire du portrait prend alors un autre sens lorsque le porteur de miroir est une femme, directement concernée par une démonstration dans laquelle elle s'inclut désormais comme preuve vivante du mérite du sexe féminin. Le recueil d'éloges ne vise plus simplement à façonner — de l'extérieur — la conduite des femmes ; il constitue une invitation à adhérer à une communauté dont le texte devient un miroir aux reflets multiples, qui fait se répondre les figures de l'auteure, de la femme louangée et de la lectrice.

La liste de femmes célèbres ou exemplaires représente un type de discours qui a connu un essor particulier avec la diffusion du *De mulieribus claris* (1360) de Boccace.[3] À la Renaissance, des listes sont parfois intégrées dans des ouvrages défendant le sexe féminin où elles servent d'illustration au discours apologétique, par exemple dans le *De nobilitate et præcellentia feminei sexus* (1509) de Cornelius Agrippa de Nettesheim.[4] Au courant du XVIIe siècle, tout en continuant à être utilisés dans des apologies comme *Le Triomphe des Dames* de François de Gerzan,[5] les éloges de femmes se développent comme un genre textuel en soi,[6] particulièrement dans les années 1640, sous la régence d'Anne d'Autriche, qui voit la parution de plusieurs recueils, tels le *Théâtre françois des seigneurs et des dames illustres* (1642) de François Dinet, la *Gallerie des dames illustres* (1642) de François de Grenaille ou les *Éloges des douze dames illustres* (1646). Sous le signe d'une mise en scène théâtrale, mais surtout picturale, comme le suggèrent leurs titres, ces ouvrages sont l'occasion de brosser le portrait — ou, à tout le moins, de tracer le profil — d'un grand nombre de figures modèles, réunies et ordonnancées grâce à la métonymie d'origine architecturale que constitue la galerie.[7] Celle-ci devient le principe de composition de

[3] À propos de l'influence de ce texte, on consultera Desrosiers-Bonin, 'De l'*exemplum* antique à l'*exemplar* vivant dans *La Cité des Dames* de Christine de Pizan', p. 299 et suiv.

[4] Sur l'importance de l'ouvrage de Corneille Agrippa, voir Breitenstein, 'Traduction, transferts culturels et construction des publics', pp. 91–92.

[5] François de Gerzan, *Le Triomphe des Dames*. Voir, en particulier, les chapitres 6 ('De la valeur & générosité des Femmes') et 7 ('De l'excellence de l'esprit des Femmes'), qui sont conçus comme des compilations d'exemples.

[6] Pascal, 'Les Recueils de femmes illustres', p. 3.

[7] Au XVIIe siècle, le terme désigne moins le lieu que constitue la galerie que les tableaux qui sont exposés dans cet espace d'étalage de la richesse ; Conroy, 'Ekphrasis, Edification, and the Iconography of Women', p. 225.

plusieurs ouvrages[8] qui se présentent comme l'*ekphrasis* d'œuvres d'art réelles ou imaginaires, dans l'esprit de la *Galeria* de Giambattista Marino.[9] Certains des ouvrages consacrés aux éloges de femmes fortes dans les années 1640 comportent d'ailleurs des gravures[10] ou renvoient explicitement à des peintures,[11] en exploitant, à divers degrés, la métaphore de la galerie comme lieu de monstration et d'héroïsation de certaines figures en tant qu'antithèses de la faiblesse du sexe féminin.[12]

Je m'attarderai tout d'abord aux deux grands recueils dédiés à Anne d'Autriche que forment, d'un côté, les *Éloges et les vies des Reynes, des Princesses et des Dames Illustres* du minime Hilarion de Coste, publié en 1630 et remanié en 1647,[13] et, de l'autre, la *Gallerie des femmes fortes* (1647) du jésuite Pierre Le Moyne.[14] Ils donnent à voir des conduites dévotes et vertueuses et, de manière plus accessoire, des qualités intellectuelles,[15] à travers une pratique épidictique du portrait dont la portée didactique repose sur des effets de miroir établis entre les dames illustres et les lectrices. Si les femmes ne sont pas explicitement désignées comme les destinataires de ces livres,[16] il est clair que l'exemplarité qui y est mise en place les désigne comme le premier public, que l'on invite à s'inscrire dans une économie de l'émulation dont la clé est Anne d'Autriche. Dans

[8] Pascal, 'Les Recueils de femmes illustres', pp. 3–4.

[9] Voir Maber, 'Poetic Picture-Galleries in Seventeenth-Century France'. Cette pratique s'associe au discours épidictique, comme dans les *Éloges des hommes illustres peints en la gallerie du Palais Roial* de B. Griguette, qui suggère — alors qu'il n'en est rien — qu'il s'agit de l'*ekphrasis* de tableaux se trouvant dans la galerie des hommes illustres de l'ancien Palais Cardinal.

[10] Conroy ('Ekphrasis, Edification, and the Iconography of Women ', p. 226) établit une liste des ces recueils avec illustrations.

[11] Pensons aux *Éloges des* XII *dames illustres, grecques, romaines, et françoyses*, censés décrire les tableaux de l''Alcove de la Reine'. À ce sujet, voir Pascal, 'Représenter la régence ?', p. 91.

[12] Conroy, 'Ekphrasis, Edification, and the Iconography of Women', p. 225.

[13] Hilarion de Coste, *Les Éloges et les vies des Reynes, des Princesses et des Dames Illustres en piété, en courage et en doctrine*. Les renvois à cet ouvrage seront indiqués par le sigle *EVR*, suivis de la page et placés entre parenthèses dans le corps du texte.

[14] Pierre Le Moyne, *La Gallerie des femmes fortes*. Les renvois à cet ouvrage seront désormais indiqués par le sigle *GFF*.

[15] Ainsi, parmi les 206 notices qui forment les *Éloges* d'Hilarion de Coste, seulement soixante-six sont consacrées à des figures associables au savoir.

[16] L'identification des destinataires réels des éloges collectifs soulève un certain nombre de problèmes sur lesquels se penche le dossier 'Publics et publications dans les éloges collectifs de femmes à la fin du Moyen Âge et sous l'Ancien Régime', préparé par Renée-Claude Breitenstein pour *Études françaises*.

leurs éloges collectifs, les deux auteurs proposent des descriptions qui évoquent à certains égards le portrait littéraire, très populaire à l'époque,[17] et ce, même si Hilarion de Coste, en disant rédiger les 'Vies' de dames illustres, situe son texte dans un cadre hagiographique où la narration, plus que la description, semble centrale. Délaissant l'exemplarité antique qu'affectionnait le XVIe siècle, l'auteur dit 'publier les faits les plus memorables de quelques-unes qui se sont rendues recommandables de nostre temps, & du temps de nos peres' (*EVR*, f. iiv°), c'est-à-dire 206 femmes correspondant à trois grandes catégories : 'femmes illustres en piété', 'en courage & valeur', et 'pour leur sçavoir'. Cette entreprise, que Coste assimile aux *Vies parallèles* de Plutarque,[18] est mise au service d'un didactisme moral perceptible dans le choix et la description de ces dames qui sont offertes aux lectrices comme des modèles spéculaires, c'est-à-dire, à l'instar des *Specula* et *Mirouers* d'origine médiévale,[19] non comme le reflet réel des lectrices mais comme l'image sublimée de ce qu'elles pourraient être, à l'imitation de Jeanne de Portugal, par exemple, qui 'a esté de son temps un vray miroir [au sens de modèle] de chasteté' (*EVR*, p. 8). Comme ces 'vies' sont avant tout des éloges, la narration y est subordonnée à la visée épidictique — c'est-à-dire à la description — de façon à constituer des portraits dynamiques. Pour des reines et princesses comme Louise de Lorraine ou Marie Stuart, l'anecdote est très présente, tandis que dans le cas de personnalités de condition plus modeste, comme les Dames Des Roches ou Marie de Gournay, la dimension descriptive l'emporte, assortie de témoignages ou d'hommages, tels des poèmes élogieux qui confirment la valeur exemplaire de la dame.[20]

Par la façon dont elle joue avec les métaphores picturales — et ce, dès son titre —, la *Gallerie des femmes fortes* offre davantage de matière de nature spéculaire. Le Moyne, préoccupé par l'expression rhétorique des rapports entre peinture et littérature, comme l'illustrent bien ses *Peintures morales* (1640–43),[21] met en place une parole imagée — conformément à l'esthétique jésuite de la

[17] Voir Plantié, *La Mode du portrait littéraire en France (1641–1681)*.

[18] À vrai dire, il affirme que son ouvrage est tout aussi justifié que la traduction qu'Amyot a procurée des *Vies* de Plutarque : '[...] j'ay autant & plus de raison de faire les Vies de ces Heroïnes qui ont vécu & sont decedées en la creance de l'Eglise Catholique [...] que le docte Amiot [...] a employé ses veilles à traduire les vies des Hommes Illustres Payens, écrites par Plutarque' (*EVR*, p. 10).

[19] Voir, par exemple, le *Mirouer des femmes vertueuses*.

[20] Comme ceux rédigés par Colletet et La Mothe Le Vayer fils, qui consolident l'éloge qu'Hilarion de Coste consacre à Gournay.

[21] Fabrizio-Costa, 'G.B. Marino et le Père P. Le Moyne', p. 19.

Contre-Réforme[22] — au service de l'héroïsation du sexe féminin, dans le but avoué de favoriser l'éducation des femmes.[23] En alléguant les autorités religieuses qui ont 'contribué [...] à l'instruction des Femmes', le jésuite dit cependant donner une 'nouvelle forme à cette matiere' en associant à la philosophie, 'la Peinture, la Poësie et l'Histoire' (*GFF*, f. aa iiiv° et aa iiiir°) pour présenter une galerie de vingt gravures de Claude Vignon, réalisées par Gilles Rousselet et Abraham Bosse, qui montrent des femmes appartenant à quatre catégories : les juives, les barbares, les romaines et les chrétiennes. L'image de la galerie s'épanouit, dans les pièces liminaires, au moyen de métaphores picturales, mais aussi théâtrales, qui annoncent la nature ekphrastique de ces éloges voulant peindre chacune de ces héroïnes, non de l'extérieur comme chez Philostrate, mais de l'intérieur, de 'tous les traits & de tous les mouvemens du cœur : de toutes les postures & de toutes les couleurs de l'ame' (*GFF*, f. a iiiiv°). Ainsi présentée comme une façon d'orienter et d'enrichir la dimension visuelle,[24] cette peinture par les mots prend la forme d'une série de six textes successifs : le premier est un commentaire en prose du portrait, complété par un sonnet qui, comme le dit l'auteur, 'est un autre tableau fait en petit'. Le troisième volet consiste en un éloge, abrégé de la vie de l'héroïne, qui est assorti d'une réflexion morale où l'auteur fait 'remarquer ce qu'il y a de plus utile & de plus instructif dans l'exemple qui a precedé' en avertissant 'les Femmes de leurs devoirs & de leurs obligations' (*GFF*, f. eer°). L'objectif didactique de chaque portrait commence à se préciser par ce resserrement des liens entre passé et présent, qui s'épanouit dans un cinquième volet : une question morale, développement plus abstrait de l'exemplarité du portrait (par exemple, dans le cas de Débora, en proposant une réflexion sur la capacité des femmes à gouverner). L'éloge se termine par un retour au portrait proprement dit, au moyen d'exemples qui établissent des parallèles entre la figure centrale et d'autres illustrations — généralement plus récentes — des mêmes dispositions.[25] Ainsi Isabelle d'Espagne et Marguerite d'Autriche, duchesse de Parme, sont-elles mises en rapport spéculaire avec Débora, comme 'preuve avantageuse & moderne pour le gouvernement des Femmes' (*GFF*, p. 22). Chaque section de la galerie exploite ainsi, dans des

[22] Conroy, 'Ekphrasis, Edification, and the Iconography of Women', p. 225.

[23] Il affirme en effet, dans sa préface, que les 'vices seroient rares parmy les Hommes, si les Femmes dont naissent les Hommes, estoient toutes sages' (*GFF*, f. aa iir°).

[24] Sur les rapports texte-image et leur visée rhétorique dans la *Gallerie*, voir Conroy, 'Description or Prescription ?', p. 2.

[25] En vertu du principe selon lequel l'exemple, 'veu de prez, il fasse plus d'impression & agisse avec plus de force' (*GFF*, f. eer°).

registres complémentaires, les données fournies par la gravure, de manière à montrer la Femme forte 'habillée à nostre mode, & parée des ornemens de nos Muses, c'est-à-dire en lui donnant un visage dans lequel les contemporaines sont plus susceptibles de se reconnaître'.[26]

La mise en rapport spéculaire des femmes fortes et des lectrices est préparée par l''Epistre panegyrique' à Anne d'Autriche, qui précède la préface auctoriale, et dans laquelle on présente la galerie à suivre comme un théâtre où les figures illustres, désignées comme 'les plus beaux spectacles de leurs Siecles', sont elles-mêmes 'descenduës de leurs Throsnes & de leurs Theatres, pour estre [les] spectatrices' de la régente (*GFF*, f. aa iiiv°). Cette rhétorique de la flatterie qui passe par l'assujettissement des dames illustres à la reine n'est guère surprenante,[27] mais elle repose ici sur un jeu de regards nous conduisant d'une femme à l'autre : de l'ancienne à la moderne, du modèle textuel à la lectrice, du singulier au collectif, à travers la figure réelle et contemporaine de la régente. En tant qu''Heroine Chrestienne',[28] Anne d'Autriche devient un miroir qui magnifie et concentre les qualités féminines (piété, prudence, justice, grâces et magnificence) décrites par ce recueil qui se présente comme une véritable galerie des glaces où se démultiplient, en se superposant, les visages féminins.

Dans leur description des dames illustres, les deux religieux que sont Hilarion de Coste et Pierre Le Moyne projettent une exemplarité féminine plutôt conservatrice,[29] fondée sur des qualités dont le savoir n'occupe pas le premier plan. Signe des temps, deux recueils d'éloges signés par des femmes paraissent dans les années 1660 : *Les Dames illustres* (1665) de Jacquette Guillaume et les *Éloges des illustres sçavantes anciennes et modernes* (1668) de Marguerite

[26] Ce faisant, l'auteur valorise lui aussi la modernité, tout en donnant un coup de griffe à la fiction romanesque, qui propose des contre-modèles d'éducation féminine ; ainsi Le Moyne rappelle-t-il 'à ceux qui courent après les Phantosmes des Romans, que la verité n'est pas seulement plus instructive, mais qu'elle est encor plus belle & plus divertissante que le mensonge' (*GFF*, f. eer°).

[27] On la retrouve dans la dédicace qu'Hilarion de Coste adresse à la régente dans ses *Éloges* : 'Ce ne sont pas uniquement des Françoises, mais des Reynes & des Princesses de toutes les nations qui vous reconnoissent pour leur Reyne' (*EVR*, p. 4).

[28] La régente illustre le mieux les vertus qui permettent aux femmes fortes de passer de la sphère domestique à la sphère publique, puisque 'toutes les victoires commencent dans [le] cabinet' de la régente, 'par le zele & par la priere' de celle-ci (*GFF*, f. er°).

[29] Conroy ('Ekphrasis, Edification, and the Iconography of Women', pp. 227–34) démontre l'existence, dans la *Gallerie* de Le Moyne, de tensions entre certaines gravures (représentant des attributs masculins plus marqués) et leur commentaire ekphrastique (qui ne s'engage pas dans la valorisation de tels attributs).

Buffet.[30] Ces auteures — dont on ne sait rien — s'éloignent quelque peu du modèle héroïque pour s'attarder aux dispositions intellectuelles féminines, qui constituent la clé de voûte de discours apologétiques où la liste exemplaire joue un rôle central.[31] Prolongeant une tradition apologétique qui ne peut établir des rapports entre sexes qu'en termes de supériorité ou d'infériorité, le titre de Guillaume indique qu'il s'agit de faire la preuve 'que le sexe féminin surpasse en toute sortes de genres le sexe masculin'. À cette fin, l'auteure convoque un cortège de dames, surtout dans la deuxième partie où elle illustre, au moyen de portraits, les réalisations féminines dans les arts et les sciences. Les *Éloges* de Buffet prennent, quant à eux, un parti-pris plus moderne, celui de l'égalité des sexes, en étalant '[c]es veritez [...] si connuës que les femmes peuvent avoir autant d'habileté que les hommes, en tout ce qui les rend habiles, qu'on ne peut rien opposer au contraire' (*EIS*, pp. 215–16). Par son efficacité et son rationalisme, l'ouvrage préfigure la démonstration que François Poullain de la Barre fera en 1673 dans son *De l'égalité des hommes et des femmes*. En dépit de leur style et de leur construction fort différents, les deux livres déclinent le portrait selon une logique qui, sans renoncer à la force du nombre (à la manière d'Hilarion de Coste) ou de l'allégorie (comme chez Le Moyne), fait valoir l'appartenance de l'auteure au sexe féminin comme un élément qui oriente fortement l'exemplarité, puisque la démonstration concerne au premier chef les auteures elles-mêmes qui offrent, par leur travail d'écriture, la preuve concrète des aptitudes de leur sexe. Tandis que la spécularité des éloges des deux religieux souligne implicitement l'écart entre les dames illustres et les lectrices par la valorisation de modèles souvent absolus, les textes de Buffet et Guillaume reposent sur un jeu de miroirs plus complexe,[32] où l'auteure, au nom d'une communauté féminine, offre au lectorat des exemples familiers, auxquels il est plus facile de s'identifier.

Après une apologie générale — et plutôt abstraite — du sexe féminin, qui occupe les 237 premières pages, Buffet fait place, dans les 105 pages restantes, à dix-neuf descriptions de femmes 'modernes' qui occupent le cœur de l'ouvrage,

[30] Jacquette Guillaume, *Les Dames illustres où par bonnes et fortes raisons, il se prouve, que le sexe Feminin surpasse en toute sorte de genres le sexe masculin* (ouvrage désigné par le sigle *DI*) ; Marguerite Buffet, *Éloges des illustres sçavantes anciennes et modernes*, publiés à la suite des *Nouvelles observations sur la langue françoise*, p. 277 (*EIS*).

[31] Sur le caractère synthétique de l'usage des pratiques apologétiques et épidictiques chez Buffet, voir Ducharme, 'Une formule discursive au féminin'.

[32] C'est-à-dire en exploitant les effets spéculaires propres aux écrits didactiques qu'évoque Winn dans 'La *Dignitas mulieris* dans la littérature didactique féminine', p. 14.

suivies de cinquante-deux très brefs éloges d'"illustres sçavantes des siecles passez'. Dans ces portraits, que l'on peut considérer comme des dessins au fusain tant ils vont à l'essentiel, Buffet fait souvent sentir sa présence au moyen d'indications à la première personne associant le 'je' auctorial au sexe féminin par l'usage d'un 'nous' — peu présent dans le volet apologétique de l'ouvrage —, qui montre que les qualités de ces femmes rejaillissent sur l'ensemble de la communauté féminine. Ainsi Christine de Suède est-elle présentée comme 'le plus bel ornement de nostre sexe',[33] tout comme Louise de Prie, maréchale de La Mothe, est 'l'honneur & l'exemple de nostre sexe, & l'admiration de ceux qui ont le bon-heur de la connoistre & de l'imiter' (*EIS*, p. 256). Buffet parsème son propos d'indications personnalisantes qui la montrent active dans l'établissement du portrait, ajoutant, par exemple, au sujet d'Anna Maria von Schurman qu'elle a '[...] appris mesmes qu'on va la consulter quand on a quelques points de Theologie difficiles à décider' (*EIS*, p. 244). Ce type d'indication contribue à façonner l'*ethos* d'une femme elle-même érudite qui défend son sexe en faisant valoir surtout son appartenance à la France et sa fréquentation des dames illustres françaises. Elle consacre en effet dix-sept de ses dix-neuf notices à des compatriotes, quelques-unes très connues, comme Julie d'Angennes, duchesse de Montausier, fille de la célèbre marquise de Rambouillet, mais d'autres moins notoires, telles Mmes de Bonnevaut, de Gredeville, ou de Miraminy. Buffet se fait le témoin de la valeur de ces dames, en affirmant que son inventaire, loin d'être exhaustif, relève d'un échantillonnage fondé sur sa propre expérience, qui devient *de facto* une garantie d'authenticité : 'Quand je ne voudrois icy parler que des habiles qui sont à Paris, elle sont dans un si grand nombre, que je serois reduite à plusieurs volumes : j'ay voulu ne parler que de celles qui sont dans ma connoissance' (*EIS*, pp. 281–82). Elle affirme fréquenter certaines de ces dames, telle Isabelle de Choiseul-Praslin, comtesse du Plessis-Guénégaud, qui tenait salon à l'hôtel de Nevers : 'Si la vertu & la modestie de Madame la Comtesse du Plexis me tire la plume des mains pour m'obliger au silence, puisque je sçay que cette illustre Dame n'aime les loüanges que pour les distribuer, ayant l'honneur de la connoistre si parfaitement que je fais ; je ne puis refuser à moy-même la satisfaction que j'ay de luy rendre justice [...]' (*EIS*, pp. 259–60). Buffet suggère ainsi l'existence d'un réseau de sociabilité, réel ou imaginé, qui donne à voir une exemplarité s'exprimant par des figures contemporaines que l'on invite à

[33] Elle relègue les anciennes à la fin, inversant l'ordre chronologique. À ce sujet, voir Beaulieu, 'Jacquette Guillaume et Marguerite Buffet'.

mieux connaître et à imiter, à l'instar de l'auteure elle-même.[34] Buffet revendique indirectement sa place parmi ces femmes en présentant, avant ses *Éloges*, des *Observations sur la langue françoyse* dont l'ambition est, selon le libellé de la page titre, 'd'enseigner aux Dames l'art de bien parler & de bien écrire sur tous sujets'. Déclarant avoir dû 'se soustenir par la profession des Lettres', elle offre à la reine, Marie-Thérèse d'Autriche, une réalisation d'importance, preuve tangible du savoir féminin que l'on soumet au lectorat avant même la démonstration apologétique.[35] Par la seule construction en deux volets de l'ouvrage, Buffet se présente, en dépit de la modestie rhétorique qu'elle affiche, comme une forme de miroir de ses dames illustres, de façon à déployer un registre assez large de stratégies — discours théorique, éloges, démonstration concrète du savoir — pour créer un ensemble dont l'auteure est en fin de compte le centre optique.

Ce dispositif par lequel la figure auctoriale rassemble modèles et lectrices en raison de la dimension spéculaire de son activité d'écriture est encore plus manifeste dans *Les Dames illustres*. Guillaume cherche elle aussi à établir la preuve du savoir féminin, mais elle intègre celui-ci dans la trame même de sa démonstration en faisant de son livre un *compendium* du savoir des femmes ayant pour but de faire taire les détracteurs du groupe auquel elle appartient : 'J'ay pour ce sujet un syllogisme convainquant, & capable d'enchaisner le reste de nos ennemis, ne leur laissant l'usage de la parole que pour confesser leur subjection' (*DI*, p. 13). En plus de mentionner ou de décrire une centaine de femmes anciennes et modernes, dans la deuxième partie de l'ouvrage, l'auteure s'attarde au portrait de vingt figures récentes ou contemporaines en attribuant à sept de celles-ci des propos ou des écrits savants qui se trouvent insérés, sur le mode de la prosopopée, dans le cadre épidictique. Très substantiels, ces développements occupent à eux seuls plus de la moitié des 262 pages de la seconde partie de l'ouvrage, nous faisant momentanément oublier la visée épidictique pour livrer de véritables leçons de théologie, d'histoire ou de géographie, dont la plus longue s'étend sur soixante-dix-sept pages. Cet usage d'un discours prétendument citationnel donne vie à un certain nombre de figures décrites, qui s'éloignent de la rhétorique picturale mais muette des femmes pratiquée par Le Moyne pour devenir de vrais portraits parlants : le miroir réfléchit la voix en plus de l'image. Parfois, ce discours citationnel est limité à quelques fragments

[34] Par exemple, elle dit, à propos de Mlle Dupré, que son 'bel esprit est si fécond' qu'il 'faut la connoître pour en pouvoir parler assez avantageusement, & luy rendre la justice qu'elle merite' (*EIS*, p. 281).

[35] À propos des *Observations*, on consultera l'étude de Meli, 'Un bien dire à l'usage des bourgeoises'.

textuels (par exemple, des poèmes, cantiques et maximes dans le cas de Thérèse d'Avila), mais le plus souvent, il occupe la majeure partie de l'éloge (ainsi, trente-sept des trente-huit pages consacrées à Christine de Suède font entendre les 'raisonnements' de la princesse sur la religion catholique). Ce sont surtout des Françaises que Guillaume fait parler, des contemporaines qu'elle évoque de manière vague ('deux sœurs') ou qu'elle désigne au moyen d'une signature abrégée (Mademoiselle ****). Si cette façon d'éviter l'identification semble relever d'une forme de bienséance, son caractère éminemment paradoxal dans un cadre épidictique suggère qu'il s'agit d'une stratégie pour donner la place centrale à des femmes qui sont peu notoires (parfois en raison de leur jeunesse),[36] en présentant des exemples dont les réalisations regardent vers l'avenir plutôt que vers le passé. De cette façon, dans le miroir de l'exemplarité que l'ouvrage donne à voir, les princesses et les femmes d'exception cèdent la place à des figures plus modestes qui font la preuve de leurs aptitudes en devenant les sujets et non seulement les objets du discours élogieux. Par ces effets de ventriloquie, Guillaume crée une sorte de compilation du savoir féminin, qui vaut moins par l'originalité de son contenu — puisque les connaissances reproduites proviennent d'ouvrages qui circulaient à l'époque[37] — que par ses modalités d'attribution de discours à des femmes au visage imprécis et avec lesquelles, par conséquent, il est plus facile de s'identifier. D'autant que ces 'illustres inconnues', si l'on peut dire, semblent appartenir au cercle de Guillaume, comme cette sienne parente, experte en philosophie, qui 'a fait quantité de Pieces sçavantes'. Les références aux relations de Guillaume s'accompagnent de mentions d'une participation de l'auteure à la discussion sur ce savoir livresque.[38] Ainsi souligne-t-elle au passage à propos de

[36] Ainsi Guillaume dit-elle au sujet de Mademoiselle **** : 'Quoy que cette jeune Damoiselle soit d'un âge si tendre, qu'elle pourroit bien passer pour enfant, elle a neantmoins tant de belles connoissances & tant de beaux secrets, que je croirois luy faire tort, si je ne luy donnois place entre les Dames Illustres' (*DI*, p. 309).

[37] À la fin de son ouvrage, Guillaume avoue d'ailleurs avoir emprunté à autrui : 'Mon cher Lecteur, je crois que vous ne vous étonnerez point de voir en quelque endroit de cét ouvrage, les termes mesmes de quelques Autheurs, ayant trouvé le langage agreable, j'ay bien voulu vous laisser vostre part de la satisfaction que j'ay receuë en les lisant, peut-estre m'en sçaurez vous gré, en le recevant, reconnoissant qu'en cela j'ay plus regardé vostre interest que le mien' (*DI*, p. 442). Bien sûr, ces savoirs sont empruntés à des ouvrages qui circulaient à l'époque. Voir Beaulieu, '"La Gloire de nostre sexe"', pp. 139–41.

[38] Guillaume reprend — mais avec une visée apologétique différente — le désir de jouer un rôle actif dans l'illustration de défense de la culture française que l'on trouve, à la fin du siècle précédent, chez Antoine du Verdier et François Grudé, sieur de La Croix du Maine, ou, plus près d'elle, dans la *Bibliothèque françoise* (1664) de Charles Sorel. De manière discrète, La Croix du

Mademoiselle ****, autorité en matière de pierres précieuses : 'Quoy qu'ailleurs j'aye dit quelque chose de la boëtte de grenade, je n'ay pas laissé de persuader nostre jeune Sçavante de nous entretenir de ses particularitez' (*DI*, p. 310).

L'un des buts de l'ouvrage semble être l'établissement d'une communauté féminine virtuelle à travers un 'nous' collectif — encore plus présent que chez Buffet — dont la figure de proue est la dédicataire, Élisabeth Marguerite d'Orléans (1646–96), duchesse d'Alençon. La rhétorique dédicatoire habituelle s'exprime à travers une métaphore astronomique qui lie les dames illustres et la cousine du roi : 'On peut les comparer [les dames illustres] à des Astres qui regardent V. A. R. comme un Soleil brillant dont ils recevront tout leur éclat ; mais qui se confondront dans sa grande lumière' (*DI*, f. aiiiir°). Grâce à cette image qui distingue et rassemble tout à la fois la princesse et les dames, l'auteure vise à illustrer la valeur du sexe féminin par la qualité et la quantité des dames méritoires convoquées pour constituer un réseau dont la dédicataire est la clé de voûte symbolique et auquel l'auteure sert littéralement de porte-parole, si l'on considère l'importance accordée au discours citationnel. Cet effet de mise en relation est formulé dès la dédicace où les femmes en jeu se trouvent associées à la visée apologétique : 'Car je ne sçaurois mieux faire voir la gloire de nostre Sexe, qu'en mettant sous vostre protection, des grandes Saintes, des Reynes, & des Princesses si recommandables, qu'elles se peuvent vanter de surpasser les hommes en tout ce qui est louable [...]' (*DI*, f. a iiiv°).

Retenons l'importance de ce 'nous' qui s'associe au 'vous' — celui de la dédicataire et des lectrices[39] — dans une dynamique spéculaire donnant au texte de Guillaume, et à un moindre degré à celui de Buffet, une résonance autre que ce que l'on trouve dans les galeries de portraits signées par des hommes. Le caractère plus familier et concret des portraits offerts au lecteur fait en sorte que les *Dames illustres* de Guillaume — probablement le recueil d'éloges collectifs le plus spéculaire de l'époque — délaissent le caractère silencieux du portrait pour devenir véritablement une galerie de voix, un peu comme les *Femmes illustres, ou les Harangues héroïques*, publiées en 1642 et 1644 sous la signature de Georges de Scudéry, mais attribuables à sa sœur Madeleine. Au contraire de cet ouvrage qui donne à entendre des femmes antiques, l'ouvrage de Guillaume

Maine fait même valoir sa connaissance des savants de l'époque : '[j'ay] eu desir d'avoir l'amitié de tant d'hommes doctes qui sont aujourd'huy vivants' (*Premier volume de la Bibliotheque du Sieur de La Croix du Maine*, f. aiiv°). Je remercie Hélène Cazes de ses observations sur cette question.

[39] Par opposition au 'ils' désignant les hommes qui méprisent le sexe féminin. Sur cet aspect, voir Beaulieu, '"La Gloire de nostre sexe"', pp. 133–35.

fait valoir une modernité encore en devenir, dont on perçoit la voix plus que le visage, vraisemblablement pour que les lectrices puissent imaginer leur propre réflexion dans ces identités protéiformes, sous la conduite d'une auteure qui met son propre savoir au service de la cause des femmes.

Ouvrages cités

Sources primaires

Bouvignes, Louis de, *Miroir de la vanité des femmes mondaines* (Namur : Adrien La Fabrique, 1675)

Buffet, Marguerite, *Éloges des illustres sçavantes anciennes et modernes*, publiés à la suite des *Nouvelles observations sur la langue françoise* (Paris : Jean Cusson, 1668)

Coste, Hilarion de, *Les Éloges et les vies des Reynes, des Princesses et des Dames Illustres en piété, en courage et en doctrine*, 2 tomes (Paris : Sébastien et Gabriel Cramoisy, 1647)

Du Salut, Nicolas, *La Vie de Mademoiselle de Neuvillars, miroir de perfection pour les femmes mariées et pour les âmes dévotes* (Paris : Sébastien Cramoisy, 1649)

Gerzan, François de, *Le Triomphe des Dames : Dédié à son Altesse Royale Mademoiselle* (Paris : François de Gerzan, 1646)

Griguette, Bénigne, *Éloges des hommes illustres peints en la gallerie du Palais Roial* (Dijon : Pierre Palliot, 1646

Guillaume, Jacquette, *Les Dames illustres où par bonnes et fortes raisons, il se prouve, que le sexe Feminin surpasse en toute sorte de genres le sexe masculin* (Paris : Thomas Jolly, 1665)

La Croix du Maine, François Grudé (sieur de), *Premier volume de la Bibliothèque du Sieur de La Croix du Maine* (Paris : Abel Langelier, 1584)

Le Moyne, Pierre, *La Gallerie des femmes fortes* (Paris : Antoine de Sommaville, 1647)

Sources secondaires

Beaulieu, Jean-Philippe, 'Jacquette Guillaume et Marguerite Buffet : Vers une historiographie du savoir féminin ?', dans *Les Femmes et l'écriture de l'histoire*, éd. par Sylvie Steinberg et Jean-Claude Arnould (Rouen : Presses universitaires de Rouen, 2008), pp. 325–39

——, '"La Gloire de nostre sexe" : Savantes et lectrices dans *Les Dames illustres* (1665) de Jacquette Guillaume', *Études françaises*, 47.3 (2011), 127–42

Breitenstein, Renée Claude, 'Traduction, transferts culturels et construction des publics dans deux éloges collectifs de femmes de la première moitié du XVIe siècle', *Études françaises*, 47.3 (2011), 91–107

——, éd., 'Publics et publications dans les éloges collectifs de femmes à la fin du Moyen Âge et sous l'Ancien Régime', *Études françaises*, 47.3 (2011), 5–142

Conroy, Derval, 'Description or Prescription ? Verbal Painting in Pierre Le Moyne's *Gallerie des femmes fortes* (1647)', *French Forum*, 36.2–3 (2011), 1–17

——, 'Ekphrasis, Edification, and the Iconography of Women : The Case of Pierre Le Moyne's *Gallerie des femmes fortes*', dans *Reverberations : Staging Relations in French Since 1500 : A Festschrift in Honour of C. E. J. Caldicott*, éd. par Phyllis Gaffney, Michael Brophy et Mary Gallagher (Dublin : University College Dublin Press, 2008), pp. 224–38

Desrosiers-Bonin, Diane, 'De l'*exemplum* antique à l'*exemplar* vivant dans *La Cité des Dames* de Christine de Pizan', dans *Les Femmes et l'écriture de l'histoire*, éd. par Sylvie Steinberg et Jean-Claude Arnould (Rouen : Presses universitaires de Rouen, 2008), pp. 299–307

Ducharme, Isabelle, 'Une formule discursive au féminin : Marguerite Buffet et la *Querelle des femmes*', *Papers on French Seventeenth-Century Literature*, 30.58 (2003), 131–55

Fabrizio-Costa, Silvia, 'G. B. Marino et le Père P. Le Moyne : Autour de la galerie littéraire', *Studi Francese*, 148 (2006), 17–34

Maber, Richard, 'Poetic Picture-Galleries in Seventeenth-Century France', dans *Texte(e)/Image*, éd par M.-A. Dutton (Durham : University of Durham Press, 1999), pp. 41–68

Meli, Cinthia, 'Un bien dire à l'usage des bourgeoises : *Les Nouvelles Observations sur la langue française* (1668) de Marguerite Buffet', dans *Femmes, rhétorique et éloquence sous l'Ancien Régime*, éd. par Claude La Charité et Roxanne Roy (Saint-Étienne : Publications de l'Université de Saint-Étienne, 2012), pp. 87–101

Pascal, Catherine, 'Les Recueils de femmes illustres au XVIIe siècle', communication donnée lors des premières rencontres de la SIEFAR 'Connaître les femmes de l'Ancien Régime' (Paris, 20 juin 2003) <http://www.siefar.org/docsiefar/file/Pascal-dicos.pdf> [page consultée le 12 juin 2010]

——, 'Représenter la régence ? Image(s) de reine(s) dans les *Éloges des douze dames illustres grecques, romaines et françoises dépeintes dans l'alcove de la reine* (1646)', dans *Les Femmes et l'écriture de l'histoire*, éd. par Sylvie Steinberg et Jean-Claude Arnould (Rouen : Presses universitaires de Rouen, 2008), pp. 89–102

Plantié, Jacqueline, *La Mode du portrait littéraire en France (1641–81)* (Paris : Champion, 1994)

Winn, Colette H., 'La *Dignitas mulieris* dans la littérature didactique féminine (du XVe au XVIIe siècle) : Les Enjeux idéologiques d'une appropriation', *Études littéraires*, 27.2 (1994), 11–24

The Reflection Sundial at Palazzo Spada in Rome: The Mirror as Instrument, Symbol, and Metaphor

Ulrike Feist

Introduction

In 1632 Cardinal Bernardino Spada (1594–1661) purchased a mid-sixteenth-century palazzo in Rome, which was subsequently known as Palazzo Spada.[1] At the time of its purchase the Cardinal's brother, Virgilio Spada (1596–1662), remarked that the palazzo was so perfect that — even if one should have wished to — it could not have been made more beautiful.[2] Despite the opinion of his brother, however, Bernardino Spada immediately began renovating and extending the palazzo, and redecorating the rooms — work that lasted more than thirty years, right up to the Cardinal's death, and cost almost twice

[1] The most important and comprehensive publication on the building history of the Palazzo Spada is still the monographic study by Neppi, *Palazzo Spada*. This foundational work continues to be referenced by publications such as Cannatà, *Palazzo Spada: Arte e Storia*; Cannatà, Vicini, and Ghetti, *Palazzo Spada: Le decorazioni restaurate*; Càndito, 'Corrispondenze ottico-prospettiche tra le opere di Maignan e di Borromini a Palazzo Spada'; Vicini, *Il Collezionismo del Cardinale Fabrizio Spada in Palazzo Spada*; Tabarrini, *Borromini e gli Spada*.

[2] 'Mi rallegro con V[ostra] E[minenza] che ha trovato palazzo così finito, che non ci potrà porre un mattone di vantaggio, quando anche volesse'; ASR, FSV 463, chap. 22, fol. 2v.

Ulrike Feist (ulrike.feist@uv.hu-berlin.de) is Advisor to the Vice President at Humboldt University of Berlin. She studied art history and history in Berlin and Rome, receiving her PhD from the University of Augsburg. She is the author of *Sonne, Mond und Venus*, on visualizations of astronomical knowledge in early modern Rome.

The Mirror in Medieval and Early Modern Culture: Specular Reflections, ed. by Nancy M. Frelick, CURSOR 25 (Turnhout: Brepols, 2016), pp. 271–286 10.1484/M.CURSOR-EB.5.111052
BREPOLS PUBLISHERS

the original purchase price.[3] One of the most famous changes to be made was the erection of the forced perspective colonnade, which he commissioned the architect Francesco Borromini to construct in the Giardino Segreto of his palace in 1652.

By far less well known, yet equally impressive, is the so-called Galleria della Meridiana (the Sundial Gallery).[4] In 1644, the Cardinal had decorated the vaulted ceiling of a twenty-two-metre-long gallery with a monumental sundial fresco painted by Giovanni Battista Magni according to a design by the French mathematician Emmanuel Maignan (Figure 12.1).[5] On the first floor, the Piano Nobile, at the end of the stairs on the left, one finds the door to the Galleria della Meridiana, and this leads directly into the Sala Grande, the most representative space in the palazzo.

This sundial is of a particular type, called a catoptric sundial ('catoptric' comes from the Greek *κατοπτρικός*, meaning 'specular') — it is also known as a mirror or reflection sundial. Here, time is not indicated with the help of a shadow cast by a stylus or gnomon, but by a point of light reflected by a mirror. The small round mirror is installed in the central window of the five mezzanine windows.[6] Apart from a small opening, the window is sealed to reduce light dispersion and to ensure the clear visibility of the reflected point of light (Figure 12.2).

In what follows, I will attempt, by analyzing the different functions of the mirror, to answer the question of why the Cardinal chose a monumental sundial as a decoration for his palazzo. It will become clear that Bernardino Spada's decision for this particular form of sundial was in fact a very conscious one. My remarks will be divided into three parts. First, I will look at the mirror's instrumental function; I will then analyze its iconographic and symbolic function; finally, I will try to reveal its metaphorical function.

[3] 'Ma à pena entrato il Card[ina]le ad habitarvi, cominciò a chimerizarvi sopra, et ha continuato trenta e più anni a farvi lavorare, a segno che tutta l'economia in haver fatta la prima compra con qualche vantaggio ha giovato poco mentre oltre li scudi 31500 si trova haver speso fino al giorno della morte altri scudi 50000 [...]'; ASR, FSV 463, chap. 22, fol. 2v.

[4] See Neppi, *Palazzo Spada*, pp. 188–204, and Feist, *Sonne, Mond und Venus*, pp. 11–73.

[5] The first invoice was issued in January of 1644; by May the painting of the gallery had been finished. Cf. the invoices issued by Andrea Battaglini and Giovanni Battista Magni in ASR, FSV 747.

[6] The original mirror has been lost. See Neppi, *Palazzo Spada*, p. 194.

Figure 12.1. Galleria della Meridiana, Palazzo Spada, Rome.
© Per gentile concessione della Soprintendenza Speciale per il Patrimonio Storico, Artistico ed Etnoantropologico e per il Polo Museale della città di Roma.

Figure 12.2. Mezzanine window and reflected point of light, Galleria della Meridiana, Palazzo Spada, Rome. Photo by author. © Per gentile concessione della Soprintendenza Speciale per il Patrimonio Storico, Artistico ed Etnoantropologico e per il Polo Museale della città di Roma.

The Mirror as Instrument

The mirror's most obvious function is its instrumental one. Logically, a mirror sundial cannot function without a mirror. Light hits the mirror and is reflected as a point of light onto the vaulted ceiling. Because the window faces south-east, the dial records only the hours from sunrise until the early afternoon. The complex network of lines is made up of numerous hour lines marked in different colours — besides the equinoctial, these include the antique, Italian, and Babylonian hour lines, among others.[7] The catoptric sundial in the Palazzo Spada is in fact one of the largest and most complicated painted networks for temporal measurement; its hour lines offer a total of twenty-two different ways of reading the time.[8] The size of the sundial's quadrant insures an impressively exact operation following the general rule: the larger an instrument is, the finer the graduation that can be carried out — and hence the greater the precision of the indications.

[7] For the functions of the individual hour lines, see Neppi, *Palazzo Spada*, pp. 194–203.

[8] Whitmore, *The Order of Minims*, p. 168.

However, the mirror does not only fulfill its instrumental function during the day; also at night moonlight is reflected onto the vaulted ceiling. On one wall of the gallery, Cardinal Spada had a moondial installed for which one only needed to know the moon's age (that is, day X after the last new moon) to read the hour displayed on the sundial.[9] By setting the values on the moondial's moveable wooden disks, it indicated the nocturnal time.

Both sundial and moondial were designed by the French mathematician and friar of the Order of Minims, Emmanuel Maignan (1601–76). The wooden moondial was constructed by the carpenter Andrea Battaglini, who in one of his accounts complains about the highly time-consuming process of its construction, in the course of which he was forced to make repeated corrections to the instrument according to Maignan's instructions.[10] In 1636 Maignan, who was born in Toulouse, had taken up residence at the Minim convent Trinità dei Monti in Rome, where in 1637 he constructed a catoptric sundial for one of the hallways. In 1642 Bernardino Spada assumed the office of Cardinal Protector of the Order,[11] and at this time at the latest must have seen the sundial in the convent building. Clearly, he immediately commissioned Maignan with the design of an even grander sundial for his palazzo, since the gallery in the Palazzo Spada was already redecorated in 1644.

As archival documents reveal, Maignan also made other optical and astronomical instruments for Spada, including a periscope. The latter was made by combination of various mirrors, with which it was possible 'to observe, from within a room with closed windows who was passing by on the street below.'[12] Clearly, the Cardinal was interested in the mathematical sciences — astronomy, perspective, and optics. Indeed, he also financed the treatise *Perspectiva horaria*, published by Maignan in 1648, which devotes over seven hundred pages to the history of time measurement, optics, catoptrics, and dioptrics; the work is dedicated to Spada.

[9] For the moondial see Neppi, *Palazzo Spada*, p. 198.

[10] '[...] ogni cosa secondo l'ordine del P[ad]re Emanuel della Trinità de Monti; d[ett]o lavor serve per un Artefitio di mattematica, qual fù di assai perdità di tempo, perche bisognò mutar hor d'un modo, hor d'un'altro, secondo che detto commandava'; ASR, FSV 747, invoice made out by the carpenter and engraver Andrea Battaglini on 24 July 1645.

[11] Neppi, *Palazzo Spada*, p. 189.

[12] '[...] un'altro artefitio di mattematica secondo l'ordine del sud[ett]o P[ad]re commandatomi da V[ostra] Em[inen]za [...] che d[ett]o serve per vedere chi passa per strada à stare in stanza con fenestre chiuse'; ASR, FSV 747, invoice made out by the carpenter and engraver Andrea Battaglini on 24 July 1645.

Figure 12.3. Galleria della Meridiana, engraving from Emmanuel Maignan, *Perspectiva horaria* (Rome: Philippi Rubei, 1648), between pp. 390 and 391.

In the treatise, the catoptric sundial in the Palazzo Spada is presented as Maignan's most important work and is illustrated in a large-format engraving (Figure 12.3). Standing in the gallery of the Palazzo Spada are four figures looking up at the sundial on the vaulted ceiling. The physiognomies of the two male figures whose faces are visible seem to be portraits of Maignan and Spada. This is suggested particularly by the hollow-cheeked face of the man on the left side of the engraving that resembles portraits of Maignan, and by the dress and beard of the man on the right side, which distinguish him as the cardinal. Together with another member of the Order, they are depicted in the act of presenting the unusual scientific artwork to a noble visitor, who, pointing with outstretched arm, is apparently inquiring with interest about the workings of the sundial.

Cardinal Bernardino Spada is only one example among many showing that the former theory of the supposed hostility of the Catholic Church toward the natural sciences is far from being valid. Indeed, for more than six centuries — from the rediscovery of ancient knowledge during the late Middle Ages to the Enlightenment — the Roman Catholic Church provided more financial and social support for astronomical research than any other institution, as has already been pointed out by John L. Heilbron.[13] Support for astronomy was certainly an attempt to solve an administrative problem: the necessity of determining the exact date of Easter for a number of years in advance.

It was two years after the death of Galileo Galilei (1564–1642), who in 1633 had been convicted and placed under house arrest by Pope Urban VIII, that Spada, who had Urban VIII to thank for his cardinalship, transformed a gallery of his own palazzo into a scientific instrument. It is not known whether Cardinal Spada, like Galileo, was convinced of the correctness of the theory of heliocentrism. What is certain, however, is that he was receptive to the new culture of knowledge of his time as defined by modern astronomy, an experimental research practice and mathematization.

The Mirror as Symbol

The sundial gallery must be considered in connection with the palazzo's Sala Grande, which only a few years earlier — in 1635 — was redecorated with a series of *quadratura* frescos depicting important figures from the history of

[13] Heilbron, *The Sun in the Church*, p. 3.

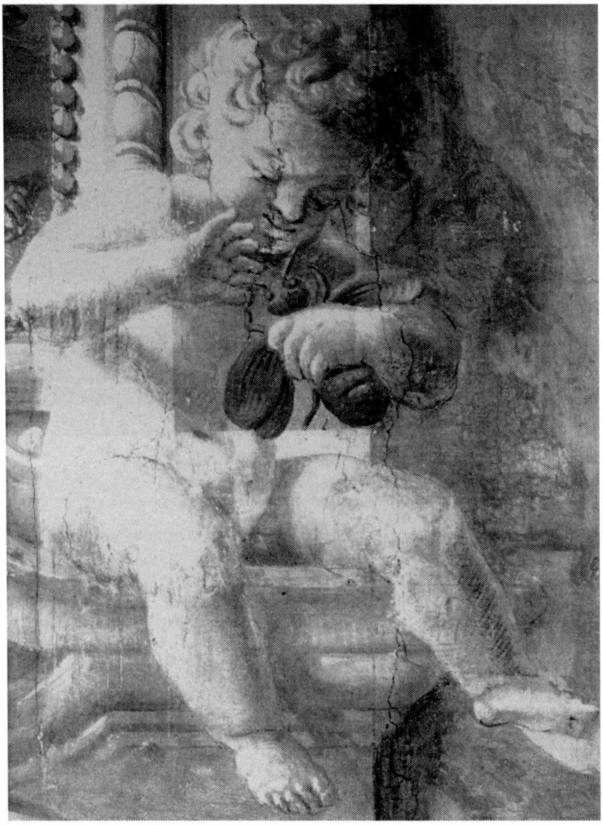

Figure 12.4.
Angelo Michele Colonna and Agostino Mitelli, Putto with bee. Fresco, 1635, Sala di Pompeo (Sala Grande), Palazzo Spada, Rome. © Per gentile concessione della Soprintendenza Speciale per il Patrimonio Storico, Artistico ed Etno-antropologico e per il Polo Museale della città di Roma.

the Papal States.[14] What is important here, however, is the iconographic foundation of the program, which makes references to Bernardino Spada and his patron Pope Urban VIII, and which becomes visible in many small details. The Cardinal's coat of arms was composed of three swords (*spada* means 'sword') accompanied by three lilies in the upper part. Pope Urban VIII had chosen the sun as his *impresa*, while his heraldic animal was the bee. All of these symbols appear in the décor of the Sala Grande: the coffered ceiling is adorned with lilies, bees, and suns. On the walls, numerous small painted angels play with swords or bees (Figure 12.4) — the pictorial program of the Sala Grande is both a self-homage on the part of Bernardino Spada and a Cardinal's homage to Urban VIII Barberini.

[14] The frescos were executed by the Bolognese artists Angelo Michele Colonna and Agostino Mitelli. See Neppi, *Palazzo Spada*, pp. 137–39.

Figure 12.5. Angelo Michele Colonna and Agostino Mitelli, Allegory of Virtue. Fresco, 1635, Sala di Pompeo (Sala Grande), Palazzo Spada, Rome. © Per gentile concessione della Soprintendenza Speciale per il Patrimonio Storico, Artistico ed Etnoantropologico e per il Polo Museale della città di Roma.

In addition, a number of medallions with allegorical figures decorate the walls of the hall. One of these allegories holds a mirror in her hand — according to Cesare Ripa's *Iconology*, the mirror was considered an attribute of Prudence. Another allegory is adopted with all its attributes from Ripa's *Iconology*: the winged female figure in a flowing robe holds a staff in her right hand and a laurel wreath in the left (Figure 12.5). Over her chest, her robe is decorated with a sun. This is an allegory of Virtue.

Mirror and sun — these were also the central elements of the catoptric sundial: the mirror as a symbol of prudence; the sun as an attribute of virtue. It was certainly not by chance that Pope Urban VIII selected the sun as his personal *impresa*. The sundial gallery was not only a scientific instrument but also, like the Sala Grande, was highly charged symbolically. And this was even increased by the Latin epigrams that could also be found in the gallery.

The Mirror as Metaphor

On the gallery's walls and between the mezzanine windows are a total of eight plaques with epigrams imaginatively extolling the exceptional nature of the sundial. One of the epigrams presents a dialogue between the sun and the mirror. Under the heading 'The Sun Complains about the Mirror (That Signals the Hours)', the sun turns indignantly to the mirror:[15]

> You shameless mirror before my eyes, product of fragile glass,
> where does this new unheard-of insolence come from, which allows you
> to reflect my countenance and my rays, which glow like fire,
> to bring me down from my pole and hold me on earth, and to
> confine me under the vault of a narrow roof?[16]

To this the mirror replies to the sun:

> Golden sun, you who light everything in your glow,
> have you not seen the inexorable edict of the highest god Jupiter:
> the wise will rule over the stars?
> For that reason the wise prince, the bright light of the scarlet senate,
> has brought you into his house.
> The hand and the wise industry of a great astronomer force me
> to serve the wishes of so great a patron.
> And it is only proper for us both to obey the wise.[17]

In this epigram the small mirror appears as master over the sun. It has the power to fetch the sun from the sky and bring it into the palazzo. This in turn is authorized by the highest Roman god, Jupiter. Following his predication, the mirror only carries out the wishes of its wise patron: the astronomer (Maignan) and above all the wise prince (Cardinal Spada), who is described as the 'bright light of the scarlet senate'. This is a reference to the College of Cardinals that

[15] All translations done by the author.

[16] 'Sol de Speculo Horoscopo conqueritur | Vile oculis Speculum fragilis labor improbe vitri | Unde tibi surgit novus hic temerarius ausus | Ut vultum radiosque meos licet igne micantes | Reddere meque polo deductum addicere terrae | Claudereque angusti tentes sub fornice tecti ?'

[17] 'Speculum Solis | Auree Sol tu cuncta tuo qui lumine lustras | Num quoque vidisti descriptum Adamante supremi | Hic Iovis edictum sapiens dominabitur Astris ? | Hinc igitur Sapiens Princeps idemque Senatus | Purpurei Lux clara suas te accersit ad aedes | Hinc manus et magni sapiens me industria cogit par | Astronomi, tanti votis servire Patroni | Et nos ergo ΣΟΦΟΙΣ par est parere duobus.'

dressed in scarlet robes. Into the slightly convex surface of the small round mirror of the sundial a radiant sun had in fact been engraved, so that the sun was also 'caught' visibly and permanently in the mirror.[18]

The mirror and the sun also appear repeatedly in the other epigrams. In one place we find the sun full of admiration for this unusual work of art, saying: 'Now I have to take a double path: one over the sky, the other through the Cardinal's palace'.[19] In another place the sun even appears as a painter:

> In the imitation of nature a hand here attempted to depict the sky;
> but to paint the sun it could not do.
> However, the sun did not want to be left out of the work,
> so it paints itself and its path with its own light.[20]

The epigrams were written by Guillaume du Planté, who, like Maignan, belonged to the Order of Minims.[21] Nevertheless, it is certain that Cardinal Spada would have given him clear instructions as to the general themes of the celebratory verses.

In addition, one of the inscriptions was composed by Bernardino Spada personally.[22] Here too, the Cardinal describes the sundial's special capacity to capture the sun's rays. The claim he makes for the work is expressed most clearly in the final lines.

> Presentation of the new sundial
> Facing east a small window opens in a gap
> and hardly allows the daylight to enter.
> The sun itself, coming from the sky, pauses at the window's threshold;
> it does not allow its rays to enter.
> If at the outer edge you place a mirror,
> it throws the rays it receives onto the ceiling.
> A round mirror also throws the rays into this vault
> so that the point of light glows in the palazzo as in the celestial axis.
> Numerous lines divide the vaulted ceiling into different hours;

[18] See Neppi, *Palazzo Spada*, p. 194.

[19] 'Ecce duplex mihi nunc est ineuda via. | Altera coelestis per nota palatia Regis, | Altera Purpurei per nova tecta Patris.'

[20] 'Aemula naturae manus hic depingere coelos | Tentavit, Solem pingere non potuit. | Ne tamen incepto desit Sol ipse labori | Seque suumque suo lumine pingit iter.'

[21] See Neppi, *Palazzo Spada*, p. 191.

[22] 'Primum Epigramma ipse Eminentissimus Dominus Cardinalis scripsit'; Maignan, *Perspectiva horaria*, p. 395.

> the lines that are lit in the ceiling display the day.
> The iron stylus and an iron shadow are banished,
> and the world has golden, not iron, days.[23]

With the notion of the Golden Age, Cardinal Spada adopted a formula that was used by classical authors such as Plato, Virgil, and Ovid to describe an original paradise-like state, which is inevitably followed by a period of decline: the Iron Age. Cardinal Spada metaphorically equated the iron stylus of the 'common' sundial with the Iron Age. However, stylus and shadow are eventually cast out by the new catoptric sundial, which captures the light through the sun's rays. As the commissioner of his mirror sundial, the Cardinal stages himself, in the truest sense of the word, in the right light, as founder of a new Golden Age.

From mirror to Golden Age — the Cardinal succeeded with great skill in fully exploiting the metaphorical dimension of a small round instrument. However, behind this extremely ambitious rhetoric, it is possible that there was a hidden attempt to compensate for his family's deficient social status. The Spada family was not part of the long-established Roman nobility; Bernardino's father was still a coal merchant in Romagna.[24] Within a single generation, however, Bernardino was able to advance to the highest ranks of Roman aristocracy. His social advancement, which was typical of early modern Rome, was made possible by the specific nature of the papacy as an ecclesiastical elective monarchy, which led to a continuous exchange of the elites and hence a competitive social climate.

The key to social advancement was the right contacts and patronage — initially in the sense of finding a patron (in this case Urban VIII), but also, above all, in the sense of becoming active as a patron oneself. Bernardino's art patronage led to impressive results, as can still be seen today when viewing his collection of paintings and sculptures at the Palazzo Spada. That he also placed great value on scientific patronage has been shown by the example of the sundial gallery and the financial support of Maignan's treatise on optics.

[23] 'Notio Novi Horologii Solaris | Excipiens Austrum modico se pandit hiatu | Et vix admittit parva fenestra diem | E coelo sol ipse ruens in limine sistit | Ad loca nec defert interiora iubar. | Si tamen extremo sternas in margine vitrum | Quos recipit radios in laqueare iacit. | Orbiculare vitrum radios quoque vibrat in orbem | Ut micet in tecto sicut in axe globus. | Linea crebra tholum varias partitur in horas | Quam globus irradiat temperat illa diem. | Ferreus hinc Gnomon et ferrea pellitur umbra | Aurea pro ferro tempora mundus habet.'

[24] For more information on Bernardino Spada's biography and his family story, see Karsten, *Kardinal Bernardino Spada*.

Figure 12.6. Giovanni Battista Magni, Allegories of Astronomy, Cosmography, Geometry and Optics. Fresco, 1644, Galleria della Meridiana, Palazzo Spada, Rome, photo by author.

Naturally, the sundial was also a work of art, and its decorative function was essential. The quadrant of the sundial is represented as a drapery, lined by a painted yellow-brown frieze as if it were a decorative textile border, and appears to be carried through the sky by seven flying putti. At the edges of the dial face are two figurative frescos, which, like the sundial, were painted by Giovanni Battista Magni. On one side one can recognize Mercury, the messenger of the gods, with the model of a sundial in his hands, while a throng of gods enthroned on the clouds above look in admiration at this invention.[25] The fresco at the other end of the ceiling shows the four allegories of Astronomy, Cosmography, Geometry, and Optics — the sciences whose mastery was necessary for the construction of a sundial (Figure 12.6). Using different mathematical and

[25] 'Altera verò constructi iam Astrolabij specimé à Mercurio in coelum deferri, caetu Deorum suspiciente, ac mirante tam praeclari inventì novitatem'; Maignan, *Perspectiva horaria*, p. 392.

astronomical instruments (armillary sphere, angle, compass, dividers, plumb line, mirror) the figures are busy constructing a catoptric sundial.[26]

While these allegories certainly have first and foremost a decorative function, they are not the only representations of artistic character. The dial face with its network of colourful hour lines, the different-coloured numbers, inscriptions, and zodiac signs is certainly also very decorative and aesthetically attractive (see Figure 12.1).

Conclusion

Finally, it can be noted that the Cardinal's decision to have a gallery of his palazzo decorated with a monumental mirror sundial was probably based on a number of different motives: the sundial was an especially ingenious form of artistic and aesthetic decoration, and it was also a functioning instrument that had a practical, daily use. The sheer monumental size of the instrument, however, already suggests that there were reasons for choosing this type of ceiling decoration other than mere aesthetic and practical functions. It was a crucial part of Spada's patronage activity and thus also a means of self-representation and a way of giving social legitimacy to his power.

It is certainly worth noting in this context that the sundial of the Palazzo Spada received only very limited public attention, restricted to contemporary optical treatises. Jean-François Niceron, Emmanuel Maignan's student and friar, mentioned it in his *Thaumaturgus opticus* of 1646.[27] Also Athanasius Kircher described it with much praise in his optical treatise *Ars magna lucis et umbrae*, published the same year.[28] Surprisingly, however, the sundial of the Palazzo Spada does not appear once in the extensive contemporary travel literature dealing with the Eternal City.[29]

The sundial's strategy of visualization therefore does not seem to have been truly successful. Searching for possible reasons, the question arises whether the work was perhaps perceived as being too complex and hence was commented on only by astronomical experts. It is equally conceivable that the specific deco-

[26] 'Perspectivam unàque Astronomiam, & cum Cosmographia Geometriam, deque construendo Astrolabio'; Maignan, *Perspectiva horaria*, p. 392.

[27] Jean-François Niceron, *Thaumaturgus opticus*, pp. 178–79. For Niceron, see also Whitmore, *The Order of Minims*, pp. 155–62 and Neppi, *Palazzo Spada*, p. 189.

[28] Athanasius Kircher, *Ars magna lucis et umbrae*, p. 649.

[29] See Feist, *Sonne, Mond und Venus*, pp. 68–70.

ration with a catoptric sundial was the result of the personal initiative of the Frenchman Maignan that did not have any imitations after his return to France, as no other mathematician in Rome was able and willing to develop a decorative scheme based on astronomical calculations. Possibly, the decoration of the Galleria della Meridiana with its celebratory epigrams was also felt to be inappropriately ambitious for a cardinal and understood as a claim of entitlement to the papal throne. This ultimate climb on the career ladder remained beyond Cardinal Spada's reach. Nevertheless, he completed a remarkable career, thanks to the patronage of Pope Urban VIII.

The fact that high-ranking representatives of the church regarded astronomical studies as extremely useful and supported them accordingly has been revealed using the example of the sundial in the Galleria della Meridiana and the collaboration between Cardinal Spada and the Minim friar Maignan. It is not inconceivable that, in commissioning a catoptric sundial, the design of which represents a perfect union of science and art, as well as a functioning and precise instrument of daily service, Spada was simply ahead of his time. No doubt, he saw in scientific patronage an effective means of augmenting his own reputation.

Above all, however, due to the sun imagery as it is expressed implicitly through the iconography as well as explicitly in the epigrams, the sundial has strong symbolic implications. In its function as instrument, symbol, and metaphor, the mirror was ultimately decisive in lending the catoptric sundial these larger dimensions of meaning. The mirror and sun symbolism surely alludes to Berdardino Spada's patron, Pope Urban VIII Barberini, who already during his time as cardinal had used the sun symbol as a personal *impresa*. Consequently, the statement that Cardinal Spada 'has brought [the sun] into his house' must also be understood as a kind of homage to Maffeo Barberini. As a cardinal and also later as pope, Barberini had entered Spada's life like a radiant sun that ignited his career and nourished its flames so they could shine brightly and untainted. The sun theme reappears in various places throughout the Palazzo Spada, for instance, in a fresco of the so-called Stanza del Sole on the ground floor as well as in the stucco of the panelled ceiling in the Sala Grande. It is in the mirror sundial of the Galleria della Meridiana, however, that this sun-theme literally found its most impressive reflection.

Works Cited

Archival Sources

Rome, State Archive (ASR), MS Fondo Spada Veralli (FSV) 463
——, MS Fondo Spada Veralli (FSV) 747

Primary Sources

Kircher, Athanasius, *Ars magna lucis et umbrae* (Rome: Scheus, 1646)
Maignan, Emmanuel, *Perspectiva horaria, sive de horographia gnomonica tum theoretica, tum practica libri quatuor* (Rome: Philippi Rubei, 1648)
Niceron, Jean-François, *Thaumaturgus opticus* (Paris: Langlois, 1646)

Secondary Works

Càndito, Cristina, 'Corrispondenze ottico-prospettiche tra le opere di Maignan e di Borromini a Palazzo Spada', *Mélanges de l'école française de Rome: Italie et Méditerranée*, 117.1 (2005), 73–89
Cannatà, Roberto, ed., *Palazzo Spada: Arte e Storia* (Rome: Bonsignori, 1992)
Cannatà, Roberto, Maria Lucrezia Vicini, and Mario Lolli Ghetti, eds, *Palazzo Spada: Le decorazioni restaurate* (Milan: Electa, 1995)
Feist, Ulrike, *Sonne, Mond und Venus: Visualisierungen astronomischen Wissens im frühneuzeitlichen Rom* (Berlin: Akademie, 2013)
Heilbron, John L., *The Sun in the Church: Cathedrals as Solar Observatories* (Cambridge, MA: Harvard University Press, 2001)
Karsten, Arne, *Kardinal Bernardino Spada: Eine Karriere im barocken Rom* (Göttingen: Vandenhoeck & Ruprecht, 2001)
Neppi, Lionello, *Palazzo Spada* (Rome: Editalia, 1975; repr. 1990)
Tabarrini, Marisa, *Borromini e gli Spada: Un palazzo e la committenza di una grande famiglia nella Roma barocca* (Rome: Gangemi, 2009)
Vicini, Maria Lucrezia, *Il Collezionismo del Cardinale Fabrizio Spada in Palazzo Spada* (Rome: Markonet, 2006)
Whitmore, P. J. S., *The Order of Minims in Seventeenth-Century France* (The Hague: Martinus Nijhoff, 1967)

Index

Cursor Mundi

All volumes in this series are evaluated by an Editorial Board, strictly on academic grounds, based on reports prepared by referees who have been commissioned by virtue of their specialism in the appropriate field. The Board ensures that the screening is done independently and without conflicts of interest. The definitive texts supplied by authors are also subject to review by the Board before being approved for publication. Further, the volumes are copyedited to conform to the publisher's stylebook and to the best international academic standards in the field.

Titles in Series

Chris Jones, *Eclipse of Empire? Perceptions of the Western Empire and its Rulers in Late-Medieval France* (2007)

Simha Goldin, *The Ways of Jewish Martyrdom* (2008)

Franks, Northmen, and Slavs: Identities and State Formation in Early Medieval Europe, ed. by Ildar Garipzanov, Patrick Geary, and Przemyslaw Urbanczyk (2008)

William G. Walker, *'Paradise Lost' and Republican Tradition from Aristotle to Machiavelli* (2009)

Carmela Vircillo Franklin, *Material Restoration: A Fragment from Eleventh-Century Echternach in a Nineteenth-Century Parisian Codex* (2010)

Saints and their Lives on the Periphery: Veneration of Saints in Scandinavia and Eastern Europe (c.1000–1200), ed. by Haki Antonsson and Ildar Garipzanov (2010)

Approaching the Holy Mountain: Art and Liturgy at St Catherine's Monastery in the Sinai, ed. by Sharon E. J. Gerstel and Robert S. Nelson (2011)

'This Earthly Stage': World and Stage in Late Medieval and Early Modern England, ed. by Brett D. Hirsch and Christopher Wortham (2011)